Computational and Network Modeling of Neuroimaging Data

Neuroimaging Methods and Applications Series
Series Editor: Peter Bandettini, PhD

Chief, Section on Functional Imaging Methods, Laboratory of Brain and Cognition
Director, Functional MRI Core Facility
National Institute of Mental Health, National Institutes of Health, Bethesda, MD, United States

About the Series:

Neuroimaging Methods and Applications is a book series that embodies the collective expertise and experience of the neuroimaging community. The books in this series cover neuroimaging approaches with an emphasis on fMRI. They include the latest insights and practical information about instrumentation, acquisition methods, processing, multimodal integration, and both clinical and neuroscience research applications. They also include such topics as neuromodulation and computational modeling. The series is intended to provide useful information, insight, and perspective to neuroimaging researchers and clinicians at all levels—from undergraduates to full professors. The topics and content are intended to complement a wide range of reader experience and expertise, including neuroscience, psychology, psychiatry, neurology, engineering, physics, statistics, mathematics, computer science, and physiology.

Titles published:

Advances in Resting-State Functional MRI—Edited by Jean Chen and Catie Chang
Connectome Analysis: Characterization, Methods, and Applications—Edited by Markus D. Schirmer, Tomoki Arichi, and Ai Wern Chung, Computational and Network Modeling of Neuroimaging Data—Edited by Kendrick Kay

Computational and Network Modeling of Neuroimaging Data

Edited by

Kendrick Kay

Center for Magnetic Resonance Research, University of Minnesota, Minneapolis, MN, United States

Academic Press is an imprint of Elsevier
125 London Wall, London EC2Y 5AS, United Kingdom
525 B Street, Suite 1650, San Diego, CA 92101, United States
50 Hampshire Street, 5th Floor, Cambridge, MA 02139, United States

Notices

Knowledge and best practice in this field are constantly changing. As new research and experience broaden our understanding, changes in research methods, professional practices, or medical treatment may become necessary.

Practitioners and researchers must always rely on their own experience and knowledge in evaluating and using any information, methods, compounds, or experiments described herein. In using such information or methods they should be mindful of their own safety and the safety of others, including parties for whom they have a professional responsibility.

To the fullest extent of the law, neither the Publisher nor the authors, contributors, or editors, assume any liability for any injury and/or damage to persons or property as a matter of products liability, negligence or otherwise, or from any use or operation of any methods, products, instructions, or ideas contained in the material herein.

ISBN 978-0-443-13480-7

For information on all Academic Press publications
visit our website at https://www.elsevier.com/books-and-journals

Publisher: Mara Conner
Acquisitions Editor: Tim Pitts
Editorial Project Manager: John Leonard
Production Project Manager: Sujithkumar Chandran
Cover Designer: Christian Bilbow

Typeset by STRAIVE, India

Working together to grow libraries in developing countries

www.elsevier.com • www.bookaid.org

Contents

CHAPTER 3   **Cognitive modeling: Joint models use cognitive theory to understand brain activations 53**
Brandon M. Turner

**CHAPTER 4 Network modeling: The explanatory power
of activity flow models of brain function 81**
Michael W. Cole

**CHAPTER 5 Biophysical modeling: An approach for
understanding the physiological fingerprint
of the BOLD fMRI signal 119**
Mario Gilberto Báez-Yáñez and Natalia Petridou

CHAPTER 6   Biophysical modeling: Multicompartment biophysical models for brain tissue microstructure imaging **159**

H. Farooq, P.K. Pisharady, and C. Lenglet

CHAPTER 7   **Dynamic brain network models: How interactions in the structural connectome shape brain dynamics** **209**
Joana Cabral and John D. Griffiths

Contributors

Mario Gilberto Báez-Yáñez
Translational Neuroimaging Group, Center for Image Sciences, University Medical Center Utrecht, Utrecht, The Netherlands

Joana Cabral
Life and Health Sciences Research Institute (ICVS), Medical School, University of Minho, Braga, Portugal

Michael W. Cole
The Center for Molecular and Behavioral Neuroscience, Rutgers University, Newark, NJ, United States

H. Farooq
Center for Magnetic Resonance Research, Department of Radiology, University of Minnesota, Minneapolis, MN, United States

John D. Griffiths
Krembil Centre for Neuroinformatics, Centre for Addiction and Mental Health; Department of Psychiatry & Institute of Medical Sciences, University of Toronto, Toronto, ON, Canada

Zijin Gu
School of Electrical and Computer Engineering, Cornell University and Cornell Tech, New York, NY, United States

Catherine Hanson
RUBIC, Psychology and CMBN, Newark, NJ, United States

Stephen José Hanson
RUBIC, Psychology and CMBN, Newark, NJ, United States

Keith W. Jamison
Department of Computational Biology, Cornell University, Ithaca; Department of Radiology, Weill Cornell Medicine, New York, NY, United States

Tomas Knapen
Department of Movement and Behavioral Sciences, Vrije Universiteit Amsterdam; Spinoza Centre for Neuroimaging; Netherlands Institute for Neuroscience, Amsterdam, The Netherlands

Amy Kuceyeski
Department of Computational Biology, Cornell University, Ithaca; Department of Radiology, Weill Cornell Medicine, New York, NY, United States

C. Lenglet
Center for Magnetic Resonance Research, Department of Radiology; Institute for Translational Neuroscience, University of Minnesota, Minneapolis, MN, United States

Martin A. Lindquist
Department of Biostatistics, Johns Hopkins University, Baltimore, MD, United States

Andre Marquand
Donders Institute for Brain, Cognition and Behaviour, Radboud University Medical Centre; Department of Cognitive Neuroscience, Radboud University Nijmegen Medical Centre, Nijmegen, The Netherlands

Elisha P. Merriam
Laboratory of Brain and Cognition, NIMH, NIH, Bethesda, MD, United States

Shinji Nishimoto
Graduate School of Frontier Biosciences, Osaka University; Center for Information and Neural Networks (CiNet), National Institute of Information and Communications Technology; Graduate School of Medicine, Osaka University, Osaka, Japan

Natalia Petridou
Translational Neuroimaging Group, Center for Image Sciences, University Medical Center Utrecht, Utrecht, The Netherlands

P.K. Pisharady
Center for Magnetic Resonance Research, Department of Radiology, University of Minnesota, Minneapolis, MN, United States

Zvi N. Roth
School of Psychological Sciences, Tel Aviv University, Tel Aviv, Israel

Saige Rutherford
Donders Institute for Brain, Cognition and Behaviour, Radboud University Medical Centre; Department of Cognitive Neuroscience, Radboud University Nijmegen Medical Centre, Nijmegen, The Netherlands

Mert R. Sabuncu
School of Electrical and Computer Engineering, Cornell University and Cornell Tech; Department of Radiology, Weill Cornell Medicine, New York, NY, United States

Yu Takagi
Research and Development Center for Large Language Models, National Institute of Informatics, Tokyo; Graduate School of Frontier Biosciences, Osaka University; Center for Information and Neural Networks (CiNet), National Institute of Information and Communications Technology, Osaka, Japan

Brandon M. Turner
The Ohio State University, Columbus, OH, United States

Thomas Wolfers
Donders Institute for Brain, Cognition and Behaviour, Radboud University Medical Centre, Nijmegen, The Netherlands; Department of Psychiatry and Psychotherapy, Tübingen Center for Mental Health, University of Tübingen; German Center for Mental Health, Tübingen, Germany

Preface

Why this book? Neuroimaging is witnessing a massive increase in the quantity, quality, and types of data that can be acquired from human brains. Yet data alone do not provide insights. What is needed is quantitative modeling in order to achieve powerful interpretations of brain data. The goal of this book is to survey, introduce, and explain model-based approaches to understanding the brain as measured through the tools of neuroimaging (e.g., MRI, EEG, MEG). A central characteristic of this book is the great diversity of approaches, targeting different brain systems, different spatial and temporal scales, and different aspects of biology and psychology. In fact, a quick glance at the chapters in this book is sufficient to realize that *neuroscience is currently not really a coherent field, but rather a collection of different subfields.* This is troubling and suggests that synthesis and integration across subfields are desperately needed.

Scope of this book. This book seeks to provide an authoritative overview of the diverse modeling approaches that have been fruitfully applied to neuroimaging data. The restriction to neuroimaging is, of course, an artificial choice. But this restriction provides two advantages: (i) we can imagine a unifying starting point in terms of the data that can be practically collected from a given human brain, and (ii) the smaller, more tractable scope helps forge cross-cutting comparisons across modeling approaches. The collection of model-based approaches included in this book is inevitably incomplete, given the huge diversity of neuroscience research. Nonetheless, I believe this book achieves a remarkable breadth and clarity of content and thus should provide the reader with deep insights into the modeling of neuroimaging data.

Design of this book. Each of the 11 chapters in this book introduces and motivates a specific modeling approach, explains how the modeling is actually conducted, and ends with a forward-looking perspective on where that type of modeling is going. Importantly, all chapters conform to a common parallel structure. This structure is designed to promote coherence and help the reader make sense of the diverse material. In addition, all chapters spend time elaborating the motivations, priorities, and assumptions that are associated with a given type of modeling. Such details are critical but are often missing from conventional research articles and may not be apparent to nonexperts.

Cross-cutting comparisons. As you read this book, I hope you discover intriguing similarities and differences across modeling approaches. Here, I comment on a few that I have observed. There are striking similarities across modeling approaches with respect to mathematical and statistical techniques and issues contained therein, such as model selection, model fitting, model interpretation, and model generalization. On the other hand, there are stark contrasts in terms of the topics and phenomena that are emphasized. Some approaches emphasize behavior and building strong theoretical principles for interpreting behavior before analyzing brain data (see Chapter 3), whereas in other approaches, behavior is barely even mentioned. Some approaches emphasize network interactions between brain regions (see Chapters 4 and 8), while others do not (see Chapter 2). While all modeling approaches connect to empirical data and are therefore necessarily grounded in statistics (see Chapter 1), it is interesting to observe the variation in the extent to which statistical issues are prioritized. Several modeling approaches emphasize biological details, especially the complexities of the microscopic features that generate the macroscopic neuroimaging data that we can measure (see Chapters 5 and 6). Biological details might be particularly critical for understanding individual differences (see Chapter 11) as well as predicting the effects of perturbations such as those achieved through stimulation or pharmacology (see Chapter 7). In contrast, other approaches adopt a less biological and more abstract view of the brain (e.g., Chapters 8 and 9) and/or focus on exploiting patterns that manifest in neuroimaging data irrespective of their biological origins (e.g., Chapters 9 and 10). Finally, it is interesting to compare the extent to which modeling approaches address the brain from general, large-scale perspectives (e.g., Chapters 7, 9, and 11) or from specific, focused perspectives (e.g., Chapters 2 and 3).

Concluding remarks. I hope you enjoy this book and appreciate the wide diversity of approaches toward quantitative modeling of the brain. Although I imagine that both modeling experts and modeling novices will benefit from the book chapters, I speculate that it is likely the newer generations of neuroscientists who will be able to connect, integrate, and extend the model-based approaches that have been developed thus far. This book, I think, is just a small step toward developing a more mature neuroscience in which diverse models are integrated with aligned priorities, goals, assumptions, and terminology. The puzzle pieces are here; it's time to put them together.

Kendrick Kay

Statistical modeling: Harnessing uncertainty and variation in neuroimaging data

Martin A. Lindquist

Department of Biostatistics, Johns Hopkins University, Baltimore, MD, United States

Introduction to statistical modeling
What is statistics?

As fields become increasingly quantitative, the proper evaluation of scientific results becomes more difficult. Neuroimaging data are increasingly being used in a wide array of fields, and it is important for researchers to be able to properly evaluate them and use them to make appropriate conclusions. Proper understanding of what data can and cannot tell us is critical. *Statistics* is an interdisciplinary field concerned with developing and studying methods for collecting, analyzing, interpreting, and presenting data. In doing so, statisticians draw on several mathematical and computational tools. Statistics is an important part of how we use data to make decisions and predictions in neuroimaging and beyond.

Two of the most important concepts underlying the field of statistics are *uncertainty* and *variability*. Uncertainty refers to either a lack of data or an incomplete understanding. This could be due to the fact that the outcome in question has not yet been determined (e.g., will it snow tomorrow?), or alternatively that the outcome has been determined but we don't yet know what it is (e.g., what is the result of a diagnostic test just taken but not yet analyzed?). We use probabilities to make statements about uncertainty. Uncertainty can be reduced or eliminated with more or better data. Variability refers to the inherent heterogeneity or diversity of data in an assessment. Variation arises because even if the same measurement was repeated, the results would likely change. Statisticians seek to understand, control, and model sources of variation. These two concepts are linked, as uncertainty emerges because of variability. But they are different concepts. For example, one may consider uncertainty in the estimate of a population average, or variation in the estimate across different samples. Importantly, statistics allows us to make decisions in the face of uncertainty.

Computational and Network Modeling of Neuroimaging Data. https://doi.org/10.1016/B978-0-443-13480-7.00012-0

Statistics in neuroimaging

Neuroimaging is a highly interdisciplinary field, and statistics and statistical modeling has long played an important role (Lindquist, 2008; Ombao et al., 2016). While in this chapter we focus primarily on problems related to the analysis of structural and functional magnetic resonance imaging (MRI) data, we note that most of the points we make are equally relevant for other imaging modalities, including positron emission tomography (PET), electroencephalography (EEG), and magnetoencephalography (MEG), among others.

In structural neuroimaging, statistics has, for example, been used to assess differences in brain development and its link to neurodevelopmental disorders (Giedd et al., 1999; Wallace et al., 2010). In addition, machine learning techniques have been used to create brain-based biomarkers, for example, of age (Cole et al., 2018). In functional neuroimaging, statistics has been successfully used to detect brain regions with task-related changes in brain activity (e.g., *brain mapping*) (Poldrack, 2018; Friston et al., 1994), to identify brain regions that exhibit similar activation patterns over time (e.g., *functional connectivity*) (Friston, 1994; Biswal et al., 1995), and to make predictions and classify participants based on brain activity (Haxby et al., 2001; Norman et al., 2006).

Indeed, it is hard to overstate the importance that statistics has played in neuroimaging research, as it underlies most findings and conclusions that have been made. For example, the area of brain mapping is largely driven by the use of the so-called *general linear model* (GLM) (Friston et al., 1995; Worsley and Friston, 1995), which encompasses classical statistical techniques such as regression, *t*-tests, and analysis of variance (ANOVA). Any time researchers threshold a brain map or a graph (i.e., make a determination whether or not a voxel or edge is significant), they use statistical techniques to derive appropriate values; these include random field theory (Worsley et al., 1996; Worsley, 1996) and permutation methods (Nichols and Holmes, 2002), among others. In addition, when one is faced with building a predictive model in settings where the number of observations (e.g., subjects) is smaller than the number of features (e.g., voxels), penalized regression models such as the LASSO or ridge regression are commonly used (Wager et al., 2013). Fig. 1 shows some examples of statistical methods in the neuroimaging literature.

Statistics helps neuroimaging researchers produce reliable results, through control of false-positive rates (Nichols and Hayasaka, 2003). It provides tools to draw conclusions about populations (e.g., people with a certain disease) based on a small sample of participants drawn from the population in question. It provides methods for effectively comparing different populations. One example is determining how a population of participants with schizophrenia compares with a population of normal controls with respect to some brain measure (Gur and Gur, 2010). It also helps make important clinical decisions (e.g., as a tool for preparing surgery) (Silva et al., 2018). Importantly, statistics allows us to harness and better understand the variability inherent in neuroimaging data. This ultimately drives our ability to compare data across space and time within an individual, compare different individuals from a

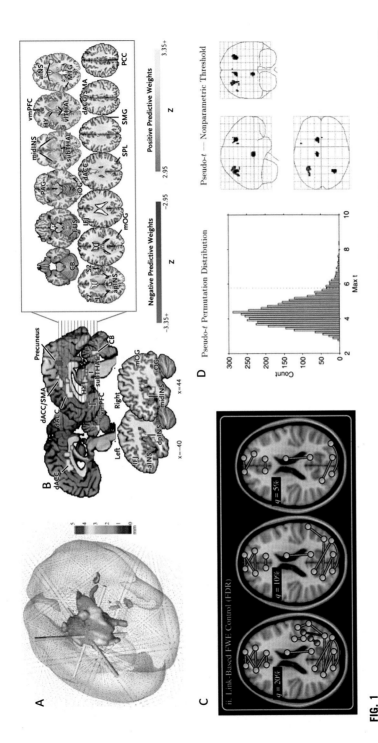

FIG. 1

Examples of statistics in neuroimaging. (A) Deformation-based morphometry of nonmissile trauma. Trauma minus control average deformations, sampled every 6 mm, with Hotelling's T^2 statistic for significant differences. (B) A neurological signature for physical pain. The map shows weights that exceed a threshold (a false discovery rate of q < 0.05). (C) The network-based statistic as well as edge-based FWE control provided by the false discovery rate (FDR) were used to detect a contrast that was simulated between two groups (q = 0.05, 0.1 and 0.2). (D) (left) Permutation distribution of maximum repeated measures t statistic. Here, the dotted line indicates the 5% level corrected threshold. (right) Maximum intensity projection of pseudo t statistic image threshold at 5% level, as determined by permutation distribution.

(A) Data from Worsley, K.J., Taylor, J.E., Tomaiuolo, F., Lerch, J., 2004. Unified univariate and multivariate random field theory. Neuroimage 23 (Suppl 1), S189–95. (B) From Wager, T.D., et al., 2013. An fMRI-based neurologic signature of physical pain. N. Engl. J. Med. 368, 1388–1397. (C) From Zalesky, A., Fornito, A., Bullmore, E.T., 2010 Network-based statistic: identifying differences in brain networks. Neuroimage 53, 1197–1207. (D) From John T. Ashburner, Karl J. Friston, Stefan J. Kiebel, Thomas E. Nichols, William D. Penny. Statistical Parametric Mapping: The Analysis of Functional Brain Images. (Elsevier (Singapore) Pte Limited, 2010)

certain population, compare populations, compare longitudinal measures from the same individual, and compare different individuals whose data come from different sites using different collection techniques. Statistics is becoming increasingly important as neuroimaging experiments become more complex and ask more difficult questions.

What is a statistical model?

A statistical model is a mathematical description of a real-world phenomenon. Statistical models can contain both random and nonrandom (or deterministic) components. The random component allows us to model different sources of variation in the response, while the nonrandom component often seeks to explain the effects of certain experimental interventions or stimuli on the response. Models generally consist of *parameters* that need to be estimated using the available data sample. Estimates of model parameters are generally referred to as *statistics*. Because the value a statistic takes varies depending on the sample, the statistic has a probability distribution associated with it, the *sampling distribution*, which quantifies what values the statistics can take and how often it takes them in repeated sampling. The sampling distribution is used to perform statistical inference and test claims about the parameter(s) of interest. It is important to note there are typically a number of ways to estimate a given parameter (i.e., multiple possible statistics that could be used), so we often choose ones with beneficial properties (e.g., small bias, small variance, certain distributional priorities, or robustness to outliers).

To illustrate, assume we have an exogenous variable of interest (e.g., an experimental manipulation) that causes a change in activation in an area of the brain. In practice, we do not observe the true activity as the signal is intermixed with noise. Further, suppose we collect data on n different individuals assumed to be randomly sampled from a population of potential participants. Our goal is to determine whether there is a true nonzero effect in the population of interest. To do so, we begin by setting up the following statistical model:

$$y_i = \theta + \varepsilon_i, \quad \varepsilon_i \sim N\left(0, \sigma^2\right), \quad i = 1, \ldots n$$

Here, y_i is the observed data for individual i. It is assumed to be a random draw from a normal distribution with mean θ and variance σ^2. This is a model of the *population distribution* from which the n individuals were drawn. Hence, the mean and variance are population parameters that describe characteristics of the larger population. These are the parameters we seek to estimate and perform inference on, as we ultimately want to make conclusions about the population. Estimation is performed by computing a sample statistic (i.e., an estimate of the parameters using data from the sample of n participants). In our example, both the sample mean and median would be reasonable estimates of the population mean. We typically choose our statistic based on its properties. The mean is often used in this context as it is the *maximum likelihood estimate*, which entails it has numerous beneficial large-sample properties (e.g., asymptotic normality) (Casella and Berger, 2021). However, if there are a lot of outliers in the data, using a more robust statistic such as the median or trimmed-mean could be an attractive alternative.

Once we estimate the model parameters, we often seek to perform a statistical test that considers the evidence for a certain effect. To infer whether the voxel shows an effect or not, one performs a hypothesis test using a *test statistic*. This is a statistic used explicitly for statistical hypothesis testing. It typically involves a measure of the magnitude of the effect (e.g., the mean) divided by the standard deviation of its sampling distribution (describing how much the observed effect is expected to vary from sample to sample). In our example, when using the sample mean as our statistic, we typically use a t-statistic as our test statistic. The *null hypothesis* is often that the effect) is zero (i.e., $\theta=0$). The corresponding *alternative hypothesis* could be that it is greater than zero (i.e., $\theta>0$). Our problem is now to determine whether there is enough information in the data (i.e., as summarized using the test statistic) to reject the null hypothesis in favor of the alternative hypothesis. This can be done by deciding whether or not the test statistic is consistent with what we would expect to observe under the null. If not, we reject the null hypothesis in favor of the alternative hypothesis.

Like other parametric statistics, if we are willing to make assumptions, then many test statistics (e.g., the t-statistic) have a known *probability density function* (PDF), stating what values it can take under repeated samples assuming the null hypothesis is true. This implies that if the true effect is zero (i.e., the null hypothesis is true), we can use it to compute the *P-value*, the probability of observing a t-statistic as or more extreme than the one observed, based on the area under the PDF. We typically compare this to a preset acceptable false-positive rate, or *alpha* (α) value (e.g., $P < .05$). If the P-value is below this value, we consider the effect *significant* and reject the null hypothesis. Otherwise, we consider it nonsignificant. In our example, if we assume the data are drawn from a normal distribution, then the test statistic will follow a t-distribution with $n-1$ degrees of freedom.

The above description, and most statistical methods covered in the introductory statistics courses, are frequentist (or classical) methods. In the frequentist point of view, probabilities describe long-run relative frequency and parameters are fixed unknown constants. Because they do not fluctuate, no useful probability statements can be made about them. Statistical procedures are instead designed to have well-defined long-run frequency properties. For example, a 95% confidence interval is designed so that if we repeat the sampling approach many times, roughly 95% of the intervals constructed would capture the true population parameter. A hypothesis test with $\alpha=0.05$ would incorrectly reject a true null hypothesis 5% of the time.

Bayesian inference is an alternative approach that provides a different perspective. It is the process of inductive learning using Bayes' rule (Hoff, 2009). Taking the Bayesian point of view, probabilities describe a degree of belief. Probability statements can be made about parameters, even though they are fixed constants. Inferences are made about a parameter θ by producing a probability distribution for it. In Bayesian analysis, one chooses a density $p(\theta)$, the *prior* that expresses our beliefs about θ before we see any data. Then, one chooses a statistical model $p(y|\theta)$, the *likelihood*, that reflects our belief about y given θ. In the previous example, the likelihood corresponds to a normal distribution with mean θ and variance σ^2. After observing y, we update our beliefs and calculate the *posterior* distribution $p(\theta|y)$. Updating is performed using Bayes' rule: $p(\theta|y) \propto p(y|\theta)p(\theta)$.

The prior distribution is a subjective distribution, based on the experimenter's belief and formulated prior to viewing the data. The choice of prior is crucial. If no prior information about the parameters are available, noninformative priors can be used. These types of priors let the data "speak for itself." Alternatively, one can choose the prior in such a way that the posterior lies in the same family of distributions as the prior (conjugate priors). For example, if the likelihood follows a normal distribution, choosing a prior that also follows a normal distribution guarantees that the posterior is normally distributed.

The posterior distribution contains all current information about the parameter θ and is used for subsequent inference. Numerical summaries (e.g., mean, median, mode) of the distribution can be used to obtain point estimates of the parameter. For example, the mode of the posterior distribution gives the maximum a posteriori (MAP) estimate. Importantly, in the Bayesian setting, one can make probability statements about the parameters based on the posterior (e.g., $P(\theta > 1|y)$). It can often be difficult to perform inference using the posterior, particularly if its distributional form is complicated. In these cases, one can use techniques such as Markov-chain Monte-Carlo (MCMC) (Robert and Casella, 2004) or variational Bayes (Fox and Roberts, 2012) to sample from a posterior distribution and estimate the distribution of interest.

Examples of successful statistical models

What makes a good statistical model? Ultimately, the answer to that question depends on your goals and on the data. A common goal is to use the model to *explain* some phenomena of interest. Here, we may seek the model with the lowest possible error, while being as parsimonious (i.e., having as few parameters) as possible. This typically provides a model that is both accurate and interpretable. In machine learning (a.k.a. statistical learning) models, the goal is to *predict* some phenomena. Here, we seek a model that generalizes well to new data (i.e., the prediction error should be as small as possible when applied to a new dataset). The appropriateness of a model of course also depends on the data and how it is collected. For example, a technique such as Granger causality might be appropriate for EEG data, but less so for fMRI data due to inherent differences in the properties of the data (David et al., 2008).

Neuroimaging data consist of both signal and noise. In the context of fMRI data, the signal of interest corresponds to the hemodynamic response (i.e., the BOLD signal that we seek to measure), and noise comes from multiple sources, including thermal noise, head motion, and physiological noise. Making a good model for neuroimaging data is nontrivial. Standard neuroimaging studies give rise to massive amounts of noisy data that exhibit a complicated spatio-temporal correlation structure (i.e., the variation in nearby timepoints and spatial locations is related to one another). While there exists a substantial literature on the analysis of time series (Shumway and Stoffer, 2017), most approaches have traditionally been designed for the analysis of a single time series and don't necessarily scale well when applied to thousands of voxel-wise time series. This necessitates taking shortcuts, and a deep understanding of the data is required to allow researchers to make an informed choice whether a particular shortcut is reasonable or not. The correlation *between* time courses extracted from different locations in the brain (i.e., the functional connectivity) adds an additional layer of complication, necessitating taking spatial correlations into consideration. In addition, fMRI has a low signal-to-noise ratio (SNR), which means the signal is weak compared to the amount of noise. This makes it difficult to detect true sources of signal and separate them from noise. In addition, there are often a number of confounders that can influence the performance of a model. For example, if participants move a lot during data collection, it is useful to include this information in the model (Johnstone et al., 2006; Ciric et al., 2017). Other possible confounders include age, body-mass index, head volume, and more (Alfaro-Almagro et al., 2021).

Statistics have long played an important role in understanding the nature of the data and obtaining relevant results that can be used and interpreted by neuroscientists. In this context, uncertainty can correspond to whether a specific task leads to activation in a region of the brain or whether two brain regions are functionally related to one another. Variation takes many forms in neuroimaging, including variation across time, space, experimental runs, and sessions, as well as between participants, populations, scanners, analysis pipelines, and collection sites; see Fig. 2 for some examples. In neuroimaging, statistical models seek to quantify these sources of variation to make claims about the brain and obtain generalizable knowledge. An additional complication is the relatively small number of participants in a given study, as well as the relatively small amount of data collected on each participant. The landscape surrounding this is rapidly changing (see the last section for more detail), but having the appropriate amount of data to appropriately characterize the different sources of variation has historically been a big problem in the field (Poldrack et al., 2016).

Arguably, the most successful statistical models in neuroimaging research have been from the class known as the *linear models*. Linear models are the cornerstone of statistical methodology and its most studied branch. They allow us to model the relationship between an outcome variable and one or more predictors and help us determine which of these variables are important for explaining the relationship. In addition, they are relatively easy to work with and are applicable to many real-world settings.

The workhorse of statistical models in neuroimaging research has long been the so-called *general linear model* (GLM) used for assessing brain activity (Worsley and Friston, 1995). It is a linear model that can be used to investigate whether the brain responds to a single type of event, compare different types of events, and assess correlations between brain activity and behavioral performance or other psychological

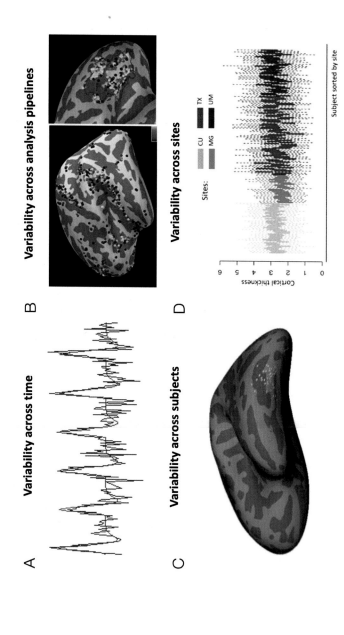

FIG. 2

Sources of variability in fMRI. (A) Variability in a voxel-wise fMRI time course from a single subject. The blue curve represents the observed time course and red the underlying signal. (B) Peak activation locations in group analyses vary widely across different preprocessing and analysis choices. Carp (2012a) analyzed nearly 7000 combinations of basic analysis choices; hotter colors reflect more frequent peak activation locations across different combinations. (C) The spatial distribution of individual subject peaks for [words > fixation] in ventral occipito-temporal cortex. (D) Variability in cortical thickness sorted by site (Fortin et al., 2018). Plots color-coded by imaging site: Columbia University (CU), University of Texas Southwestern (TX), Massachusetts General Hospital (MG), and University of Michigan (UM).

(C) From Duncan, K.J., Pattamadilok, C., Knierim, I., Devlin, J.T., 2009. Consistency and variability in functional localisers. Neuroimage 46, 1018–1026.

variables. The GLM is particularly useful when multiple predictors can be used to explain the variability in a single continuous outcome variable. In a typical neuro-imaging experiment, the outcome is typically a time series extracted from either a brain voxel or region of interest. The predictors are generally signals corresponding to a task of interest. The GLM approach treats the data as a linear combination of model functions (predictors) and noise. Though we assume the model functions have known shapes, we use the GLM to model their unknown amplitudes and test whether that particular function is present in the data; see Fig. 3A. Researchers typically include additional predictors to account for other sources of variation in the data, including ones that model signal changes related to drift, head movement, and phys-iological (e.g., respiration and cardiac pulsation) artifacts. These are referred to as nuisance covariates because they are model factors associated with known sources of variability but are not directly related to the task or to the experimental hypothesis.

The analysis is typically performed in the *massive univariate* setting, which implies that a separate GLM is fit at every single voxel in the brain and the results of the subsequent statistical tests are saved in a *statistical map* depicting the values of the test statistic at each voxel; see Fig. 3B. Determining which voxels are active (i.e., whether activation is significantly different from 0) is nontrivial. In the massive-univariate setting, thousands of decisions are required to determine which voxels are significant. Thus, researchers are faced with a large multiple comparisons prob-lem. Here, statistics has played a critical role, and techniques such as random field theory (RFT) (Worsley et al., 1996; Worsley, 1996; Brett et al., 2003) have had an enormous impact on the field through their implementation in the major software packages (e.g., SPM, FSL, and AFNI). Much of the early work on RFT was driven by statistician Keith Worsley and provides an example of a case where statistics played a large role in the manner in which neuroimaging data were analyzed. At the same time, it provides an example of a case where neuroimaging drove statistical research in exciting new directions.

The use of statistical models of course extends far beyond brain mapping and the GLM. For example, they play an important role in the study of effective connectivity (Friston et al., 2003; Ramsey et al., 2010), functional connectivity (Rissman et al., 2004; Marrelec et al., 2006; Varoquaux et al., 2010), and network analysis (Rubinov and Sporns, 2010). In each of these examples, statistical models are used to explain variation, make decisions about whether an effect (e.g., connection between two regions) is significant, and compare populations. In addition, it is becoming increas-ingly common to perform prediction and classification using machine learning (Norman et al., 2006; Wager et al., 2013). Here, researchers seek to develop tech-niques for analyzing information represented in brain activity patterns. In contrast to the mass-univariate approaches described above, these techniques generally take into consideration relationships between multiple variables (e.g., activity in multiple voxels). Examples include *decoding* models (Pereira et al., 2009; Haynes, 2015) that seek to determine whether class-specific information is present in the data; *encoding* models (Naselaris et al., 2011) that describe how information about stimuli are represented in brain activity; and *representational similarity analysis* (RSA)

(Kriegeskorte and Kievit, 2013) that measures the representational similarity of two stimuli based on the distance between their respective response patterns. Much of the important research in this area takes place in a gray zone between the disciplines of computer science and statistics. Clearly, many techniques with roots in statistics have played an important role in machine learning, including linear regression, penalized regression, logistic regression, and random forests (Hastie et al., 2009). However, that said, there are fundamental differences between how statisticians and computer scientists approach data; see, for example, Breiman (2001) for a detailed discussion.

Assumptions of statistical models

Every statistical model makes some form of assumption. For example, when using the GLM to construct brain maps, we rely on several statistical assumptions. Ultimately, any conclusions about which voxels are active are only as good as the validity of these assumptions. Standard assumptions include that the (i) variation across time unaccounted for by the model is independent from one another and come from a single population (i.e., independent, identically distributed or IID); (ii) errors are independent of the experimental effects; (iii) errors are normally distributed with equal variance across experimental conditions; (iv) the model captures experimental effects accurately (i.e., we have correctly specified the predictors); and (v) responses are linear with respect to the predictors and do not depend on the history of task demands or ongoing spontaneous processes. If these assumptions are valid, the map will likely provide an accurate picture of whether or not the identified brain regions contain at least some true effect. If not, then inferences may be imprecise and possibly even invalid. Luckily, linear models are relatively robust to violations of these assumptions providing some security against minor violations.

Sometimes, we relax these assumptions or make alternative ones. For example, instead of assuming the fMRI time series data are IID, it is common to instead use a time series model that allows one to model the autocorrelation between the error in adjacent time points (e.g., an AR(p) process) (Purdon et al., 2001; Lund et al., 2006). This necessitates making a stationarity assumption, which implies the mean, variance, and autocorrelation structure does not change over time. Further, if we use RFT to threshold the brain maps, we need to make additional assumptions (Worsley et al., 1996). These include that the spatial smoothness of the fMRI signal is constant over the brain and that the spatial autocorrelation function has a squared exponential shape. There have been some concerns about the validity of these assumptions (Eklund et al., 2016) which we discuss further below.

Bayesian versions of the GLM inherit all of the assumptions made in the frequentist model above, as those are assumptions about the likelihood function. In addition, we make additional assumptions about the distributions of any parameters that appear in the model. For example, we make distributional assumptions about the prior. In addition, if one chooses a prior with small variance, this will lead to a situation where the data (through the likelihood) have less impact on the resulting posterior. Thus, it is important to feel comfortable with the choice of priors when evaluating the results of a Bayesian model.

So far, the models we have discussed have been parametric, implying that they are based on parameterized families of probability distributions (e.g., the normally distributed model considered earlier). Nonparametric statistics is another class of methods where the data are not assumed to come from a prescribed model determined by a small number of parameters. Instead, they can be distribution-free or flexible in other ways. Nonparametric statistics are often used when the assumptions of parametric tests are violated. The control of multiple comparisons is one such example, where nonparametric permutation tests have proven to be an attractive alternative to parametric alternatives (e.g., RFT) (Nichols and Holmes, 2002; Nichols and Hayasaka, 2003; Eklund et al., 2016).

Building, testing, interpreting statistical models
Model building and estimation

When applying statistics to real-world problems, we need to separate between the *model* used to describe the data, the *method* of parameter estimation, and the *algorithm* used to obtain them. The model uses probability theory to describe the parameters of the distribution thought to be generating the data, the method defines the loss function that is minimized to find the unknown model parameters, and the algorithm defines the way the chosen loss function is minimized.

To date, most neuroimaging studies have made use of classical statistical methods, such as linear models and null-hypothesis testing (e.g., *t*-tests). Therefore, let us begin by focusing on the classical brain mapping approach and walk through a number of modeling choices. Consider, for example, fitting a GLM to an fMRI time series y from a single voxel. The activity in that voxel is modeled as the linear combination of a number of independent predictors related to task conditions and other nuisance covariates of no interest (e.g., head movement estimates and drift). The analysis includes, for each task condition or event type of interest, construction of a time series of the predicted shape of the signal response, which uses prior information about the shape of the vascular response to a brief impulse of neural activity. Most often, a canonical hemodynamic response function (HRF) is used for this purpose (Friston et al., 1998; Lindquist et al., 2009). Though we assume the predictors have known shapes, we need to estimate their unknown amplitudes. The noise portion of the model accounts for extra variation not explained by the predictors, and distributional assumptions are typically made about its form. Fig. 3A shows an example of a simple GLM with predictors corresponding to two different conditions.

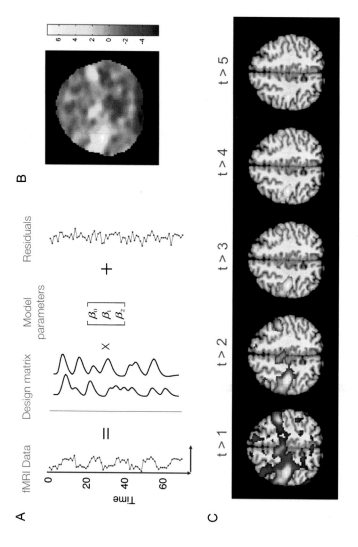

FIG. 3

Example of the GLM approach. (A) The fMRI data from a single voxel are the outcome. The design matrix has three columns, an intercept and two predictors, and three corresponding model parameters (beta values). (B) An example of a statistical image. Separate hypothesis tests are performed at each voxel of the brain to determine whether activation in that voxel is significantly different from zero. The results are summarized in an image showing the value of the test statistic at each voxel. (C) Examples of thresholded statistical images. The statistical image from panel B is thresholded using five different possible values, and voxels deemed significant are color-coded and superimposed onto an anatomical image. Clearly, the choice of threshold has a large impact on which voxels are deemed active, and subsequent interpretations.

The GLM can now be expressed as:

$$y = X\beta + \varepsilon, \varepsilon \sim N(0, I\sigma^2)$$

In this formulation, y is an $n \times 1$ vector containing the observed time series data observed at n time points, X is an $n \times k$ design matrix containing the predefined predictors (e.g., task-related signal and nuisance components), β is a $k \times 1$ vector of unknown parameters depicting the contribution of each predictor, and ε is an $n \times 1$ vector of unexplained error terms. The model describes how the data y depend on the unknown parameters β and σ^2. This rather simple model makes a number of assumptions, including that y is normally distribution with mean $X\beta$, and the elements of ε (i.e., the error at different time points) are IID with constant variance σ^2.

Next, we seek to determine the parameter estimates that "best" fit the observed data. This necessitates first defining what is meant by best, which is done through the choice of *loss function*. For example, when using the GLM, we often define best as the estimates of the unknown parameters that minimize the least-squares criteria:

$$f(\beta) = (y - X\beta)^T (y - X\beta).$$

In contrast, when using ridge regression, we define best as the parameter values that minimize the following penalized least-squares criteria:

$$f(\beta) = (y - X\beta)^T (y - X\beta) + \lambda \beta^T \beta$$

Note the additional penalty term ($\lambda \beta^T \beta$) shrinks the β value toward zero, and λ is a parameter that controls the amount of regularization. Another variant using $\lambda \sum_{i=1}^{p} |\beta_i|$ as the penalty term provides us with the least absolute shrinkage and selection operator (LASSO) solution.

For many models, minimizing the loss function necessitates using an algorithm. However, in the case of the model above, a closed-form result exists when minimizing the least-squares criteria, namely,

$$\widehat{\beta} = (X^T X)^{-1} X^T y$$

The parameter σ^2 is typically estimated as follows:

$$\widehat{\sigma} = \frac{1}{n - p} \left(y - X\widehat{\beta} \right)^T \left(y - X\widehat{\beta} \right).$$

Ridge regression also has a closed-form solution. In contrast, no such closed-form solution exists for the LASSO. Hence, an optimization algorithm such as coordinate descent is instead used to minimize the loss function in this setting (Hastie et al., 2009).

The estimate $\widehat{\beta}$ is often referred to as the ordinary least-squares (OLS) solution. This has the property of being the best linear unbiased estimate (BLUE) of β. This implies that in the class of all linear (in y) unbiased estimators, it has the smallest variance. Unbiasedness (i.e., the property that the expected value of $\widehat{\beta}$ in repeated samples equals β) and small variance are both desirable properties in an estimator. The sampling distribution for the OLS estimate is given by:

$$\widehat{\beta} = N \left(\beta, (X^T X)^{-1} \sigma^2 \right),$$

and this result is the basis for subsequent inference. It should be noted that it may be possible to find estimates with smaller variances by allowing for some bias, as when using penalized least-squares. This is referred to as a *bias-variance trade-off*.

A key assumption of the OLS approach is that the error terms are considered independent. In practice, this assumption is violated for fMRI data as it tend to exhibit autocorrelation. There are a number of problems that may arise if the autocorrelation is ignored. First, while the estimated regression coefficients will still be unbiased, it will no longer have minimum variance. This implies that it is no longer the BLUE, and we should be able to find a better unbiased estimator. Second, the estimates of σ^2 will be biased. Third, the variance of β will be underestimated and resulting t-statistics will be inflated. Fourth, tests using the t and F distributions may not be applicable. For these reasons, it is critical for the autocorrelation to be accounted for in order to conduct valid single-subject inference.

Thus, in many situations, we assume that $\varepsilon \sim N(0, V\sigma^2)$. Here, the term V models the correlation between the error at different time points. There are many ways of choosing the value of V, ranging from formats that make strong simplifying assumptions about the data's form, such as the one-parameter AR(1) model (Purdon et al., 2001), to forms which use many parameters (Woolrich et al., 2001). As with any model fitting procedure, a trade-off exists in the number of included parameters. Models with many parameters generally produce close fits to the observed data. However, models with few parameters, if chosen carefully, can produce more accurate estimates as they are less susceptible to random noise patterns present in the data (Olszowy et al., 2019). An important note when choosing V is that the time between adjacent time points (or TR) matters. Most early models were derived assuming a TR of 2 s or more. These models are not expected to carry over exactly when analyzing modern fMRI data with subsecond resolution (Bollmann et al., 2018).

When assuming $V \neq I$, the least squares solution is written as follows:

$$\hat{\beta} = \left(X^T V^{-1} X\right)^{-1} X^T V^{-1} y.$$

This is sometimes referred to as the *generalized* least-squares (GLS) solution. The GLS estimate is the BLUE in the setting where V is known. Here, the sampling distribution is given by:

$$\hat{\beta} = N\left(\beta, \left(X^T V^{-1} X\right)^{-1} \sigma^2\right).$$

As an alternative, we could take a Bayesian view and place a prior on the unknown model parameters. Consider the standard GLM, and for simplicity, assume that σ^2 is known. Further, assume the following prior $\beta \sim N(0, \tau^2 I)$ where τ is known. The posterior distribution will also follow a normal distribution as the normal prior is a conjugate prior for the normal likelihood. The mean of the posterior distribution will correspond to the ridge regression solution, illustrating the links between penalized regression and Bayes linear models. If instead we had used a Laplace prior, the mean would have been equivalent to the LASSO solution.

The GLM can be extended in various directions depending on the variance components included. The classical single-subject GLM includes a term modeling variation across time. Multilevel GLM exist that also includes a term used to model variation across subjects (Lindquist et al., 2012). Additional terms can be included to model variation across space (spatial GLM) (Penny et al., 2005; Mejia et al., 2020), variation across sessions (longitudinal models) (Guillaume et al., 2014), and variation across sites (multisite GLM); see Fig. 2 for example of some of these sources of variation.

Statistical testing

Once we have estimated the parameters of interest, focus shifts from estimation to performing statistical testing. This can be performed by computing a *test statistic* which is often a measure of the magnitude of the effect, divided by the standard deviation of the sampling distribution for the effect. The latter quantifies how much the observed effect is expected to vary from sample to sample. This is the basis for the well-known t-statistic used for statistical inference:

$$t = \frac{\widehat{\beta} - \beta}{SE\left(\widehat{\beta}\right)}.$$

To infer whether a voxel shows an effect a hypothesis test is performed. The *null hypothesis* is that the effect is zero (i.e., $\beta = 0$). Correspondingly, the alternative hypothesis could be that it is greater than (or different from) zero (i.e., $\beta > 0$). The problem is now to determine whether there is significant information in the data to reject the null hypothesis in favor of the alternative hypothesis. This will be true if the test statistic is inconsistent with what we would observe under the null, which can be measured using the *P*-value (small *P*-values correspond to inconsistent behavior).

An important statistical concern when analyzing neuroimaging data is that we perform a large number of tests, often one for each voxel in the brain. Hence, we are faced with a large multiple testing problem. The need for appropriate multiple comparisons correction when performing statistical inference is not a new problem. However, it has come to the forefront in many new modern data-intensive disciplines. Researchers in areas such as imaging and genetics are routinely required to simultaneously perform thousands of statistical tests. Ignoring this multiplicity leads to problems with false positives, thereby introducing nonreproducible results into the literature (Lindquist and Mejia, 2015). Fig. 3B shows an example of an unthresholded statistical map, while Fig. 3C shows the thresholded maps at a variety of different values.

A standard approach for handling multiple comparisons is to simultaneously control the probability of obtaining a false positive for every statistical test (i.e., in every voxel) in the brain by adjusting the threshold for significance. In neuroimaging, researchers have used several different approaches for controlling the false-positive rate. Here, one typically differentiates between methods that control the family-wise error rate (FWER) (Nichols and Hayasaka, 2003), which is the probability of obtaining any false positives in the brain, or for the false discovery rate (FDR) (Benjamini and Hochberg, 1995; Genovese et al., 2002), which is the proportion of false positives among all rejected tests. Methods that control the FWER have long been the standard in neuroimaging research and include Bonferroni correction and RFT. A shortcoming of these methods is that they are known to be conservative (i.e., the probability of getting a false positive is less than expected) (Nichols and Hayasaka, 2003). In contrast, methods that control the FDR (e.g., the Benjamini–Hochberg method (Benjamini and Hochberg, 1995)) tend to be more lenient (Genovese et al., 2002). Whether to choose an FWER or FDR-controlling procedure ultimately depends on the problem. In certain cases, obtaining a false positive may be considered a more serious error than a false negative. In others, it may be the opposite. If the "cost" of obtaining a single false positive is high, then FWER controlling procedures provide the best protection and should be used. If instead one were willing to accept a certain number of false positives to guard against excessive false negatives, then FDR-controlling procedures are more appropriate.

The earlier description relates to voxel-level corrections for multiple comparisons. However, the most popular approach for dealing with multiple comparisons is to use the so-called cluster-extent based thresholding (Woo et al., 2014; Carp, 2012b). This procedure consists of two steps (Nichols and Hayasaka, 2003). First, an arbitrary voxel-level *primary threshold* is used to define clusters of suprathreshold voxels. Second, a cluster-level *extent threshold*, measured in units of contiguous voxels, is determined based on the estimated distribution of cluster sizes under the null hypothesis of no activation in any voxel in that cluster. The cluster-level extent threshold that controls FWER can be obtained from the sampling distribution of the largest null hypothesis cluster size among suprathreshold voxels within the brain, which can be estimated using RFT (Worsley et al., 1996; Worsley, 1996) or nonparametric methods (Nichols and Holmes, 2002). Much of its popularity is due to its high sensitivity to weak and diffuse signals (Smith and Nichols, 2009). In contrast, voxel-level corrections for multiple comparisons are so stringent that they can dramatically increase the risk for false negatives without extremely large sample sizes. However, cluster-level correction provides low spatial specificity when clusters are large, as researchers can only infer there is signal *somewhere* within a significant cluster and cannot make inferences about the statistical significance of specific locations within the cluster. This can be problematic if a liberal cluster-defining primary threshold is used, as this can create large clusters spanning several anatomical regions (Woo et al., 2014).

When using standard statistics (e.g., *t*-statistics) and comparing them with their assumed distributions, we use parametric statistics. When we use the data itself to compute *P*-values, we use nonparametric statistics. This generally involves making fewer statistical assumptions. Permutation tests and the bootstrap are popular ways of using the data, rather than distributions based on assumed normality, to estimate variability (Nichols and Holmes, 2002; DiCiccio and Efron, 1996). However, in practice, parametric methods can be quite robust and practical.

Instead of performing hundreds of thousands of tests and correcting for multiple comparisons, one could instead imagine using a multilevel approach with voxels grouped within regions of interest. If properly constructed, this type of model could circumvent many of these issues, at the cost of increased computational demands. It is often argued that performing analysis in the Bayesian framework allows one to circumvent the multiple comparisons problem. While a Bayesian model that models all the parameters of interest simultaneously (as described previously) would be a step in the right direction, simply fitting a voxel-wise multiple regression in a Bayesian framework does not necessarily alleviate these concerns. Simply switching between statistical paradigms, but still performing the same procedure (i.e., thresholding), does not properly address these issues. Ultimately, the amount of protection provided by Bayesian methods depends strongly on the choice of prior distribution, and exactly quantifying the chance of obtaining false positives can be difficult in practice (Lindquist et al., 2013).

Creating brain maps

Brain maps are constructed depicting the results of hypothesis tests across brain voxels, color coded by the strength of the evidence that there is a nonzero response to a task or a nonzero correlation between brain activity and some other external condition or outcome. If data from a single individual are used to construct the brain maps, they are referred to as *single-subject* maps. These are common in vision science and increasingly used in clinical (e.g., surgery preparation) and legal applications. These maps are generally constructed by comparing data from one condition (e.g., an experimental task) with another across repeated measurements and testing for statistical significance in each voxel. However, single-subject maps cannot be used to make *population inferences*, which are claims about how the brain functions across individuals in a population. To do this, we must take into consideration the variation across subjects. If data from a group of participants are used to construct the brain maps, they are referred to as *group-level maps*. These maps allow us to conduct statistical tests to evaluate how well the observed effects generalize to the broader population of participants. These maps are constructed by comparing experimental conditions of interest for each individual and then combined in a group analysis that considers the variability across individuals. To perform these types of analysis, it is important to include variance components that model the way participants are drawn from the population (Lindquist et al., 2012) (i.e., both model variation *within* and *across* participants). In the neuroimaging literature, this has generally been referred to as random effects models, in contrast to fixed effects models that don't include this variance components and only generalize to the participants of the study (Penny and Holmes, 2007; Mumford and Nichols, 2009).

Multilevel models have been a mainstay in the social, behavioral, and agricultural sciences for several decades and are steadily gaining in popularity in the neuroimaging community. As in single-subject analysis, group analysis is typically performed using the massive univariate approach with a single voxel analyzed at a time. However, this is more complicated in the group setting because to effectively compare data for a specific voxel across subjects, it is critical that all subjects are normalized to a stereotaxic space prior to comparison. Thus, we assume throughout that the voxels have been aligned and are comparable across subjects (an assumption that is violated in practice). Multisubject fMRI data are hierarchical in nature, with lower-level observations nested within higher levels (i.e., subjects nested within groups). Taking the multilevel nature of the data (typically described using a two-level hierarchical model) into consideration can provide the means to generate population level statistics. In the *first level*, we analyze within-subject effects for each individual in the study. In the *second level*, we perform analyses across subjects or alternatively across groups of subjects (e.g., patient vs control). Treating the variation across participants as an error term in a group statistical analysis allows the results to be generalized to new participants drawn from the same population. Researchers typically perform this analysis either in stages or combined into a single integrated model (Mumford and Poldrack, 2007). Performing a one-sample *t*-test on first-level effects is a simplified version of a full mixed-effects analysis. This is valid if the standard errors of first-level estimates are equal for all subjects. This is rarely true in practice, which has led to the development and increased popularity of full mixed-effects models. Though more computationally expensive, such models incorporate both first-level and second-level effects into a single unified model. This relaxes the assumption of homogenous standard errors by appropriately weighting, in proportion to their precision, the contribution of each subject's estimates to the group estimates. This is based on the idea that the larger a subject's standard error, the less reliable the estimate is, so the less that subject should contribute to the group results.

Pitfalls in statistical modeling

In the past couple of years, there have been several debates in the neuroimaging literature related to the appropriate use of statistics. In the following sections, we outline a few of these, as they illustrate potential pitfalls that can arise when applying statistics to neuroimaging data.

Multiplicity issues

An issue that can always arise is, as described earlier, the violation of key modeling assumptions. In general, techniques such as the GLM approach are fairly robust to such violations, but in practice, investigations of assumptions are rarely performed (though exceptions exist (Luo and Nichols, 2003; Loh et al., 2008)), and most standard statistical methods have not been validated using real data. One major violation of the assumptions in a commonly used method was discovered in work by Eklund and colleagues (Eklund et al., 2016) where they showed some of the key assumptions underlying RFT were violated in practice.

In their work, they used resting-state fMRI data from a number of healthy controls to conduct millions of "task-based" group analyses. As the resting-state data was not actually collected while performing the task in question, this could be considered as null data. Using this approach under different experimental designs, the authors estimated the false-positive rate while controlling for multiple testing using standard approaches. While theoretically they should have found 5% false positives, they instead found that the standard fMRI software packages had false-positive rates

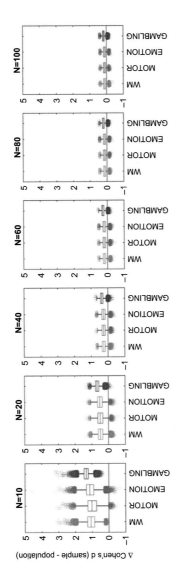

FIG. 4

Effect size bias. For each sample of size N (10, 20, 40, 60, 80, 100), the empirical effect size was computed based on the average within each significant cluster. The population effect size for that cluster was then subtracted from this value. Boxplots show the distributions across all clusters from all 100 samples for each task and sample size for four tasks from the HCP dataset. The results illustrate the effects of circular analysis on effect size estimation.

*Results taken from Geuter, S., Qi, G., Welsh, R.C., Wager, T.D., Lindquist, M.A., 2018. Effect size and power in fMRI group analysis. bioRxiv 295048. https:/doi.org/10.110**/295043.*

of up to 70%. One reason for this was the violation of certain model assumptions that underlie RFT, including that the spatial smoothness of the fMRI signal is constant over the brain and that the spatial autocorrelation function has a squared exponential shape. In contrast, nonparametric permutation tests, which did not make such assumptions, were found to perform much better.

In addition, the use of cluster-extent based thresholding has been criticized, as it can lead to problems with low spatial specificity when clusters are large. To circumvent some of these shortcomings, Woo and colleagues (Woo et al., 2014) provide recommendations to avoid these pitfalls, including using a primary threshold of $P < .001$, using more stringent primary thresholds or voxel-wise correction methods for well-powered studies, and making the level of spatial precision more transparent.

Voodoo-correlations and circularity

As an alternative toward performing whole-brain analyses, and dealing with multiplicity issues, one could imagine taking a more targeted approach toward testing. In many cases, the scientific question of interest dictates that a certain subset of the tests performed is particularly important. For example, we may be interested in focusing our analysis on certain prespecified regions of the brain. By focusing entirely on voxels that lie within these regions, one can substantially reduce the burden of multiple comparison correction. However, it is critical that the voxels to be tested be specified before actually viewing the data. Otherwise, one risks "data dredging" and uncovering biased results that will not be reproducible (Kriegeskorte et al., 2009, 2010).

This was one of the key issues behind the so-called voodoo correlations debate. In a hotly debated article, Vul and colleagues (Vul et al., 2009) pointed out that the correlations reported in many fMRI studies were commonly overstated because researchers tend to report only the highest correlations, or only those correlations that exceed some threshold. They suggested that these statistical problems are leading researchers to overstate links between behavioral variables and brain data. In particular, they argued against the practice of using a two-stage analysis procedure where the method used to select which voxels should be further investigated is not independent of the analysis performed on the resulting selected regions.

It is clearly unwise to perform a two-stage analysis that tests the significance of nonzero correlation on voxels that were chosen simply due to the fact that they exhibited high correlation in the data. However, there was significant debate about how common this type of analysis was in practice (Lieberman et al., 2009). Instead, it was argued that most studies first conduct a test of significance and thereafter simply report an aggregate correlation value for each region deemed significant in the first test. With proper control for multiple comparisons, this second procedure would not change the fact that certain voxels exhibited significant nonzero correlation in the hypothesis testing framework; however, the reported correlation will still be radically inflated (Lindquist and Gelman, 2009); see Fig. 4 for an illustration. So, while this solution solves one of the problems raised, it retains the problem of biased correlations. In general, it is unlikely that true correlations between brain activity and measures of emotion, personality, and social cognition are as high as those reported in the papers surveyed in Vul et al., which in many cases were above 0.90. Indeed, a recent study on data with large sample sizes (see below) has estimated correlations between brain function measured using task-based fMRI and complex cognitive or mental health phenotypes closer to 0.1–0.2 (Marek et al., 2022).

Too few or too many subjects?

Another topic that caused significant discussion in the field is the sample size. For example, Button and colleagues stated that studies with small sample sizes undermine the reliability of neuroscience research (Button et al., 2013; Munafò et al., 2014). The argument is that small-sample studies both have low statistical power to detect true effects and a reduced likelihood for statistically significant results to actually be true effects. Despite this growing body of evidence, most functional magnetic resonance imaging (fMRI) studies continue to be small (Poldrack et al., 2016), presumably due to the excessive costs of imaging. This is problematic, as this leads to a situation where a large number of published results are likely not replicable and perhaps even false.

In an alternative take, Friston raised concerns with increasing sample sizes, stating that a significant result obtained with a small sample size is "stronger" than if it had been obtained with a larger sample size, because small sample tests cannot detect trivial or uninteresting effects (Friston, 2012). This assertion was based on a phenomenon that he calls the "fallacy of classical inference," which is related to the fact that the null hypothesis is generally false in a strict sense (i.e., the effect is never *exactly* zero). As a consequence, the null can always be rejected with sufficient sample size. In other words, if you have enough statistical power even very small effects can be detected. While it is clear that researchers should be aware of these issues when performing null hypothesis testing, it is hard to advocate for smaller sample sizes, as this leads to less power to detect effects, less accurate parameter estimates, and wider confidence intervals (Lindquist et al., 2013). Statistical tests are used to determine whether a parameter is significantly different from some prespecified value. As the sample size increases, it becomes easier to detect such differences. If researchers find this problematic, they should ask more appropriate questions instead of reducing the sample size. After all, performing hypothesis tests is not the only type of statistical inference (Lindquist and Gelman, 2009).

If the experimental design is good and the data follow the specified model, then a statistically significant effect found in a study with a small sample size is important and likely to remain significant even as the sample size increases. However, one problem with undersampled studies is that there is limited opportunity for exploring confounding variables (e.g., gender, age). This is a variable that is related to both the independent and dependent variables in the model. For example, age could be related to both the image data and disease status. Problems can occur when the distribution of the confounding variable differs between the groups being compared. It is well known in the epidemiology literature that many large signals disappear or are attenuated when explored in a larger follow-up study where appropriate confounder correction was introduced. For these reasons, having larger sample sizes is preferred. The commonly made arguments against larger sample sizes typically regard diminished statistical returns in the form of power when factoring in important nonstatistical considerations (e.g., cost).

Recent work by Marek and colleagues (Marek et al., 2022) explored the relationship between the sample size and the effect size in detail. They find that as sample sizes grew into the thousands, replication rates began to improve and effect size inflation decreased. The largest replicable brain-wide associations for univariate and multivariate methods were $r = 0.14$ and $r = 0.34$, respectively. A take-home message is that when analyzing large datasets, researchers should expect smaller, but more reliable, associations. Most real-world associations are probably small; this doesn't necessarily make them uninteresting.

It should be noted that there are alternative ways to obtain robust and reproducible findings without the need for thousands of participants. For example, precision mapping studies, where extended data acquisition (often more than 10 h of functional data) from a few participants, allow for reliable individualized mapping of brain organization. These studies have generated individual-specific functional connectomes that have revealed important new information about spatial and organizational variability in brain networks (Laumann et al., 2015; Gordon et al., 2017).

Pipeline variability

The choice of pipeline can have a significant impact on the outcome of an fMRI study. As an illustration, Carp (2012b) applied 6912 unique analysis pipelines to the same dataset. The main finding was that while many pipelines gave rise to activation in the same general regions, peak locations varied widely as a function of analysis pipeline. Botvinik-Nezer and colleagues (Botvinik-Nezer et al., 2020) explored this issue further by providing the same dataset for independent analysis by 70 different independent research teams. This allowed for the evaluation of the actual impact of analytic flexibility on fMRI results "in the wild," illustrating that significant variation in the results of fMRI analyses are difficult to avoid in practice. An even more subtle statistical question is whether or not analyses using different pipelines are even testing the same question, as including additional covariates changes the interpretation of the estimated model parameters (Lindquist, 2020).

While this has a significant impact on the standard practice of thresholding statistical maps to make binary decisions regarding significance, it is encouraging that the underlying spatial patterns appear to be stable across pipelines. This points toward problems related to using high thresholds within a null hypothesis significance testing framework in settings where tests across many variables are performed (Lindquist, 2020). Increased sample sizes will help but are not always feasible. An alternative solution would be to use multivariate pattern-based approaches that directly utilize the spatial patterns to analyze the data rather than continuing to focus on tests of individual voxels, and a de-emphasis of testing in favor of effect size estimation and building computational models. For example, the results of Marek and colleagues (Marek et al., 2022) indicate that the largest replicable brain-wide associations were twice as high for multivariate methods compared to univariate methods.

Open challenges and future directions

The data collected in neuroimaging studies are constantly evolving and with them new statistical models will be needed. We anticipate an increase in longitudinal studies, multisite studies, large n studies, small precision mapping studies with several hours of data on a few participants, translational neuroimaging, and brain stimulation. In the following sections, we discuss a few of these topics.

Large n studies

We anticipate that a lot of research will be driven by the substantial increase in the availability of large diverse lifespan datasets consisting of more than 1000 subjects (Van Essen et al., 2013; Miller et al., 2016). With this increase, there is a need for principled statistical methods to analyze the resulting data and identify meaningful effects. Some commonly used statistical methods will carry over to this new paradigm, while others will need refinement. Uncovering reproducible associations between brain structure/function and behavioral phenotypes is difficult using standard sample sizes. As discussed earlier, as the sample size increases, even small effects are well powered and easier to detect. Detecting which of these significant effects are important will require more domain knowledge. In the future, we anticipate that alternative approaches (e.g., Bayesian methods and multivariate predictive models) and metrics (e.g., effect sizes) will find more widespread use.

There are multiple statistical benefits associated with large datasets. First, they may provide sufficient power to investigate higher-order interactions and perform data fusion across modalities (e.g., imaging and genetics). Second, they may provide the diverse demographics needed to estimate flexible models and study covariation with demographic factors with sufficient precision. Third, they allow researchers to perform analysis on subsets of data with a particular set of characteristics. Fourth, they can be used to validate results from smaller studies. Fifth, they can be used to determine priors for the analysis of smaller-sample studies. Sixth, they can be used to test the reproducibility of new methods in smaller-sample studies. Large-scale datasets allow for population inference. However, there is a need for scalable statistical inference to uncover important patterns in the data.

Translational neuroscience

Another area we expect to be important is *translational neuroscience*, which lies in the cross section of basic neuroscience and clinical applications. The arrival of functional neuroimaging generated optimism that it would both improve our understanding of the mind and deliver clinically useful tools. To date, it has not had as large of an impact on clinical practice and public health as hoped. Most early translational neuroimaging efforts used the brain mapping approach, as this was consistent with lesion studies and theories of modularity. Here, the goal is to understand what functions and processes are encoded in an isolated, target brain region of interest. These studies helped identify brain features (i.e., measures of activity in specific brain regions) that are predictive of health-related outcomes. However, clinical trials targeting these features have generally failed (Dougherty et al., 2015; Morishita et al., 2014).

One issue for these failures is that the brain mapping approach was never explicitly designed with translational goals in mind. The main goal is to test hypotheses about structure–function associations, i.e., assess if there is a nonzero effect in a voxel. They are designed for inference that brain region **B** is active conditional on stimulus (or symptom) **S** (i.e., $P(\mathbf{B}|\mathbf{S})$), and do not allow for "reverse inference" (i.e., $P(\mathbf{S}|\mathbf{B})$). In addition, they are focused on isolated brain regions, while many features of disorders are likely encoded in distributed neural systems involving networks of many regions. To overcome these shortcomings, researchers have increasingly begun to investigate alternative analysis strategies. In particular, the use of machine learning (or predictive modeling) has shown a great deal of promise. Here, the direction of inference is reversed relative to conventional brain mapping. Brain features comprise a set of predictors, and behavioral or clinical variables comprise one or more outcomes. Thus, the models integrate all available brain data into a single prediction about the outcome. This is important as this can provide focused tests that avoid multiple comparisons and increase statistical power. Instead, diagnostic value is assessed by evaluating their performance in new, out-of-sample individuals. The increased use of predictive models gives valid estimates of effect size and clinical significance, opening the door for the development of clinical biomarkers (Woo et al., 2017).

Causal inference and brain stimulation

Neuroscientists have become increasingly interested in studying effective connectivity (Friston, 1994), that is, the direct influence of one brain region on others. While researchers using the methods developed for this type of analysis often interpret their findings as indicative of causation, they generally do not provide a statement of what is meant by this term. In addition, they do not provide explicit definitions of the effects they purport to be estimating or statement of the conditions under which their interpretations are valid.

There is an increasingly active subfield in statistics referred to as *causal inference* that seeks to address these issues. In particular, the *potential outcomes* framework

(Rubin, 1974) has been widely used to define causal relationships using counterfactual conditional statements. This notation allows researchers to make explicit the assumptions involved when various experimental designs, imaging modalities, and modeling procedures are used to make causal inferences (Lindquist and Sobel, 2011, 2016; Sobel and Lindquist, 2014). These new statistical methods have the potential to make important contributions to the further development of neuroscience by determining the assumptions needed to make causal conclusions using neuroimaging data.

In the future, the combination of fMRI and transcranial magnetic stimulation (TMS) promises to integrate the ability of neuroimaging to observe brain activity with the ability of TMS to manipulate brain function (Siebner et al., 2009). Using this technique, one can infer the effect of brain activity on task performance by comparing subjects randomized to receive/not receive stimulation. Ultimately, we believe that techniques such as TMS, which allow for direct manipulation of intermediate outcomes, provide the most promising approach for studying effective connectivity. That is because such procedures rely on research designs with known properties in lieu of assumptions that are neither testable nor known to be correct.

Take-home points

- Statistics is the study of *uncertainty* and *variability*. It allows us to harness and better understand the variability that is inherent in neuroimaging data.
- Statistical modeling has played an important role in neuroimaging research. It helps produce reliable results, provides tools to draw conclusions about populations of participants, and allows one to make important clinical decisions.
- Statistical models make assumptions that it is important to understand and determine whether or not they are valid. Ultimately, any conclusions are only as good as the validity of these assumptions, and any violations can potentially lead to spurious findings.
- When developing statistical models, it is important to include all sources of variation. This allows us to compare data across space and time within an individual, compare different individuals from a population, compare populations, compare longitudinal measures from the same individual, and compare different individuals whose data come from different sites acquired using different collection techniques.
- In the past couple of years, there have been a number of debates in the neuroimaging literature related to the appropriate use of statistics. These include issues related to multiple comparisons correction and sample size, as well as controversies related to circularity and pipeline variability.
- The data collected in neuroimaging studies are constantly evolving, and this requires new statistical models. We anticipate an increase in longitudinal studies, multisite studies, large *n* studies, precision mapping studies, translational neuroimaging, and brain stimulation.

References

Alfaro-Almagro, F., et al., 2021. Confound modelling in UK Biobank brain imaging. Neuroimage 224, 117002.

Benjamini, Y., Hochberg, Y., 1995. Controlling the false discovery rate: a practical and powerful approach to multiple testing. J. R. Stat. Soc. Series B Stat. Methodology 57, 289–300.

Biswal, B., Yetkin, F.Z., Haughton, V.M., Hyde, J.S., 1995. Functional connectivity in the motor cortex of resting human brain using echo-planar MRI. Magn. Reson. Med. 34, 537–541.

Bollmann, S., Puckett, A.M., Cunnington, R., Barth, M., 2018. Serial correlations in single-subject fMRI with sub-second TR. Neuroimage 166, 152–166.

Botvinik-Nezer, R., et al., 2020. Variability in the analysis of a single neuroimaging dataset by many teams. Nature 582, 84–88.

Breiman, L., 2001. Statistical modeling: the two cultures (with comments and a rejoinder by the author). SSO Schweiz. Monatsschr. Zahnheilkd. 16, 199–231.

Brett, M., Penny, W., Kiebel, S., 2003. An Introduction to Random Field Theory. http:/cda. psych.uiuc.edu/web_407_spring_2014/random_field_theory.pdf.

Button, K.S., et al., 2013. Power failure: why small sample size undermines the reliability of neuroscience. Nat. Rev. Neurosci. 14, 365–376.

Carp, J., 2012a. On the plurality of (methodological) worlds: estimating the analytic flexibility of FMRI experiments. Front. Neurosci. 6, 149.

Carp, J., 2012b. The secret lives of experiments: methods reporting in the fMRI literature. Neuroimage 63, 289–300.

Casella, G., Berger, R.L., 2021. Statistical Inference. Cengage Learning.

Ciric, R., et al., 2017. Benchmarking of participant-level confound regression strategies for the control of motion artifact in studies of functional connectivity. Neuroimage 154, 174–187.

Cole, J.H., et al., 2018. Brain age predicts mortality. Mol. Psychiatry 23, 1385–1392.

David, O., et al., 2008. Identifying neural drivers with functional MRI: an electrophysiological validation. PLoS Biol. 6, 2683–2697.

DiCiccio, T.J., Efron, B., 1996. Bootstrap confidence intervals. SSO Schweiz. Monatsschr. Zahnheilkd. 11, 189–228.

Dougherty, D.D., et al., 2015. A randomized sham-controlled trial of deep brain stimulation of the ventral capsule/ventral striatum for chronic treatment-resistant depression. Biol. Psychiatry 78, 240–248.

Eklund, A., Nichols, T.E., Knutsson, H., 2016. Cluster failure: why fMRI inferences for spatial extent have inflated false-positive rates. Proc. Natl. Acad. Sci. U. S. A. 113, 7900–7905.

Fortin, J.-P., et al., 2018. Harmonization of cortical thickness measurements across scanners and sites. Neuroimage 167, 104–120.

Fox, C.W., Roberts, S.J., 2012. A tutorial on variational Bayesian inference. Artif. Intell. Rev. 38, 85–95.

Friston, K.J., 1994. Functional and effective connectivity in neuroimaging: a synthesis. Hum. Brain Mapp. 2, 56–78.

Friston, K., 2012. Ten ironic rules for non-statistical reviewers. Neuroimage 61, 1300–1310.

Friston, K.J., et al., 1994. Statistical parametric maps in functional imaging: a general linear approach. Hum. Brain Mapp. 2, 189–210.

Friston, K.J., et al., 1995. Analysis of fMRI time-series revisited. Neuroimage 2, 45–53.

Friston, K.J., et al., 1998. Event-related fMRI: characterizing differential responses. Neuroimage 7, 30–40.

Friston, K.J., Harrison, L., Penny, W., 2003. Dynamic causal modelling. Neuroimage 19, 1273–1302.

Genovese, C.R., Lazar, N.A., Nichols, T., 2002. Thresholding of statistical maps in functional neuroimaging using the false discovery rate. Neuroimage 15, 870–878.

Giedd, J.N., et al., 1999. Brain development during childhood and adolescence: a longitudinal MRI study. Nat. Neurosci. 2, 861–863.

Gordon, E.M., et al., 2017. Precision functional mapping of individual human brains. Neuron 95, 791–807. e7.

Guillaume, B., et al., 2014. Fast and accurate modelling of longitudinal and repeated measures neuroimaging data. Neuroimage 94, 287–302.

Gur, R.E., Gur, R.C., 2010. Functional magnetic resonance imaging in schizophrenia. Dialogues Clin. Neurosci. 12, 333–343.

Hastie, T., Friedman, J., Tibshirani, R., 2009. The Elements of Statistical Learning. Springer New York.

Haxby, J.V., et al., 2001. Distributed and overlapping representations of faces and objects in ventral temporal cortex. Science 293, 2425–2430.

Haynes, J.-D., 2015. A primer on pattern-based approaches to fMRI: principles, pitfalls, and perspectives. Neuron 87, 257–270.

Hoff, P.D., 2009. A First Course in Bayesian Statistical Methods. Springer New York.

Johnstone, T., et al., 2006. Motion correction and the use of motion covariates in multiple-subject fMRI analysis. Hum. Brain Mapp. 27, 779–788.

Kriegeskorte, N., Kievit, R.A., 2013. Representational geometry: integrating cognition, computation, and the brain. Trends Cogn. Sci. 17, 401–412.

Kriegeskorte, N., Simmons, W.K., Bellgowan, P.S.F., Baker, C.I., 2009. Circular analysis in systems neuroscience: the dangers of double dipping. Nat. Neurosci. 12, 535–540.

Kriegeskorte, N., Lindquist, M.A., Nichols, T.E., Poldrack, R.A., Vul, E., 2010. Everything you never wanted to know about circular analysis, but were afraid to ask. J. Cereb. Blood Flow Metab. 30, 1551–1557.

Laumann, T.O., et al., 2015. Functional system and areal organization of a highly sampled individual human brain. Neuron 87, 657–670.

Lieberman, M.D., Berkman, E.T., Wager, T.D., 2009. Correlations in social neuroscience aren't voodoo: commentary on Vul et al. (2009). Perspect. Psychol. Sci. **4**, 299–307.

Lindquist, M.A., 2008. The statistical analysis of fMRI data. SSO Schweiz. Monatsschr. Zahnheilkd. 23, 439–464.

Lindquist, M., 2020. Neuroimaging results altered by varying analysis pipelines. Nature 582, 36–37.

Lindquist, M.A., Gelman, A., 2009. Correlations and multiple comparisons in functional imaging: a statistical perspective (commentary on Vul et al., 2009). Perspect. Psychol. Sci. 4, 310–313.

Lindquist, M.A., Mejia, A., 2015. Zen and the art of multiple comparisons. Psychosom. Med. 77, 114–125.

Lindquist, M.A., Sobel, M.E., 2011. Graphical models, potential outcomes and causal inference: comment on Ramsey, Spirtes and Glymour. Neuroimage 57, 334–336.

Lindquist, M.A., Sobel, M.E., 2016. Effective connectivity and causal inference in neuroimaging. In: Handbook of Neuroimaging Data Analysis, p. 419.

Lindquist, M.A., Meng Loh, J., Atlas, L.Y., Wager, T.D., 2009. Modeling the hemodynamic response function in fMRI: efficiency, bias and mis-modeling. Neuroimage 45, S187–S198.

Lindquist, M.A., Spicer, J., Asllani, I., Wager, T.D., 2012. Estimating and testing variance components in a multi-level GLM. Neuroimage 59, 490–501.

Lindquist, M.A., Caffo, B., Crainiceanu, C., 2013. Ironing out the statistical wrinkles in 'ten ironic rules'. Neuroimage 81, 499–502.

Loh, J.M., Lindquist, M.A., Wager, T.D., 2008. Residual analysis for detecting mis-modeling in fMRI. Stat. Sin. 18, 1421–1448.

Lund, T.E., Madsen, K.H., Sidaros, K., Luo, W.-L., Nichols, T.E., 2006. Non-white noise in fMRI: does modelling have an impact? Neuroimage 29, 54–66.

Luo, W.-L., Nichols, T.E., 2003. Diagnosis and exploration of massively univariate neuroimaging models. Neuroimage 19, 1014–1032.

Marek, S., et al., 2022. Publisher correction: reproducible brain-wide association studies require thousands of individuals. Nature 605, E11.

Marrelec, G., et al., 2006. Partial correlation for functional brain interactivity investigation in functional MRI. Neuroimage 32, 228–237.

Mejia, A.F., Yue, Y.R., Bolin, D., Lindgren, F., Lindquist, M.A., 2020. A Bayesian general linear modeling approach to cortical surface fMRI data analysis. J. Am. Stat. Assoc. 115, 501–520.

Miller, K.L., et al., 2016. Multimodal population brain imaging in the UK Biobank prospective epidemiological study. Nat. Neurosci. 19, 1523–1536.

Morishita, T., Fayad, S.M., Higuchi, M.-A., Nestor, K.A., Foote, K.D., 2014. Deep brain stimulation for treatment-resistant depression: systematic review of clinical outcomes. Neurotherapeutics 11, 475–484.

Mumford, J.A., Nichols, T., 2009. Simple group fMRI modeling and inference. Neuroimage 47, 1469–1475.

Mumford, J.A., Poldrack, R.A., 2007. Modeling group fMRI data. Soc. Cogn. Affect. Neurosci. 2, 251–257.

Munafò, M., et al., 2014. Scientific rigor and the art of motorcycle maintenance. Nat. Biotechnol. 32, 871–873.

Naselaris, T., Kay, K.N., Nishimoto, S., Gallant, J.L., 2011. Encoding and decoding in fMRI. Neuroimage 56, 400–410.

Nichols, T., Hayasaka, S., 2003. Controlling the familywise error rate in functional neuroimaging: a comparative review. Stat. Methods Med. Res. 12, 419–446.

Nichols, T.E., Holmes, A.P., 2002. Nonparametric permutation tests for functional neuroimaging: a primer with examples. Hum. Brain Mapp. 15, 1–25.

Norman, K.A., Polyn, S.M., Detre, G.J., Haxby, J.V., 2006. Beyond mind-reading: multi-voxel pattern analysis of fMRI data. Trends Cogn. Sci. 10, 424–430.

Olszowy, W., Aston, J., Rua, C., Williams, G.B., 2019. Accurate autocorrelation modeling substantially improves fMRI reliability. Nat. Commun. 10, 1220.

Ombao, H., Lindquist, M., Thompson, W., Aston, J., 2016. Handbook of Neuroimaging Data Analysis. CRC Press.

Penny, W., Holmes, A., 2007. Random effects analysis. In: Statistical Parametric Mapping: The Analysis of Functional Brain Images, pp. 156–165.

Penny, W.D., Trujillo-Barreto, N.J., Friston, K.J., 2005. Bayesian fMRI time series analysis with spatial priors. Neuroimage 24, 350–362.

Pereira, F., Mitchell, T., Botvinick, M., 2009. Machine learning classifiers and fMRI: a tutorial overview. Neuroimage 45, S199–S209.

Poldrack, R.A., 2018. The New Mind Readers: What Neuroimaging Can and Cannot Reveal About Our Thoughts. Princeton University Press.

Poldrack, R., et al., 2016. Scanning the Horizon: challenges and solutions for neuroimaging research. bioRxiv. https://doi.org/10.1101/059188.

Purdon, P.L., Solo, V., Weisskoff, R.M., Brown, E.N., 2001. Locally regularized spatiotemporal modeling and model comparison for functional MRI. Neuroimage 14, 912–923.

Ramsey, J.D., et al., 2010. Six problems for causal inference from fMRI. Neuroimage 49, 1545–1558.

Rissman, J., Gazzaley, A., D'Esposito, M., 2004. Measuring functional connectivity during distinct stages of a cognitive task. Neuroimage 23, 752–763.

Robert, C.P., Casella, G., 2004. Monte Carlo Statistical Methods. Springer New York.

Rubin, D.B., 1974. Estimating causal effects of treatments in randomized and nonrandomized studies. J. Educ. Psychol. 66, 688.

Rubinov, M., Sporns, O., 2010. Complex network measures of brain connectivity: uses and interpretations. Neuroimage 52, 1059–1069.

Shumway, R.H., Stoffer, D.S., 2017. Time Series Analysis and Its Applications. Springer International Publishing.

Siebner, H.R., et al., 2009. Consensus paper: combining transcranial stimulation with neuroimaging. Brain Stimul. 2, 58–80.

Silva, M.A., See, A.P., Essayed, W.I., Golby, A.J., Tie, Y., 2018. Challenges and techniques for presurgical brain mapping with functional MRI. Neuroimage Clin. 17, 794–803.

Smith, S.M., Nichols, T.E., 2009. Threshold-free cluster enhancement: addressing problems of smoothing, threshold dependence and localisation in cluster inference. Neuroimage 44, 83–98.

Sobel, M.E., Lindquist, M.A., 2014. Causal inference for fMRI time series data with systematic errors of measurement in a balanced on/off study of social evaluative threat. J. Am. Stat. Assoc. 109, 967–976.

Van Essen, D.C., et al., 2013. The WU-Minn Human Connectome Project: an overview. Neuroimage 80, 62–79.

Varoquaux, G., Gramfort, A., Poline, J.-B., Thirion, B., 2010. Brain covariance selection: better individual functional connectivity models using population prior. In: Lafferty, J.D., Williams, C.K.I., Shawe-Taylor, J., Zemel, R.S., Culotta, A. (Eds.), Advances in Neural Information Processing Systems 23. Curran Associates, Inc, pp. 2334–2342.

Vul, E., Harris, C., Winkielman, P., Pashler, H., 2009. Puzzlingly high correlations in fMRI studies of emotion, personality, and social cognition. Perspect. Psychol. Sci. 4, 274–290.

Wager, T.D., et al., 2013. An fMRI-based neurologic signature of physical pain. N. Engl. J. Med. 368, 1388–1397.

Wallace, G.L., Dankner, N., Kenworthy, L., Giedd, J.N., Martin, A., 2010. Age-related temporal and parietal cortical thinning in Autism Spectrum Disorders. Brain 133, 3745–3754.

Woo, C.-W., Krishnan, A., Wager, T.D., 2014. Cluster-extent based thresholding in fMRI analyses: pitfalls and recommendations. Neuroimage 91, 412–419.

Woo, C.-W., Chang, L.J., Lindquist, M.A., Wager, T.D., 2017. Building better biomarkers: brain models in translational neuroimaging. Nat. Neurosci. 20, 365–377.

Woolrich, M.W., Ripley, B.D., Brady, M., Smith, S.M., 2001. Temporal autocorrelation in univariate linear modeling of FMRI data. Neuroimage 14, 1370–1386.

Worsley, K.J., 1996. The geometry of random images. Chance 9, 27–40.

Worsley, K.J., Friston, K.J., 1995. Analysis of fMRI time-series revisited—again. Neuroimage 2, 173–181.

Worsley, K.J., et al., 1996. A unified statistical approach for determining significant signals in images of cerebral activation. Hum. Brain Mapp. 4, 58–73.

Sensory modeling: Understanding computation in sensory systems through image-computable models

Zvi N. Roth[a], Elisha P. Merriam[b], and Tomas Knapen[c,d,e]

[a]*School of Psychological Sciences, Tel Aviv University, Tel Aviv, Israel,* [b]*Laboratory of Brain and Cognition, NIMH, NIH, Bethesda, MD, United States,* [c]*Department of Movement and Behavioral Sciences, Vrije Universiteit Amsterdam, The Netherlands,* [d]*Spinoza Centre for Neuroimaging, Amsterdam, The Netherlands,* [e]*Netherlands Institute for Neuroscience, Amsterdam, The Netherlands*

Introduction to image-computable models

In this chapter, we focus on the application of image-computable models to fMRI. To clarify the scope of this chapter, we will first explain our definition of an image-computable model. A *model* is a formal, falsifiable instantiation of a theory often involving an explicit computational implementation. An *image-computable model* receives arbitrary visual stimuli as input, and outputs a transformation of those stimuli that can be compared to brain measurements. One of the most important characteristics of image-computable models is, as the name suggests, that they do not require any information in addition to the images themselves (Hermes et al., 2019). That is, they give rise to an end-to-end transformation from stimulus to measurement without reference to learning, annotations, or labels. Image-computable models can be productively combined with models of cognitive concepts such as attention (Klein et al., 2014; Kay et al., 2015; van Es et al., 2018), but here we eschew discussing those more elaborate models, and instead focus on the use of image computable models for studying the early visual cortex.

We also limit our treatment to one specific flavor of image-computable model; *mechanistic, neurally inspired processing models* that parsimoniously explain

Computational and Network Modeling of Neuroimaging Data. https://doi.org/10.1016/B978-0-443-13480-7.00007-7

fMRI data with as few parameters as possible. These models' operations resemble operations hypothesized to be carried out by the brain, with the individual model components performing biologically plausible computations. For instance, this definition excludes artificial neural network (ANN) models, for two reasons. First, due to their millions of parameters, ANNs have an extraordinarily high degree of complexity. So while such models produce impressive behavior, their complexity can often make it difficult to relate the computations they perform to the visual system. Second, although the overarching architecture and learning principles of ANN models can be regarded as neurally inspired, the specific parameter tunings that give rise to their operations are not intended to be biologically interpretable.

Another defining feature of image-computable models that we discuss in this chapter is that they are mechanistic; the goal is to understand local computations and the modeling is not explicitly geared toward statistical hypothesis-testing inference, as is discussed, for example, in Chapter 1. Moreover, image-computable models are used to explain the responses at the single-voxel level, as opposed to regions of interest. This means these models explicitly capitalize on the interpretation of voxel-level BOLD responses as sampling the joint activity of a population of some thousands of neurons. Fitting with this level of description, in David Marr's conceptualization of a hierarchy of models, these models live at the junction of the computational and algorithmic levels.

The reason we adhere to these specific scope-defining limitations here is that thusly defined image-computable models make for highly valuable scientific models. There are a number of reasons for this utility, centering on their transparency: not only can their sequence of operations be fully understood, their parameters are also interpretable both computationally and, often, biologically. This set of qualities makes image-computable models an important and highly successful tool to understanding how the living human brain's computations represent visual information from measured BOLD responses.

Examples of successful image-computable models

Two broad classes of image-computable model have been used to model responses in primary visual cortex: the population receptive field (pRF) model and the Gabor filter bank model.

pRF model

The pRF model starts with a simplifying assumption that the BOLD activity in a voxel represents the sum of the receptive fields of all of the neurons in the cortical tissue sampled by that voxel. Different neurons within a voxel's neural population will have a range of RF centers (i.e., RF location scatter), orientations, and spatial frequencies (REFs, Hubel and Wiesel). Responses specific to orientation, eye-of-origin, spatial frequency, etc. are relatively diverse within the population of neurons sampled by a single voxel at standard resolution. But responses selective for spatial location are more consistent within this population: the dominant component of a voxel's population response is related to visual location. Thus, by disregarding the stimulus properties other than visual–spatial location, the pRF model is focused explicitly on mapping the retinotopic structure of the visual system and understanding the visual system's spatial computations. The main assumption here is that the population response may be well approximated by a circularly symmetric 2D Gaussian function defined in visual space. The main challenge is to find the position and size of the Gaussian that best approximates the population response (Fig. 1).

The pRF model is the primary successor to older techniques for mapping the retinotopic structure of the human visual cortex. These earlier methods used periodic stimulus designs to estimate the location in the visual field that produces an optimal response. Periodically rotating wedges and expanding/contracting rings would be used as stimuli to estimate voxels' polar angle and eccentricity preferences, respectively, based on the phase of periodic responses to visual stimulation (Sereno et al., 1995; Engel et al., 1997). The pRF modeling approach confers a number of advantages over these older techniques.

1. It is an explicit, image-computable model of visual responses in the spatial dimension: its estimated parameters are directly interpretable as relating to neural tuning inside our voxels.
2. Consequently, whereas these older techniques generally produced a point-estimate of spatial tuning, the pRF model also estimates the spatial extent of spatial tuning via its size parameter (the standard deviation of the circular Gaussian function).
3. The pRF modeling approach can simultaneously fit the shape of the hemodynamic response function as just an additional (set of) model parameter, which may do a better job of fitting population receptive fields in different tissue types (large veins, capillaries, different layers) (Kay et al., 2020).
4. Stimulus flexibility. Early retinotopic mapping techniques were limited to rings and wedges used to separately map out radial and angular components of the retinotopic map because of their dependence on a periodic stimulus design. Because the pRF model is image computable, it is more flexible and can run on arbitrary sequences of stimuli.

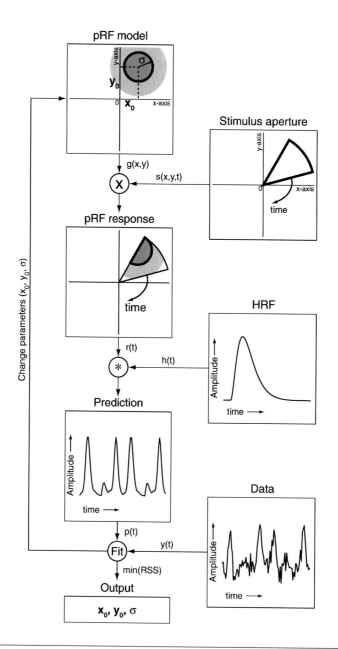

FIG. 1

pRF model and fitting procedure. The simple pRF model is defined by a Gaussian with three parameters: a center (x and y coordinates) and a standard deviation. The pRF fitting procedure finds values for these three parameters that best explains the response to the stimulus.

Stimulus designs

This stimulus flexibility freed researchers to use more irregular experimental designs, for example, stimuli with blank periods. Blank periods are effective when attempting to demonstrate spatially selective responses in brain regions with very large receptive fields (since stimuli without blanks can saturate responses in such areas) (Dumoulin and Wandell, 2008). The typical pRF mapping stimulus design is to scan a fixating subject while they view bar stimuli that traverse the visual field, in much the same way that Hubel and Wiesel used bars of light to map neuronal receptive fields in visual cortex (Hubel and Wiesel, 1962). This primary stimulus design of bars traveling across the screen combines sparseness of responses (for a given voxel, the bar only occupies its pRF for a relatively brief period, making response timing highly informative) with strength of responses (whenever the bar stimulus occupies a pRF, this is likely for a duration longer than 4 s—producing response of block-design strength). Generally, the textures inside these bars are optimized to evoke maximal response at a given location in the visual field. The texture in the bar can additionally be chosen to optimally activate brain regions tuned to specific visual categories, such as faces, bodies, scenes, etc. This has resulted in the use of bar apertures filled with full contrast counter-phase flickering checkerboard patterns as a generalized broadband stimulus (Dumoulin and Wandell, 2008; Amano et al., 2009; Fracasso et al., 2016), fast-alternating naturalistic visual stimuli, or even cartoon stimuli tailored to evoke stronger responses in high-level visual cortex (Silson et al., 2015; Benson et al., 2018; Kim et al., 2023).

Extensions to the linear pRF model

Early extensions to the basic, linear Gaussian pRF model captured two separate response signatures that the basic model cannot account for. These are surround suppression (depression of BOLD responses below empty-screen baseline whenever a stimulus impinges on a voxel's inhibitory penumbra), and response compression (sublinear increases of BOLD responses with increasing stimulus strength). Separate models were created to account for these different aspects of measured BOLD responses. Time courses displaying suppressive surrounds, predominantly appearing in V1, V2, and V3, are adequately modeled using a difference-of-Gaussians model. In this model, a larger, negative Gaussian's response is subtracted from the positive response of a smaller Gaussian with the same center location (Zuiderbaan et al., 2012). This subtractive response signature is still a linear one. However, BOLD responses also increasingly show nonlinear, compressive, response signatures as we move up the visual hierarchy. That is, as we increase the strength of a given pRF's stimulation, the response to this stimulation increases sublinearly, plateauing at higher stimulus strengths. The compressive spatial summation (CSS) model (Kay et al., 2013) produces this type of response signature by raising the response predicted by a linear model to a

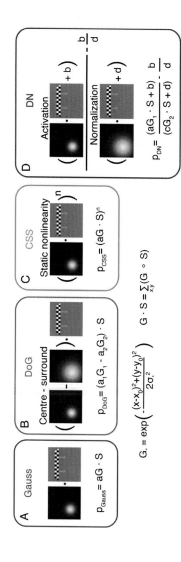

FIG. 2

Simple pRF model and elaborations. (A) The simplest pRF model is a Gaussian with three parameters (see Fig. 1). This model can be easily extended by adding additional parameters to describe phenomena such as (B) surround suppression (DoG) and (C) nonlinearities (CSS) in the neural response. (D) The DN model unifies both extensions by incorporating divisive normalization.

subunity power, generating a saturating, power-law input-response curve. A more recent model has attempted to unify these different models using a single parsimonious mechanism: divisive normalization (DN) (Heeger, 1992; Carandini and Heeger, 2011). The core concept of this canonical computational motif is that local, finely tuned responses are divided by the response of a more broadly tuned "normalization" pool of neurons. The computational motif of DN is able to explain a host of different findings across a range of different domains, from cross-orientation suppression to value-based decision-making. By implementing a divisive interaction between a larger "normalization" Gaussian pRF and a smaller "activation" Gaussian pRF, both centered on the same location, the DN pRF model is able to generate both suppressive and compressive response signatures (Aqil et al., 2021) (Fig. 2). Similar nonlinear mechanisms have been proposed to play a role in the temporal domain (Stigliani et al., 2017; Zhou et al., 2018). The normalization pool in this case has broader tuning in time: it contains a memory trace of past inputs. This allows this type of model to explain temporal subadditivity and repetition suppression. An additional testament to the parsimonious power of this approach is that the pRF modeling framework has proven effective as a scaffold for relating BOLD responses to presumed underlying electrophysiological responses, as measured using intracranial EEG (Harvey et al., 2013; Hermes et al., 2019; Groen et al., 2022). Another indication of the biological relevance of this type of image-computable modeling is that the distribution of DN pRF model parameters across the visual cortex is correlated with those of specific serotonergic (5HT) and inhibitory (GABA) receptor densities, as measured using positron emission tomography (Aqil et al., 2024).

But the pRF method also has its drawbacks. As the model is a purely spatial model without regard for additional stimulus features, one chief drawback is the need to convert the stimulus into an "energy" image. This image is created by estimating the local strength of responses to a given image, regardless of stimulus features. But the natural scenes we encounter in everyday life carry information about a multitude of cues: color, orientation, motion direction, etc. Capturing the feature-tuned responses to these types of stimuli requires going beyond spatial pRF models.

Gabor filter bank model

Although the dominant component of a voxel's response may be spatial selectivity, voxels in visual cortex, particularly in V1, have additional selectivity to cues such as spatial frequency and orientation. In the Gabor filter bank model, the image is sampled by oriented filters. The Gabor bank consists of layers of filters, each filter responding to a particular range of spatial frequencies. Within a layer, filters measure the energy at (or around) a particular orientation. In other words, the Gabor bank decomposes an image into spectral components, similar to a Fourier transform. Yet unlike a Fourier transform, this spectral decomposition is local, reflecting the amount of energy at each spatial frequency and orientation at each position in the image: the Gabor bank performs wavelet decomposition on the image.

In addition to responding to a particular combination of spatial frequency and orientation, each Gabor filter is also centered on a particular location in the image. So in a sense, this model is an extension of the pRF model to include stimulus features beyond spatial position. The Gabor filters operate on any arbitrary image, without the need to first convert the image into an "energy" image. Each layer will essentially convert the original image into an energy image at that layer's spatial frequency and orientation. The main downside of such a model is that there is a large number of parameters to fit to each voxel. Each voxel is assumed to respond not only to a particular region of the image but also to a particular spatial frequency and orientation (Fig. 3). To speed up the fitting process without requiring a prohibitive amount of data, one can use regularization, such as lasso regression, ridge regression, or early stopping (Kay et al., 2008). Alternatively, one may choose to separate between fitting each voxel's spatial selectivity and spectral selectivity. For example, the spatial selectivity may be determined based on independent pRF estimates from a separate scanning session, while spatial frequency and orientation tuning may be fitted based on responses to natural scenes (Roth et al., 2022).

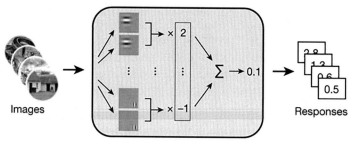

Stage 1: model estimation
Estimate a receptive-field model for each voxel

Images   Responses

Receptive-field model for one voxel

FIG. 3

The Gabor bank model. Each image is decomposed into Gabor filters at a range of orientations and spatial frequencies. Each voxel has weights for the Gabor filters, reflecting the voxel's orientation and frequency tuning. Predictions of a voxel's response amplitudes are generated by multiplying the filter weights with an image's decomposition into filter responses.

Extensions to the Gabor bank model

The Gabor bank includes filters to specific orientations and spatial frequencies, "low-level" features that V1 neurons are selective to. Extensions to the Gabor bank model include additional, more complex visual features. For example, a recent study (Henderson et al., 2023a,b) extended the Gabor bank to include mid-level visual features, computed as correlation within and between different filters (following Portilla and Simoncelli, 2000). Such mid-level features may underlie selectivity in extrastriate cortical region V2 (Freeman et al., 2011, 2013; Ziemba et al., 2016). Another study replaced the entire Gabor bank and spatial selectivity components with a combination of both low- and mid-level visual properties (color, curvature, and texture components) and semantic object-identity labels (e.g., faces, places, bodies, words, and food) to model higher-order visual areas (Khosla et al., 2022). In contrast to image-computable models, semantic encoding models require semantic labels to be assigned to each image (see Box 1).

Box 1

Approaches related to, but distinct from, image-computable models

Here, we want to highlight additional families of models that share the approach and conceptual paradigms used in fitting image-computable models. We distinguish two subtypes: encoding models, and models of functional connectivity.

Encoding models. Encoding models describe how information is encoded by a population of neurons measured in single voxels using fMRI but that are not constrained by the demand that the model's predictions are image-computable. This loosened constraint widens the applicability of these models, while sacrificing end-to-end transparency. Powerful examples are encoding models based on the semantic information occurring in images or video material. One can then estimate semantic model parameters by fitting them to fMRI data (Huth et al., 2012). Research has shown that this type of modeling can reveal structure in the brain's high-level responses to semantic information and that this structure is shared across sensory modalities, i.e., it is similar between video-based and auditory narratives (Deniz et al., 2019). The transcription of semantic information may be done manually or using neural networks, but in both cases, the process that distills the semantic information from the visual primitives remains a black box, and hence, do not meet the requirements that we set forth in this chapter.

Functional connectivity. Similar modeling strategies can also be used to attempt to explain ongoing fMRI responses (the target) as a function of responses in a source region. This means that the input to these models is not a stimulus, but rather a pattern of BOLD responses. The so-called connective field models attempt to explain target voxels' responses as a function of a "connective field" defined on the surface of a source region, in direct analogy to the population receptive field model of visual-space processing (Haak et al., 2013). These models leverage the topographic structure of the source region to condition the model of functional connectivity. They have been used to discover the topographic structure across the brain in experimental paradigms that resists image-computable approaches, such as resting-state or naturalistic movie watching experiments (Gravel et al., 2014; Knapen, 2021).

Other studies have extended the Gabor bank approach to decompose images using the filters of deep neural networks (St-Yves and Naselaris, 2018). In a similar fashion to what we described earlier, each voxel can then be modeled as a weighted combination of the different filters. An additional pRF component may be added to sample all filters according to each individual voxel's spatial preference.

In principle, the entire visual system, i.e., both early visual cortex and high-level brain areas, can be modeled with image-computable models, which in turn can also take into account task-effects (see below). Indeed, we have a solid understanding of the relevant filters with which V1 decomposes an image, and much progress has been made in determining the filters underlying V2 responses. But as we move up the visual hierarchy beyond V1 and V2, less is known about the computations being performed. Hence, we do not currently have an appropriate image-computable model for high-level visual cortex. Why is it so difficult to model high-level visual responses? Two reasons seem most likely. The first is that the relevant visual features become increasingly complex in higher order visual areas. In fact, artificial neural networks exhibit similar complexity in deep layers of the network, but we have yet to build image-computable models that successfully capture this complexity. While ANNs can capture this high-level visual complexity, they do it in an opaque manner, not by relying on explicit computations that are designed a priori by the researcher. Second, responses become increasingly influenced by task effects that are missed by models that focus on the stimulus alone.

Assumptions of image-computable models

Image-computable models in general depend on several fundamental assumptions, and each individual image-computable model has its own specific set of assumptions. This is common to all explicit modeling endeavors: formalizing a theory into a concrete set of equations and/or algorithms forces a researcher to be highly explicit about the assumptions involved, and the domains in which the model is applicable. Model comparisons and simulations can help to investigate the limits of the assumptions used in a specific model. This aspect of image-computable modeling is a core strength of the approach, in that it is key in generating parsimonious explanations of computational principles. The assumptions shared across image-computable models of fMRI voxel-based responses are important too, because they are fundamental to how we may use fMRI to investigate neural computation. So, here we highlight several assumptions of this nature.

One fundamental assumption regards how fMRI voxels sample neural responses. Activity in fMRI voxels reflects the pooled activity of a large and diverse population of neurons. The fact that fMRI voxels are a pooled measure has several important implications for how fMRI responses can be used to infer properties of activity at the neural level. In order for voxels to exhibit any sort of tuning to stimulus properties, neurons must be organized in the brain in clusters, gradients, or some other form of coarse-scale pattern that will create a reasonably homogeneous pattern of activity over the spatial scale sampled by a voxel. For example, if neurons with some particular stimulus preference were scattered randomly throughout visual cortex and intermixed with other neurons with other preferences, each voxel would sample from a large number of neurons with different tuning properties, which could, in principle, cancel each other out. The resulting voxel response should have little or no tuning to the visual feature. It is interesting to consider how tuning at the voxel level depends on the chosen voxel size, and what this voxel size dependence implies about tuning scatter in the sampled neural populations. In terms of position tuning, neurons in early visual cortex have clear position tuning (conceptualized as receptive fields) and are organized in a coarse-scale retinotopic map. In other words, nearby neurons respond to nearby locations in the visual field. As a result of a summing the activity of many nearby neurons, voxels in early visual cortex have clear position tuning as well. Similarly, nearby neurons in early visual cortex have similar spatial frequency tuning, and therefore, voxels in early visual cortex exhibit clear spatial frequency tuning (Aghajari et al., 2020; Broderick et al., 2022). But occasionally, different theories make different assumptions regarding how voxels sample neuronal activity, and these assumptions are reflected in details of the model used.

For example, the basic pRF model assumes the pRF is shaped as a symmetric Gaussian. This should be the case if neuron RFs sampled within a voxel are scattered homogeneously throughout the visual field. It is theoretically possible that neuronal RFs are scattered more in one direction (such as the radial direction, relative to fixation) than in the other perpendicular direction (the tangential direction), and this issue is currently debated (Silson et al., 2018; Lerma-Usabiaga et al., 2021).

Orientation preference in primary visual cortex is organized in cortical columns at a much finer scale than the retinotopic map (Hubel and Wiesel, 1968; Grinvald et al., 1986). Following the logic we laid out earlier, voxels may sample from neurons preferring the entire range of possible orientations, which should, in principle, result in voxels with no orientation tuning (Gardner and Merriam, 2021). Yet, fMRI decoding studies have repeatedly demonstrated the presence of orientation information in V1 voxels (Haynes and Rees, 2005; Kamitani and Tong, 2005). What is the source of decodable orientation information in these studies? One possibility, termed the "random bias" account, posits that each V1 voxel, despite sampling from columns of all orientations, happens to randomly sample from slightly more columns preferring a particular orientation. This would result in each voxel having a slightly larger response to a particular orientation (Boynton, 2005). This was an attractive conjecture because it suggested that fMRI could be used to study tuning properties of heterogenous populations of neurons within a voxel, allowing researchers to, in effect, peer past the spatial resolution limits of the fMRI measurements.

But, it turns out, this conjecture has been difficult to fully support or fully refute. Subsequent studies demonstrated that successful decoding of orientation relied on a coarse-scale map of orientation preference in V1 that corresponds to the retinotopic map, resulting in a "radial bias" across the cortical surface (Freeman et al., 2011). This finding highlights how difficult it can be to interpret the results of any decoding analysis: the mere ability to decode information from fMRI responses does not speak directly to the nature of the underlying neural computation.

Attempts to understand the neural computations that give rise to orientation decoding were further complicated by a theoretical paper that suggested that the orientation map was in fact a consequence of a subtle confound in the stimulus structure (Carlson, 2014), a finding confirmed by our lab in a subsequent empirical study (Roth et al., 2018). It turned out that stimulus edge effects have a powerful influence on the nature of the orientation map that can completely obfuscate measurement of the underlying orientation tuning; orientation selectivity was found to be determined by the spatial position of the stimulus edges, and changing the edge can completely flip the orientation map (Roth et al., 2018). According to these findings, one might conclude that each voxel samples homogeneously from columns with all orientation preferences, resulting in voxels with no reliable tuning (Carlson, 2014). But if the edge effect simply obscures columnar orientation tuning, is it possible to measure orientation tuning with fMRI?

This question led our group to develop an approach for characterizing both edge effects and orientation selectivity using a pair of image computable models and statistical model comparison. Specifically, one model was built with the assumption that no orientation tuning was present in the fMRI measurements (beyond the edge effect described earlier), and a second model was built with the assumption of orientation selectivity, above and beyond what would have been induced by the stimulus edge. We fit both models to fMRI data and found that although they both perform well, the second model, which allowed for orientation selectivity, explained more variance in the data—variance that reflected true orientation tuning (Roth et al., 2022). This study demonstrates that once we are aware of the model's assumptions, we build that assumption into the image computable model to test whether those assumptions are correct.

A second major assumption is that image-computable models assume that neural responses depend entirely on the visual stimulus. This means that it should be possible, for example, to train a model on responses recorded in the morning and accurately predict responses recorded in the afternoon, tomorrow morning, or several months from now. The viewer, or participant, may be in an entirely different state of mind after a few hours, perhaps hungrier and grumpier, or less alert, but the way in which neurons respond to visual stimuli is assumed to be unchanged. According to this assumption, we can increase the signal-to-noise ratio of the measurements by averaging over responses measured on different days, since any difference between responses measured on different days is attributed to noise. The natural analogy is a computer program, which processes an image the exact same way, any time any day, regardless of any other processes the computer is executing (or has executed) at the time.

However, recent studies suggest that this is not the case that brains are not like computers in this way. Visual responses may be influenced by the internal brain states, which reflect a wide range of factors not directly linked to sensory input, and which may change dynamically over a range of time scales. In rodents, changes in arousal have been shown to modulate visual responses throughout the visual system (Niell and Stryker, 2010; Aydın et al., 2018; Savier et al., 2019; Schröder et al., 2020). Changes in arousal also affect primate (Sirotin and Das, 2009; Cardoso et al., 2012, 2019) and human visual cortex (Roth et al., 2020; Burlingham et al., 2022), and are likely to impact visual responses as well.

Even when the internal brain state is not experimentally manipulated, visual responses may change nonetheless. A phenomenon termed "representational drift" has been demonstrated in rodents (Deitch et al., 2021; Marks and Goard, 2021), and more recently also in humans (Roth and Merriam, 2023). In representational drift, neural responses to the same stimulus exhibit systematic changes over time that are not linked to any change in the stimulus. Such "drift" implies that an image-computable model trained on data collected on day 1 will be suboptimal at predicting responses on day 2, and the difference in responses between data collected at different time points will grow monotonically with time.

Changes in neural tuning have been demonstrated with changes in attention (Womelsdorf et al., 2006), but neural tuning can also change with task (Kay et al., 2023). In a simple case, a change in task will modulate some parameters of the neural tuning, such as a shift in the response gain, or a change in the preferred stimulus. In theory, it is possible that in some cases, neural tuning changes completely with the task, such that completely different models are needed to explain the data when subjects are engaged in different tasks. This is likely the case, for example, when trying to model responses in frontal cortex (Mante et al., 2013). This type of tuning conflicts with the assumption of image-computability, as it indicates that neural responses are best explained by a combination of both image-based computation and task.

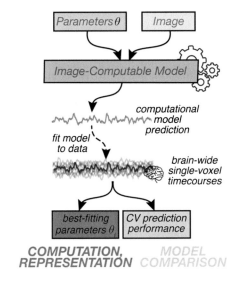

FIG. 4

The analysis paradigm used for image-computable modeling. A model converts an image and some set of model parameters into an explicit model prediction of BOLD responses. This means we can estimate optimal parameter values for all voxels in parallel, and interpret these best-fitting parameters as reflecting computational and representational properties of neural responses. Moreover, by performing cross-validation, models can be compared in terms of their ability to predict left-out data, regardless of the specific parametrizations of the models in our comparison.

Building, testing, interpreting image-computable models

An image-computable model needs to go through multiple steps to produce response predictions (Fig. 4). The initial stage of many models is to posit a filter or set of filters, and taking the dot product of the filter(s) and the stimulus. This produces an idealized linear response. The simplest models use this linear response as a signal prediction; however, more complex models apply additional steps such as rectification, exponentiation, division, etc. to perform a desired response transformation. The model-design choices made here are usually informed by mathematical considerations and relevant results from animal electrophysiology. The predicted neural signal that the model produces is then usually convolved with a hemodynamic response function to produce a prediction at the level of the measured signal.

Fitting image-computable models

There are several complementary strategies for fitting image-computable models to our neuroimaging data. We here separate these into two broad categories; direct optimization of interpretable parameters, and (penalized) regression that delivers parameter values as beta-weights. In both strategies, we are interested in two outcomes: (1) the values of best-fitting model parameters, and (2) the fit quality as assessed by cross-validated fraction of variance explained.

In direct parameter optimization, one specifies parameter values that dictate the behavior of the model and then uses the presented stimulus as input to the model to produce a predicted response—a simulation. By varying the values of the model parameters, one can produce different predicted responses. For instance, in the pRF model, these parameters are horizontal and vertical position in the visual field, and the pRF's sampling extent (the standard deviation of the pRF's Gaussian window). We can then employ the arsenal of curve-fitting routines to find the optimal parameter values, that is, the parameter values that best explain the measured signal. Thus, researchers are required to make many choices concerning whether to use grid-fitting or iterative fitting, what the appropriate cost function is, what search algorithm is appropriate, and possible combinations thereof. For instance, in many cases, parameters are fit using a coarse-to-fine approach starting with a grid search, simulating many different model predictions and using them to explain the measured signals in turn. For every voxel, the best-fitting combination of parameter values from the grid stage is then used as a starting point for an iterative search procedure. This combination of approaches ensures both that a full parameter domain is sampled and global minima within this domain are likely found while also allowing detailed search for the final parameter values. The choice of numerical optimization procedure (or sequential combination thereof) is dependent on things such as a researcher's speed-accuracy trade-off, a fitting problem's tendency to get stuck in local minima, our knowledge about the parameter space, implementational details, etc.

When fitting models using regression, one constructs a design matrix composed of separate model predictions. For instance, in the Gabor wavelet model, one prediction would be generated for each filter in the filter bank. This design matrix is then used to estimate beta weights for each of the regressors using gradient descent or general linear model fitting (see Chapter 1). Depending on the structure of the model, design matrix, and data quality and amount, one can use ordinary least squares regression (in the case of large amounts of high-quality data and/or a high-rank design matrix) or move to regularization to avoid overfitting (for example, when there are many regressors relative to the number of data points). In the case of regularized fitting, cross-validation is required to set the appropriate value for the penalization hyperparameter(s), and a separate validation set is generally used to compute explained variance of the optimized model.

The two fitting strategies outlined here both require cross-validation to compare different models, especially when these differ in their number of parameters. Their differences are mainly in the way by which the researcher imposes constraints on the model's ability to explain data. For instance, the Gaussian pRF model limits the set of possible spatial patterns with which a stimulus is compared to a very specific set: coherent, unitary, and smooth patches following a Gaussian profile defined in visual space. This very reasonable restriction of the possible outcome space can be seen as a form of regularization. In comparison, it is also possible to fit a pRF model using penalized ridge regression, using all screen pixels' time courses, convolved with an HRF, as regressors (Lee et al., 2013). In this case, the ridge penalization implements a smoothness constraint on the shape of the resulting spatial filter. This example serves to illustrate that both strategies can be used to constrain the space of possible outcomes; by a priori definition in the case of parameter optimization and based on statistical considerations in the case of penalized regression.

Additional considerations for choosing a Gabor bank model

When building a Gabor bank model, one can tweak the specific model by adjusting the architecture and the parameter values. For example, the model will have a specific number of orientations and of spatial frequency levels, and the filters will have specific orientation and spatial frequency bandwidths. These values are dependent on each other, so once a choice of filters bandwidth has been made, the number of filters will be determined by the image size.

One elegant version of Gabor bank is the Steerable Pyramid model (Simoncelli et al., 1992; Simoncelli and Freeman, 1995). This model consists of shiftable filters, similar to Gabors, that are convolved with the image. The model is designed such that the (squared) filters span Fourier space uniformly. This eliminates subtle artifacts that can be produced by a Gabor bank (Clifford and Mannion, 2015; Roth et al., 2018).

Pitfalls in image-computable modeling

As with all modeling approaches, image-computable modeling is subject to several potential pitfalls. Here, we discuss a nonexhaustive list of common sorts of errors.

Pitfall #1: A realistic appreciation of fit quality. If the model is fit using an insufficient amount of data, or if the data signal-to-noise ratio is not sufficiently high, the model predictions will be inaccurate, and hence explain only a small proportion of the total variance in the signal. To determine the reliability of the model fits, it is common to use cross-validation: we use one portion of the data to fit the model, and test how well the fitted model predicts or corresponds to the held-out portion. This is a good procedure with which to guard against overfitting and to compare models against one another, but it is important to compare their performance to the noise ceiling, which specifies the amount of explainable variance in the signal (for instance, as gauged by test–retest reliability of responses) (David and Gallant, 2005). When we observe clear divergence, this might mean there are additional aspects of the data that need modeling and indicate that one may expand either data acquisition and/or the complexity of the models used. As this type of data-model mismatch will be evident in the structure of the model fit residuals, we strongly advocate detailed inspection of model residuals as an integral stage in model development. This process will, over the course of model development, enable one to tune model complexity and generally lead to models that more parsimoniously explain data.

Pitfall #2: Implicit biases in stimuli. Because image-computable models operate on the stimulus, such approaches are susceptible to small but systematic biases in the stimuli themselves. This has been appreciated in electrophysiological approaches to receptive field mapping for some time and was part of the motivation for developing white noise RF mapping through methods such as m-sequences (Reid et al., 1997).

Here, we list a number of biases identified in image-computable models that may have varying degrees of impact on results. First, the standard traveling bar pRF design, despite its evident benefits, imposes certain appreciable biases on model outcomes. First, the standard experimental design shows bars in a circular aperture, which causes edge effects. This aperture interacts with the very predictable spatiotemporal stimulation inherent in these designs. These effects are minimized by using the so-called multifocal mapping paradigms that present multiple stimuli at different locations and spatial extents in a trial-based format, analogous to white noise mapping stimuli used in electrophysiology (Binda et al., 2013), although these designs suffer in terms of effective signal to noise and have biases of their own—for instance, in estimates of pRF size (Infanti and Schwarzkopf, 2020). As mentioned earlier, the content of the traveling bar stimuli biases in which visual regions responses will be maximal. For instance, filling bar apertures with scenes will evoke stronger responses in scene-selective cortex, allowing researchers to research spatial selectivity in those regions (Silson et al., 2015). Similar reasoning holds for different visual categories, with generalized naturalistic bar content used to evoke stronger visual–spatial responses throughout high-level visual cortex (Benson et al., 2018; Kim et al., 2023). Yet, for any stimulus category, the image statistics can potentially bias the model fits. For example, spectral statistics of natural images may bias models fit to responses to natural scenes, particularly when regularization is used to improve the model fits.

Pitfall #3: Overestimating specificity of localization. Many early efforts at building image-computable models of the visual system specify a restricted area of visual/image space for which a voxel's response is evoked. This has led to a common understanding of responses as highly localized. We want to point to two specific, recently developed reasons that call this understanding into question. First, some researchers have assumed that a voxel responds only in the area of the Gaussian standard deviation that was used to model the response. However, a Gaussian response profile implies that the voxel responds beyond the standard deviation border, albeit at a lower response amplitude. For example, if a voxel with a pRF size of 2.5 degrees is centered 3 degrees away from a stimulus, then the voxel will respond to the stimulus at approximately 50% of its maximal response amplitude. This property of pRFs has implications for stimulus vignetting: all voxels that respond to the edge of a grating will have orientation information, even if the distance from stimulus to pRF center is greater than the pRF standard deviation. Second, the results from the divisive normalization pRF model show that the normalization penumbra of a V1 voxel's response extends six- to eightfold further than its "classical," linear Gaussian pRF's extent (Aqil et al., 2021). These two examples show that responses, even at the level of V1, are much more spatially integrated than local models assume, with important implications for experimental paradigms and analyses that depend on locality. For example, selecting voxels whose responses to a specific spatial stimulus are significantly positive excludes many voxels that indeed carry information about that stimulus, for example, in their negative response deflections.

Pitfall #4: Code and method complexity. Image-computable models place a high burden on the researcher in terms of their technical process and ability to develop and maintain complicated processing pipelines. The codebase associated with an image-computable modeling project is likely to span thousands of lines of code, behooving the researcher to invest in the adherence to principled coding standards. But inherently, all modeling code of realistic complexity is likely to include bugs. Without performing additional checks, such errors in the model's implementation can lead to wrong conclusions. These additional checks should consist of detailed simulations of model predictions, the interactions between model parameters and noise levels, and analyses centering on parameter recovery (Wilson and Collins, 2019). Luckily, the field has recently started moving toward reproducible pipelines for model comparisons (Lerma-Usabiaga et al., 2020) meant to combat these types of implementation-based errors. In general, the move toward open datasets (Benson et al., 2018; Chang et al., 2019; Allen et al., 2021) and open software/methodologies promises to lead to improvements in common practice in the field.

Open challenges and future directions

In this chapter, we have described how image-computable models embody explicit, testable hypotheses regarding the neural computations carried out by the visual system. Image-computable models place a large burden on the researcher to formalize and implement the models, in contrast to an alternative approach involving artificial neural networks. In this manner, image-computable models exemplify a specific and highly valuable approach to fMRI studies of both the visual system specifically and sensory systems more generally. Task effects are increasingly seen as an interesting component of stimulus-evoked activity and highlight the inherently limited scope of image-computable models. Future generations of models should combine image-computable models with explicit models of task effects to explain brain responses in a more general domain. This will allow us to make increasingly more general models of processing: trickling the value of image-computable models up into the brain.

Take-home points

- Image-computable models are transparent: the user defines the underlying computations.
- Image-computable models require no labeling.
- Image-computable models facilitate model comparison.
- Components of image-computable models may correspond directly to neural properties.
- Image-computable models are (currently) limited to stimulus processing, excluding task effects.

References

Aghajari, S., Vinke, L.N., Ling, S., 2020. Population spatial frequency tuning in human early visual cortex. J. Neurophysiol. 123, 773–785.

Allen, E.J., St-Yves, G., Wu, Y., Breedlove, J.L., Prince, J.S., Dowdle, L.T., Nau, M., Caron, B., Pestilli, F., Charest, I., Hutchinson, J.B., Naselaris, T., Kay, K., 2022. A massive 7T fMRI dataset to bridge cognitive neuroscience and artificial intelligence. Nat. Neurosci. 25, 116–126.

Amano, K., Wandell, B.A., Dumoulin, S.O., 2009. Visual field maps, population receptive field sizes, and visual field coverage in the human MT+ complex. J. Neurophysiol. 102, 2704–2718.

Aqil, M., Knapen, T., Dumoulin, S.O., 2021. Divisive normalization unifies disparate response signatures throughout the human visual hierarchy. Proc. Natl. Acad. Sci. U. S. A. 118 (46), e2108713118.

Aqil, M., Knapen, T., Dumoulin, S.O., 2024. Computational model links normalization to chemoarchitecture in the human visual system. Sci. Adv. 10, eadj6102.

Aydın, Ç., Couto, J., Giugliano, M., Farrow, K., Bonin, V., 2018. Locomotion modulates specific functional cell types in the mouse visual thalamus. Nat. Commun. 9, 4882.

Benson, N.C., Jamison, K.W., Arcaro, M.J., Vu, A.T., Glasser, M.F., Coalson, T.S., Van Essen, D.C., Yacoub, E., Ugurbil, K., Winawer, J., Kay, K., 2018. The human connectome project 7 Tesla retinotopy dataset: description and population receptive field analysis. J. Vis. 18, 23.

Binda, P., Thomas, J.M., Boynton, G.M., Fine, I., 2013. Minimizing biases in estimating the reorganization of human visual areas with BOLD retinotopic mapping. J. Vis. 13, 13.

Boynton, G.M., 2005. Imaging orientation selectivity: decoding conscious perception in V1. Nat. Neurosci. 8, 541–542.

Broderick, W.F., Simoncelli, E.P., Winawer, J., 2022. Mapping spatial frequency preferences across human primary visual cortex. J. Vis. 22, 3.

Burlingham, C.S., Ryoo, M., Roth, Z.N., Mirbagheri, S., Heeger, D.J., Merriam, E.P., 2022. Task-related hemodynamic responses in human early visual cortex are modulated by task difficulty and behavioral performance. Elife 11, e73018.

Carandini, M., Heeger, D.J., 2011. Normalization as a canonical neural computation. Nat. Rev. Neurosci. 13, 51–62.

Cardoso, M.M., Sirotin, Y.B., Lima, B., Glushenkova, E., Das, A., 2012. The neuroimaging signal is a linear sum of neurally distinct stimulus- and task-related components. Nat. Neurosci. 15, 1298–1306.

Cardoso, M.M.B., Lima, B., Sirotin, Y.B., Das, A., 2019. Task-related hemodynamic responses are modulated by reward and task engagement. PLoS Biol. 17, e3000080.

Carlson, T.A., 2014. Orientation decoding in human visual cortex: new insights from an unbiased perspective. J. Neurosci. 34, 8373–8383.

Chang, N., Pyles, J.A., Marcus, A., Gupta, A., Tarr, M.J., Aminoff, E.M., 2019. BOLD5000, a public fMRI dataset while viewing 5000 visual images. Sci. Data 6, 49.

Clifford, C.W., Mannion, D.J., 2015. Orientation decoding: sense in spirals? Neuroimage 110, 219–222.

David, S.V., Gallant, J.L., 2005. Predicting neuronal responses during natural vision. Network 16, 239–260.

Deitch, D., Rubin, A., Ziv, Y., 2021. Representational drift in the mouse visual cortex. Curr. Biol. 31, 4327–4339. e4326.

Deniz, F., Nunez-Elizalde, A.O., Huth, A.G., Gallant, J.L., 2019. The representation of semantic information across human cerebral cortex during listening versus reading is invariant to stimulus modality. J. Neurosci. 39, 7722–7736.

Dumoulin, S.O., Wandell, B.A., 2008. Population receptive field estimates in human visual cortex. Neuroimage 39, 647–660.

Engel, S.A., Glover, G.H., Wandell, B.A., 1997. Retinotopic organization in human visual cortex and the spatial precision of functional MRI. Cereb. Cortex (New York, NY: 1991) 7, 181–192.

Fracasso, A., Petridou, N., Dumoulin, S.O., 2016. Systematic variation of population receptive field properties across cortical depth in human visual cortex. Neuroimage 139, 427–438.

Freeman, J., Brouwer, G.J., Heeger, D.J., Merriam, E.P., 2011. Orientation decoding depends on maps, not columns. J. Neurosci. 31, 4792–4804.

Freeman, J., Heeger, D.J., Merriam, E.P., 2013. Coarse-scale biases for spirals and orientation in human visual cortex. J. Neurosci. 33, 19695–19703.

Gardner, J.L., Merriam, E.P., 2021. Population models, not analyses, of human neuroscience measurements. Annu. Rev. Vis. Sci. 7, 225–255.

Gravel, N., Harvey, B., Nordhjem, B., Haak, K.V., Dumoulin, S.O., Renken, R., Curčić-Blake, B., Cornelissen, F.W., 2014. Cortical connective field estimates from resting state fMRI activity. Front. Neurosci. 8, 339.

Grinvald, A., Lieke, E., Frostig, R.D., Gilbert, C.D., Wiesel, T.N., 1986. Functional architecture of cortex revealed by optical imaging of intrinsic signals. Nature 324, 361–364.

Groen, I.I.A., Piantoni, G., Montenegro, S., Flinker, A., Devore, S., Devinsky, O., Doyle, W., Dugan, P., Friedman, D., Ramsey, N.F., Petridou, N., Winawer, J., 2022. Temporal dynamics of neural responses in human visual cortex. J. Neurosci. 42, 7562–7580.

Haak, K.V., Winawer, J., Harvey, B.M., Renken, R., Dumoulin, S.O., Wandell, B.A., Cornelissen, F.W., 2013. Connective field modeling. Neuroimage 66, 376–384.

Harvey, B.M., Vansteensel, M.J., Ferrier, C.H., Petridou, N., Zuiderbaan, W., Aarnoutse, E.J., Bleichner, M.G., Dijkerman, H.C., van Zandvoort, M.J., Leijten, F.S., Ramsey, N.F., Dumoulin, S.O., 2013. Frequency specific spatial interactions in human electrocorticography: V1 alpha oscillations reflect surround suppression. Neuroimage 65, 424–432.

Haynes, J.D., Rees, G., 2005. Predicting the orientation of invisible stimuli from activity in human primary visual cortex. Nat. Neurosci. 8, 686–691.

Heeger, D.J., 1992. Normalization of cell responses in cat striate cortex. Vis. Neurosci. 9, 181–197.

Henderson, M.M., Tarr, M.J., Wehbe, L., 2023a. Low-level tuning biases in higher visual cortex reflect the semantic informativeness of visual features. J. Vis. 23, 8.

Henderson, M.M., Tarr, M.J., Wehbe, L., 2023b. A texture statistics encoding model reveals hierarchical feature selectivity across human visual cortex. J. Neurosci. 43, 4144–4161.

Hermes, D., Petridou, N., Kay, K.N., Winawer, J., 2019. An image-computable model for the stimulus selectivity of gamma oscillations. Elife 8, e47035.

Hubel, D.H., Wiesel, T.N., 1962. Receptive fields, binocular interaction and functional architecture in the cat's visual cortex. J. Physiol. 160, 106–154.

Hubel, D.H., Wiesel, T.N., 1968. Receptive fields and functional architecture of monkey striate cortex. J. Physiol. 195, 215–243.

Huth, A.G., Nishimoto, S., Vu, A.T., Gallant, J.L., 2012. A continuous semantic space describes the representation of thousands of object and action categories across the human brain. Neuron 76, 1210–1224.

Infanti, E., Schwarzkopf, D.S., 2020. Mapping sequences can bias population receptive field estimates. Neuroimage 211, 116636.

Kamitani, Y., Tong, F., 2005. Decoding the visual and subjective contents of the human brain. Nat. Neurosci. 8, 679–685.

Kay, K.N., Naselaris, T., Prenger, R.J., Gallant, J.L., 2008. Identifying natural images from human brain activity. Nature 452, 352–355.

Kay, K.N., Winawer, J., Mezer, A., Wandell, B.A., 2013. Compressive spatial summation in human visual cortex. J. Neurophysiol. 110, 481–494.

Kay, K.N., Weiner, K.S., Grill-Spector, K., 2015. Attention reduces spatial uncertainty in human ventral temporal cortex. Curr. Biol. 25, 595–600.

Kay, K., Jamison, K.W., Zhang, R.Y., Uğurbil, K., 2020. A temporal decomposition method for identifying venous effects in task-based fMRI. Nat. Methods 17, 1033–1039.

Kay, K., Bonnen, K., Denison, R.N., Arcaro, M.J., Barack, D.L., 2023. Tasks and their role in visual neuroscience. Neuron 111, 1697–1713.

Khosla, M., Ratan Murty, N.A., Kanwisher, N., 2022. A highly selective response to food in human visual cortex revealed by hypothesis-free voxel decomposition. Curr. Biol. 32, 4159–4171. e4159.

Kim, I., Kupers, E.R., Lerma-Usabiaga, G., Grill-Spector, K., 2023. Characterizing spatiotemporal population receptive fields in human visual cortex with fMRI. J. Neurosci. 44 (2), e0803232023.

Klein, B.P., Harvey, B.M., Dumoulin, S.O., 2014. Attraction of position preference by spatial attention throughout human visual cortex. Neuron 84, 227–237.

Knapen, T., 2021. Topographic connectivity reveals task-dependent retinotopic processing throughout the human brain. Proc. Natl. Acad. Sci. U. S. A. 118 (2), e2017032118.

Lee, S., Papanikolaou, A., Logothetis, N.K., Smirnakis, S.M., Keliris, G.A., 2013. A new method for estimating population receptive field topography in visual cortex. Neuroimage 81, 144–157.

Lerma-Usabiaga, G., Benson, N., Winawer, J., Wandell, B.A., 2020. A validation framework for neuroimaging software: the case of population receptive fields. PLoS Comput. Biol. 16, e1007924.

Lerma-Usabiaga, G., Winawer, J., Wandell, B.A., 2021. Population receptive field shapes in early visual cortex are nearly circular. J. Neurosci. 41, 2420–2427.

Mante, V., Sussillo, D., Shenoy, K.V., Newsome, W.T., 2013. Context-dependent computation by recurrent dynamics in prefrontal cortex. Nature 503, 78–84.

Marks, T.D., Goard, M.J., 2021. Stimulus-dependent representational drift in primary visual cortex. Nat. Commun. 12, 5169.

Niell, C.M., Stryker, M.P., 2010. Modulation of visual responses by behavioral state in mouse visual cortex. Neuron 65, 472–479.

Portilla, J., Simoncelli, E.P., 2000. A parametric texture model based on joint statistics of complex wavelet coefficients. Int. J. Comput. Vis. 40, 49–70.

Reid, R.C., Victor, J.D., Shapley, R.M., 1997. The use of m-sequences in the analysis of visual neurons: linear receptive field properties. Vis. Neurosci. 14, 1015–1027.

Roth, Z.N., Merriam, E.P., 2023. Representations in human primary visual cortex drift over time. Nat. Commun. 14, 4422.

Roth, Z.N., Heeger, D.J., Merriam, E.P., 2018. Stimulus vignetting and orientation selectivity in human visual cortex. Elife 7, e37241.

Roth, Z.N., Ryoo, M., Merriam, E.P., 2020. Task-related activity in human visual cortex. PLoS Biol. 18, e3000921.

Roth, Z.N., Kay, K., Merriam, E.P., 2022. Natural scene sampling reveals reliable coarse-scale orientation tuning in human V1. Nat. Commun. 13, 6469.

Savier, E.L., Chen, H., Cang, J., 2019. Effects of locomotion on visual responses in the mouse superior colliculus. J. Neurosci. 39, 9360–9368.

Schröder, S., Steinmetz, N.A., Krumin, M., Pachitariu, M., Rizzi, M., Lagnado, L., Harris, K.D., Carandini, M., 2020. Arousal modulates retinal output. Neuron 107, 487–495. e489.

Sereno, M.I., Dale, A.M., Reppas, J.B., Kwong, K.K., Belliveau, J.W., Brady, T.J., Rosen, B. R., Tootell, R.B., 1995. Borders of multiple visual areas in humans revealed by functional magnetic resonance imaging. Science 268, 889–893.

Silson, E.H., Chan, A.W., Reynolds, R.C., Kravitz, D.J., Baker, C.I., 2015. A retinotopic basis for the division of high-level scene processing between lateral and ventral human occipitotemporal cortex. J. Neurosci. 35, 11921–11935.

Silson, E.H., Reynolds, R.C., Kravitz, D.J., Baker, C.I., 2018. Differential sampling of visual space in ventral and dorsal early visual cortex. J. Neurosci. 38, 2294–2303.

Simoncelli, E.P., Freeman, W.T., 1995. The steerable pyramid: a flexible architecture for multi-scale derivative computation. In: Proceedings International Conference on Image Processing. IEEE, pp. 444–447.

Simoncelli, E.P., Freeman, W.T., Adelson, E.H., Heeger, D.J., 1992. Shiftable multiscale transforms. IEEE Trans. Inf. Theory 38, 587–607.

Sirotin, Y.B., Das, A., 2009. Anticipatory haemodynamic signals in sensory cortex not predicted by local neuronal activity. Nature 457, 475–479.

Stigliani, A., Jeska, B., Grill-Spector, K., 2017. Encoding model of temporal processing in human visual cortex. Proc. Natl. Acad. Sci. U. S. A. 114, E11047–e11056.

St-Yves, G., Naselaris, T., 2018. The feature-weighted receptive field: an interpretable encoding model for complex feature spaces. Neuroimage 180, 188–202.

van Es, D.M., Theeuwes, J., Knapen, T., 2018. Spatial sampling in human visual cortex is modulated by both spatial and feature-based attention. Elife 7, e36928.

Wilson, R.C., Collins, A.G., 2019. Ten simple rules for the computational modeling of behavioral data. Elife 8, e49547.

Womelsdorf, T., Anton-Erxleben, K., Pieper, F., Treue, S., 2006. Dynamic shifts of visual receptive fields in cortical area MT by spatial attention. Nat. Neurosci. 9, 1156–1160.

Zhou, J., Benson, N.C., Kay, K.N., Winawer, J., 2018. Compressive temporal summation in human visual cortex. J. Neurosci. 38, 691–709.

Ziemba, C.M., Freeman, J., Movshon, J.A., Simoncelli, E.P., 2016. Selectivity and tolerance for visual texture in macaque V2. Proc. Natl. Acad. Sci. U. S. A. 113, E3140–E3149.

Zuiderbaan, W., Harvey, B.M., Dumoulin, S.O., 2012. Modeling center-surround configurations in population receptive fields using fMRI. J. Vis. 12, 10.

Cognitive modeling: Joint models use cognitive theory to understand brain activations

Brandon M. Turner

The Ohio State University, Columbus, OH, United States

Introduction to joint modeling

If psychology is the study of the human mind and its functions, what is a mind? Oxford Languages defines the mind as an element of a person that enables them to be aware of the world and their experiences, to think, and to feel. This definition implies that the mind is an internal component that enables perception, forms a representation (of the world), and facilitates interactions with the environment. The mind is a psychological construct that allows us to understand how thoughts are produced, and we cannot directly observe it. Yet, we can indirectly observe the functions of the mind through manifest variables of behavior, such as choices, response times, or manifest variables of the brain, such as the blood oxygenation level dependent (BOLD) response in functional magnetic resonance imaging (fMRI). With a carefully designed experiment, we can systematically study how different types of stimulation produce differences in these manifest variables. It follows then that a greater understanding of the mind can come with a better understanding of how stimulation produces simultaneous changes in large sets of manifest variables, including both behavioral and brain-related variables.

However, understanding the interactions among large sets of variables is not as easy as it may seem. One productive route forward has been to define quantitative links that define how physiological variables are related to psychological processes, known as *linking propositions* (Brindley, 1970; Teller, 1984; Schall, 2004; Turner et al., 2019). Early versions of linking propositions specified sets of strict axiomatic equality statements that would all need to be satisfied to ensure a link between psychological and physiological variables (Teller, 1984). For example, one logical set of linking propositions called the identity set offered by Teller (1984) requires the proposition that a particular characteristic of a physiological state must always imply a particular characteristic of a psychological state. However, as noted by Schall (2004), empirically testing for such linking propositions is problematic: what exactly

Computational and Network Modeling of Neuroimaging Data. https://doi.org/10.1016/B978-0-443-13480-7.00003-X

does "identical" mean in the context of measurement noise? Indeed, "...identity cannot refer to constancy within an individual over time; you cannot step twice in the same river" (Schall, 2004, p. 27). To make use of empirical tests, Schall (2004) broadened the definition of linking propositions to accommodate uncertainty both in the manifest variables and the link specification itself. When probability distributions are used to define these linking propositions, we refer to those propositions as *statistical linking functions*. Statistical linking functions have become a popular way to characterize the link between physiological and psychological variables because they allow researchers to quantify evidence through hypothesis tests.

One particularly relevant line of work first uses cognitive models to propose psychological processes, and then statistically links aspects of the cognitive models to the physiological variables (e.g., BOLD response). Because this line of work uses cognitive models to understand neuroscientific data, it is now referred to as *model-based cognitive neuroscience*. Cognitive models represent psychological processes by instantiating a theory about the mind through a set of mathematical functions. In this approach, mathematical relations govern interactions among theoretical mechanisms that are assumed to underlie psychological operations (i.e., cognition). Usually, these mechanisms have corresponding parameter values and, like knobs on a machine, changes in the parameters create changes in the dynamics of the model and subsequent predictions for manifest variables. Parameters can be estimated by "fitting" the cognitive model to data containing the manifest variables of behavior, and those parameters can then be correlated with the aspects of physiology (Forstmann et al., 2008, 2010). Hence, the statistical link is between the physiological variables and the parameters intended to characterize the psychological processes.

Although there now exist many techniques for imposing statistical linking functions (Turner et al., 2017a), the focus of this chapter is on a technique we refer to as "joint modeling" (Turner et al., 2013a), which is illustrated in Fig. 2. Joint modeling is similar to the statistical linking approach described earlier: both techniques use cognitive models to explain the psychological processes in the behavioral manifest variables, and both techniques assume a type of linking function between those processes and the physiological manifest variables. However, joint modeling distinguishes itself based on its commitment to imposing a *reciprocal* link between the variables, meaning that the statistical model allows both manifest variables to affect all of the latent parameters. To do this, joint models assume a hierarchical modeling structure: one "submodel" is devised to characterize physiological manifest variables (e.g., brain data), another submodel is instantiated via a cognitive model, and finally the two submodels are joined by a linking function, such as a multivariate Gaussian distribution or a simple regression equation. In what follows, we use a neural submodel to refer to the submodel that explains the physiological variables, behavioral submodel to refer to the submodel that explains the behavioral variables, and cognitive model to refer to the overarching psychological theory that drives the analysis and interpretation. Although typically the cognitive model is just the behavioral submodel, this may not always be the case.

In this type of hierarchical model structure, parameters that explain physiological variables inform the parameters that create the linking function, and then the linking function itself subsequently informs the parameters of the psychological variables through the cognitive model parameters (Turner et al., 2019). This mutual passing of information between the manifest variables of the brain and behavior to the latent parameters of the cognitive model is why the link is characterized as reciprocal.

Examples of successful joint models

Since the formal specification of joint models, there have been several interesting methodological developments that have created new opportunities for understanding the links between the brain and behavior. In this section, we review a few successful applications, each with slightly different priorities in terms of what the link connecting the neural data to the cognitive model is designed to do.

Linking to explain trial-to-trial fluctuations in the mind

Although the first application of joint modeling by Turner et al. (2013b) used structural data to explain differences in the speed accuracy tradeoff across participants, the very next application (Turner et al., 2015) expanded this linking structure to the individual trial level where the single-trial BOLD response was used to predict the decisions made on each trial. In this application, Turner et al. (2015) first processed the neural data using standard fMRI processing pipelines (Eichele et al., 2008) and then parameters corresponding to the strength of the hemodynamic response were extracted for each region of interest (ROI) on each trial. For example, each participant had a single-trial estimate for each of 34 ROIs over 180 trials. For the cognitive model, Turner et al. (2015) used the popular diffusion decision model (DDM; Ratcliff, 1978), which separates the decision dynamics into components consisting of the speed of information processing, the amount of bias present in the decision, the amount of information a participant required to make a response, the time required for encoding the stimulus, and the motor time to execute a response (i.e., press a button). By separating out these various decision dynamics, the strategy was to allow the pattern of neural activations to correlate with trial-to-trial fluctuations in each cognitive model component, thereby allowing researchers to interpret the activity in each brain region in terms of cognitively meaningful mechanisms. However, to exploit the covariation of cognitive components and brain activation, a linking function was required so that patterns could be inferred across all trials. To this end, Turner et al. (2015) assumed a multivariate Gaussian linking function to connect all the neural activations to all the cognitive model parameters, which created a mean vector consisting of 37 (34 ROIs and 3 cognitive model parameters) means, and a 37×37 variance-covariance matrix. Although Turner et al. (2015) could have simply used a regression model to correlate each of the 34 ROIs with each of the three model parameters, the multivariate Gaussian allowed for complex ROI-to-ROI covariation that might jointly contribute to cognitive modeling component.

After fitting the model to data, Turner et al. (2015) performed several analyses to illustrate the benefits of using a cognitive model to guide the analyses of neural data. First, Turner et al. (2015) performed a cross-validation analysis to show that a joint model that used neural data to predict the behavioral response performed better than a cognitive model that did not use the neural data to predict the same responses, which suggested that some element of the neural data was allowing the joint model to enhance its predictions for the choice and response time. Fig. 1 shows the predictions from the neurally informed DDM (NDDM; green line) and the DDM (red line) overlayed on the (log) response time data (gray data) from the test trials, which have been smoothed to better illustrate the trial-to-trial temporal dependencies (10 trial moving average; black line). Because the standard DDM has only the training data to facilitate its predictions, it predicts (in the limit) the same response time for all test trials. By contrast, because the NDDM has the training data and the neural data at test to aid in its predictions, the NDDM was much more sensitive to local changes in the response time, which improved its predictions (shown in green) relative to the standard DDM.

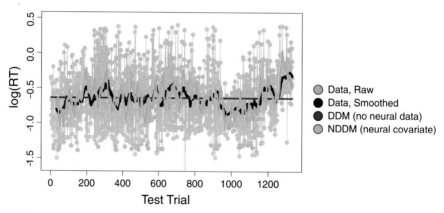

FIG. 1

Results of a cross-validation analysis. In a cross-validation analysis, Turner et al. (2015) used one-half of the choice response time data to predict the second half (i.e., termed the "test" data); here, the test data are shown on the log scale as the *gray dots*. After fitting a standard DDM and a neurally informed NDDM to the training data, they made predictions for the test data. Because the standard DDM has only the training data to facilitate its predictions, it predicts a stationary line shown in *red*. By contrast, because the NDDM has the training data and the neural data at test to aid in its predictions, the NDDM was much more sensitive to local changes in the response time, which improved its predictions (shown in *green*) relative to the standard DDM. *DDM*, diffusion decision model; *NDDM*, neural diffusion decision model; *RT*, response time.

Second, Turner et al. (2015) showed that by using the neural data, they could predict trial-to-trial fluctuations in the response time that were unaccounted for by the standard DDM. Third, Turner et al. (2015) showed that they could parcel patterns of neural activity according to cognitively meaningful regions of the parameter space. For example, by first defining a "high drift rate regime" in which choices were made quickly and accurately, they could examine patterns of neural activity that facilitated those types of responses. Similarly, they could identify patterns of neural activity that corresponded to slower, less accurate responses, and patterns that corresponded to slower motor responses.

Linking to explain relationships between EEG, fMRI, and behavior

Another application expanded the joint modeling structure to link multiple sources of brain data to the cognitive model. In Turner et al. (2016), a joint model was formed to connect distributions of fMRI and EEG data to the parameters of the linear ballistic accumulator (LBA; Brown and Heathcote, 2008), a cognitive model for perceptual decision-making that is similar to the DDM (Donkin et al., 2011a, b). In their study, Turner et al. (2016) were interested in understanding how different brain regions were involved in the subjective valuation of different delayed rewards when given two options. Turner et al. (2016) created three conditions that varied in the differences between the subjective values of the two options on each trial (e.g., an indifference condition, a small subjective difference, and a large subjective difference). As in any neuroscientific study, they first processed both the EEG and fMRI data to reduce the dimensionality to important electrodes and regions of interest, respectively. In processing the EEG data, Turner et al. (2016) used a simple event-related potential (ERP) analysis to identify regions over the ventral medial prefrontal cortex (vmPFC) that expressed noticeable differences across the three conditions during the decision time window. In processing the fMRI data, Turner et al. (2016) used a standard mask to define an ROI that was consistent with the vmPFC. In some preliminary analyses, they first confirmed that these preprocessing steps resulted in data that expressed differences across the three conditions. In addition, Turner et al. (2016) showed that the choice response time distributions were significantly altered across the three conditions; together, these preliminary analyses all suggested that their construction of the three conditions was appropriately tailored to most participants in the group.

To illustrate the utility of the neural data in constraining the cognitive model, Turner et al. (2016) developed a total of four joint models to test various links between the fMRI, EEG, and behavioral data. As a baseline, the first model completely ignored the neural data and used only the LBA model to describe the choice response time distribution across the three conditions. The next two models either used the EEG or the fMRI data to link the effect parameters in the neural modality to the drift rate parameter in the LBA. The rationale here was that larger activations of the vmPFC would be associated with more difficult decisions (i.e., where choices were less consistent and took longer), whereas smaller activations would be associated with easier decisions (i.e., where choices were more consistent and typically faster). The final model was a "trivariate" model that connected both EEG and fMRI data to the drift rate parameter. By comparing the three joint models to the behavioral-only model, Turner et al. (2016) were able to quantify the degree to which both neural measures contributed to enhancing the cognitive model's predictions in several leave-one-out cross-validation analyses. In addition, Turner et al. (2016) compared the full trivariate joint model to each of the two "bivariate" joint models (i.e., the two models containing one neural measure or the other, but not both), and found some evidence that having both EEG and fMRI provided slightly better cross-validation performance than either measure in isolation. Hence, the benefit of using a joint model is that one could infer the psychological state of an individual based only on the observed patterns of brain activity, and predicting the psychological state enables better predictions for many types of behavior.

Linking to explain nonstationary changes within a trial

One of the limitations of single-trial or across-modality linking is that at its core, the linking function is still assumed to be independent and identically distributed throughout the experiment. This can create an issue when a participant is in a different state of mind for a brief number of trials because the small number of trials gets swamped by the many other trials throughout the experiment, potentially causing some misfit or underappreciation of these critical minority trials. In the most recent application of joint modeling, Bahg et al. (2020) expanded the linking structure to build dynamic, moment-by-moment correlations between brain and behavior. Here, the linking function no longer assumed a common structure across all trials, but instead used a Gaussian process model to link the neural and behavioral data to the shared dimension of time. In other words, because both the neural and behavioral data are locked to a particular time point, any expressed deviations could be modeled uniquely because they were only expressed in one region of time rather than appealing to some global statistical structure (e.g., the multivariate Gaussian linking function).

By expanding the joint model to use Gaussian processes, Bahg et al. (2020) could visualize deviations in brain states in a continuous, multidimensional space. Although this creates an interesting visualization, it can sometimes be difficult to interpret what the deviations mean in terms of something psychologically meaningful because the Gaussian process joint model (GPJM) is less tied to a specific cognitive theory. Instead, Bahg et al. (2020) provided an analysis of the latent topology that emerged from fitting the GPJM to fMRI data using a continuous motion tracking task. To do this, Bahg et al. (2020) used multidimensional scaling to identify deviations from the most common brain states, and then subsequently examined the changes in the functional coactivation matrices of these deviated brain states from the common brain states. Here, Bahg et al. (2020) showed significant changes in the functional coactivations, with one set of deviated brain states potentially implicating an inhibitory process, and another exhibiting increases in the coactivations across the whole brain. Although more work is needed to draw stronger conclusions, the GPJM framework presents a potentially exciting new direction for linking brain to behavior on a moment-by-moment basis.

Assumptions of joint models
Assumption 1: The cognitive model well describes the cognitive process(es) used in the task

The key component of joint modeling is the behavioral submodel: although it is primarily intended to describe the behavioral data, there is also considerable emphasis on relating the processes in the model to the neural data. Speaking generally, the purpose of a model is to simplify complex patterns in data to enable a more general understanding of the essential components in said data. Engaging in computational modeling necessitates that one must be comfortable with an inherent tradeoff between the model's complexity and the model's ability to capture patterns in the data. As a general rule, when a model is more complex, it has more degrees of freedom that will allow it to capture even idiosyncratic patterns in data, and when a model is less complex, it will tend to capture fewer and fewer of the essential patterns in data (Myung, 2000; Pitt and Myung, 2002). There is considerable research and discussion trying to quantify how to achieve this theoretical "sweet spot," and although we cannot do that literature justice here, we can recommend a few great articles to initiate the reader (Myung, 2000; Myung et al., 2000; Pitt and Myung, 2002; Roberts and Pashler, 2000; Shiffrin and Nobel, 1997).

Beyond the issue of finding the right balance between complexity and generalizability is the issue of what the behavioral model is intended to help us explain: the mind. Unlike the neural submodel, which primarily is designed to describe the statistical patterns in the neural data, the behavioral submodel instantiates cognitive constructs to capture the behavioral data. In so doing, the behavioral submodel is attached to a theory about cognition, or more specifically, a theory about which set of cognitive components should be used when a person performs the task. Clearly, theories, and even the precise way in which a theory is implemented, can be wrong or misspecified. When a theory is wrong, it is usually unlikely that the parameters will exhibit any statistical relationship with the neural data, but it can happen if, for example, the theory assumes cognitive components that are similar to or mimic the "true" cognitive components used during the task. When this occurs, there is a danger that a researcher will come to incorrect or imprecise interpretations of what the correlation between a parameter of the behavioral model and the neural data actually means in terms of understanding cognition.

Unfortunately, there is no silver bullet to developing the right theory of cognition. As new theories develop, researchers seek out new experiments and new data that uniquely test for the consistency of that theory with manifest variables, and this is how progress is made. Along the way, it can be useful to engage in comparisons among sets of theories to provide direction and evidence for future researchers. To this end, joint modeling can be helpful in that different behavioral submodels can easily be swapped out and refit to data. For example, Turner (2019) showed how the same neural and behavioral data can be fit with two different joint models that differed only in the behavioral submodel. Once fit to data, researchers can engage in model selection, where the degree of fit relative to the complexity of the model can be assessed. Ideally, the model that is just complex enough to capture the data but not so complex that it could capture any data should be preferred. If both models explain the data equally well, then there is a consensus that the simpler model should be preferred, an attitude that is often referred to as the principle of parsimony because it applies a similar logic to Occam's razor (Myung and Pitt, 1997).

Assumption 2: A statistical relationship exists between the neural and behavioral data

The second assumption that should be satisfied is that there should be some statistical relationship between the neural data of interest and the cognitive model parameters. This assumption rests squarely on the linking propositions discussed earlier: if the psychological variables describe the behavioral data accurately and the neural data are manifestations of the psychological variables, then there should be some mathematical or statistical connection between them. In other words, if the psychological variables within the cognitive model are accurate reflections of the cognitive processes used during the task, then the cognitive model should accurately capture the behavioral data. However, it is also possible that the neural data are not accurate depictions of the cognitive processes. For example, if the neural data are processed incorrectly or are severely distorted by artifacts or measurement noise, then we should not expect any useful relationship between these neural measures and the parameters of even the best possible cognitive model.

Fortunately, the second assumption does not necessarily need to be satisfied in order for a researcher to use joint modeling. In fact, joint modeling is designed to evaluate the degree to which the psychological variables are related to the physiological variables (i.e., neural data). After fitting a joint model to data, if one finds a posterior estimate of the correlation parameter that is near zero, this would indicate that either the neural data are not reflective of the cognitive processes of interest, or that the cognitive model is not appropriately suited to interface with neural data. For example, if a cognitive model is flexible enough, it could capture the behavioral data correctly, but if there are too many parameters in the model, they may all fail to correlate with the neural data due to multicollinearity, which we will discuss more here. Again, we cannot emphasize enough the importance of a good cognitive model that is appropriately matched to the data of interest.

Building, testing, interpreting joint models

In this section, we will discuss some of the practical steps in building and testing joint models. As a working example, Fig. 2 illustrates a joint model for a generic fMRI experiment involving a multialternative choice task. The first row shows a time series of the neural stream of data, along with parameters δ of the neural submodel, whereas the second row shows a time series of the behavioral stream of data along with parameters θ of the behavioral submodel, where the model predictions for choice are arranged in a time series similar to the fMRI experiment. On the right, the two sets of parameters are controlled by a set of hyperparameters Ω, which comprise the linking function. In the sections that follow, we will discuss each of these components of the joint model in turn.

Select the neural data

One of the first two steps in building a joint model is to decide which aspects of the neural data are likely to reflect interesting aspects of the cognitive process. Deciding which of these aspects can be driven more by theory, such as in a confirmatory test of a link between a brain region and a model parameter, or it could be more driven by data, such as in an exploratory search of which brain areas are related to which model parameters. As we will discuss here, a practical consideration is to decide how many "features" of the neural data you are interested in studying; although significant progress has been made in enhancing the scalability of joint models, the complexity of the linking function always depends on the number of neural features that are being linked to the model parameters. Specifically, increasing the number of neural features increases the difficulty of inferring the number of significant relationships between the brain and the cognitive theory because (1) there are both technical and computational costs to estimating more parameters and (2) the parameters may trade off with one another, potentially obscuring strong inferential links.

What is a neural feature? In our view, a neural feature is simply a subset of the neural data that contains some potentially valuable information about the cognitive process under investigation. Selecting subsets of the data allows for a more pointed examination, allowing researchers to connect the eventual results to the broader scientific community. For example, if one were interested in examining the role that the dorsal-lateral prefrontal cortex (dlPFC) plays in a decision-making task, one could use an agreed upon mask to define the set of voxels in the brain that correspond to the dlPFC region. With this definition in place, one can extract the time series of those voxels and average them to create a time series that corresponds to an ROI, namely, the dlPFC. Then, if any significant relationships are found between components of the cognitive model and this constructed time series, it would substantiate the conclusion that the dlPFC is performing some function that is similar to the component of the cognitive model.

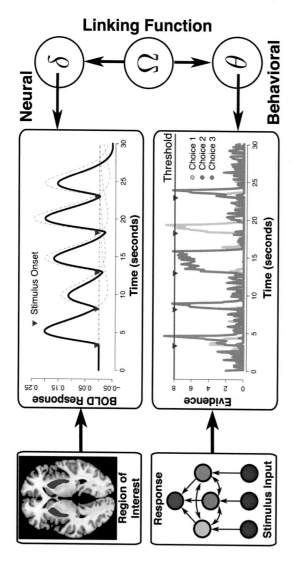

FIG. 2

Example structure of a joint model. The figure shows a hypothetical example consisting of 30 seconds worth of an experiment involving a decision among three alternatives. For neural data, regions of interest are defined (*left*) and the blood oxygenation level dependent (BOLD) response can be extracted. Statistical models can be fit to the observed BOLD time course (*middle*), and parameters δ for say, neural activation, can be estimated. For behavioral data, a cognitive model is developed (*left*) with mechanisms that are cognitively meaningful. The model can then be fit to data (*middle*), and parameters θ for say, drift rate, can be estimated. Finally, joint models specify how the neural parameters δ are related to the cognitive model parameters θ through a linking function. In each model schematic, *red triangles* indicate stimulus presentations.

Reproduced with permission from Turner, B.M., Palestro, J.J., Miletic, S., Forstmann, B.U., 2019. Advances in techniques for imposing reciprocity in brain-behavior relations.

Neurosci. Biobehav. Rev. 102, 327–336.

The top row of Fig. 2 illustrates how an ROI can be selected (left; basal ganglia), from which an average time series can be extracted (middle). In the middle panel, the red triangles correspond to the time at which the trial starts, and the dashed curves show hypothetical predictions for how activated the ROI is on that particular trial. Here, the degree of activation on each trial corresponds to a parameter that we generically refer to as δ. In this example, the δ parameter effectively scales the theoretical prediction for the BOLD response (a function called the hemodynamic response function [HRF]), such that larger values of δ produce larger BOLD responses. Finally, the solid black line represents the average of each of these predicted BOLD responses over time. The averaging in this scenario is to accommodate for the fact that the BOLD response often has carryover effects such that activation from the stimulus on trial t can still be present in the BOLD signal after the stimulus on the next trial is presented. In the end, this solid black line is the final predicted BOLD response from the neural submodel.

Because we typically do not know how the brain will react to the stimuli a priori, our goal is to estimate each of the single-trial parameters δ by fitting the neural submodel to the data. In a typical fMRI experiment, these parameters are often referred to as β. Indeed, one could simply estimate the single-trial activation parameters and simply correlate those parameter estimates with any aspect of the behavioral data (e.g., response time) or parameters of the cognitive model. However, there are good statistical and theoretical reasons to be cautious with the conclusions that can be drawn from these types of analyses (e.g., see Turner et al., 2013a, 2017a, 2018a, for detailed discussions). Perhaps the most problematic issue arises when the uncertainty of the model parameters is not accounted for when correlating (i.e., regressing) the parameter estimates against another variable: by assuming no noise in the parameter estimates, researchers risk attenuating the statistical significance (Ly et al., 2017). By contrast, the goal of joint modeling is to use both streams of data to boost the inferential power of the parameter estimation, and as a consequence, the ability of the researcher to evaluate the strength of the association between brain and theory. To guide this evaluation, we use a cognitive model.

Select the cognitive model(s)

The next step is to decide on a cognitive model to explain the behavioral data. What is a cognitive model? In our view, a cognitive model is a description of how an input (e.g., a stimulus) is mapped to an output (e.g., a response) through a set of assumed cognitive processes. The description will need to consist of a set of mathematical and statistical definitions so that the steps involved in mimicking cognition can be programmed into a function that will formally instantiate the model. Cognitive models are typically derived from overarching cognitive theories: there in fact may be several cognitive models that are derived from similar theories but are instantiated in slightly different ways. For example, an overarching theory called sequential sampling suggests that information is extracted from the environment at each moment in time and added up to create a representation of the overall information quality for a particular response. However, there are many ways in which the information could be extracted (e.g., prone to errors in encoding or not), added up (e.g., containing perceptual noise or not), maintained over time (e.g., subject to the passive loss of information or not), or even interact with other acquired information (e.g., competitive or not). Cognitive models take specific positions on each of these choices, and so they form specific instantiations of overarching theories.

In deciding which cognitive model to use, we do not need to worry whether or not we are choosing the "right" model. Instead, we can select a few different models, plug them into the joint model (creating several different joint models), and compare the results. Of course, there are practical considerations to weigh, but the main point is that joint modeling creates a platform in which different cognitive theories (or instantiations of those theories) can be formally evaluated and compared in their ability to explain how the brain produces behavior. So, although we want to select a cognitive model that will be able to fit the behavioral data well, what is also important is that the model has components that correspond to theoretical ideas that we wish to substantiate.

The bottom row of Fig. 2 illustrates a cognitive model. On the left, a diagram is used to convey that information (red) for each of three options is integrated to produce a response (pink), and this integration is competitive among the options (curved arrows). In the middle, we simulated choice dynamics from the model for a set of stimuli that were aligned with the neural time series in the top row. At each trial onset, the model accumulates information over time until eventually one of the three response alternatives gathers enough evidence to surpass a threshold amount of information (pink), which produces a response. As with the neural submodel, each trial has a parameter θ to describe how the stimuli are mapped to a response. Because we do not know the values of these parameters a priori, our goal is to estimate each parameter by fitting the joint model to data. To maximize our estimation ability, we need to specify the linking function.

Select the linking function

The last step in specifying the joint model is to specify the linking function. In a joint model, the linking function is simply the statistical specification of how the parameters of the neural submodel δ are related to the parameters of the behavioral submodel θ. Of course, there are many ways to specify the linking function, and our choice will depend on a number of practical and theoretical factors. For example, one common decision point is whether or not we wish to incorporate correlations that might exist between ROIs. If we suspect that there are no meaningful correlations between ROIs, we could use a simple general linear model to specify the linking function, such as

$$\theta_j = b_0 + b_1 \delta_j + \eta_j, \tag{1}$$

$$\eta_j \sim \mathcal{N}(0, \epsilon), \tag{2}$$

where b_0 is an intercept term, b_1 is a regression coefficient, and η_j is a noise term centered at zero and with standard deviation ϵ. Here, the neural submodel parameters are simply regressed onto the behavioral submodel parameter probabilistically, such that there is some degree of noise in this mapping process. The parameter b_1 would tell us about the strength of association between the neural and behavioral model parameters.

A very similar linking function is the multivariate normal distribution, which is written as

$$(\theta_j, \delta_j) \sim \mathcal{N}(\mu, \Sigma),$$

where μ now controls the center of the distribution, and Σ represents a variance-covariance matrix. The mean parameter vector μ will work identically to b_0 in the linking function described earlier, and Σ will quantity the strength of association between θ and δ just as b_1 did before. One big difference between these two linking functions is our ability to partition the variance-covariance matrix, and realize the conditional distribution of subsets of parameters (see Turner, 2015, for elaboration). As it is often the case that we have many ROIs, δ_j will actually a vector of parameters, one for each ROI. Similarly, we might have a vector of cognitive model parameters such that θ_j is also a vector. One of the benefits of the multivariate normal distribution as a linking function is that we can also estimate every possible correlation that might exist among the ROIs or between each model parameter and each ROI through Σ.

However, estimating every possible association will come with a potentially very high computational cost. In some cases, it will be prohibitively expensive to estimate all possible associations, so we will need to either be more selective with our choices of ROIs or reduce the number of parameters we want to estimate. Another benefit of using the multivariate normal distribution as a linking function is that we can constrain elements of the variance-covariance matrix Σ to be equal to zero. Applying constraints will reduce the flexibility of the linking function, but has the added benefits of creating a more confirmatory analysis and reducing the number of model parameters we need to estimate. As we will discuss here, there are other more strategic ways to specify Σ that will allow for scalable joint modeling analyses.

Coming back to our working diagram, Fig. 2 shows the linking function through the hyperparameter set Ω where, in the case of a multivariate normal linking function, $\Omega = \{\mu, \Sigma\}$. Here, Fig. 2 shows that the hyperparameters control the distribution of θ and δ through a hierarchical structure. This specification creates a massive inferential advantage if there are actual links between the cognitive model and the neural submodel (see Turner et al., 2013b; Turner, 2015, for elaboration). The primary benefit is through the hierarchical model structure: the information from the behavioral data informs the estimate of θ_j, which in turn informs the estimate of the parameters of the linking function Ω, which in turn inform the estimate of δ_j even before any neural data on the jth trial have been considered. This type of a priori specification provides a compelling advantage because it uses information from other trials and data streams to improve our ability to estimate the single-trial model parameters.

Fitting the model to data

With the model fully specified, our next task is to fit the model to data. This process involves estimating the model parameters at the group and participant level, and to estimate parameters, we must take one of two statistical perspectives: frequentist or Bayesian. A frequentist approach would be to find the very best value of every parameter by maximizing the "likelihood" function $\mathcal{L}(\theta, \delta, \Omega | N, B)$, which describes how likely the neural and behavioral data are given the model parameters. In other words, the parameter estimates $[\theta, \delta, \Omega]^*$ would be found by solving the following equation:

$$[\theta, \delta, \Omega]^* = \max_{\theta, \delta, \Omega} \{\mathcal{L}(\theta, \delta, \Omega | N, B)\}.$$

This approach, while technically allowed, can be very difficult when fitting a hierarchical model because the participant-level parameters depend on the group level parameters, and that dependency makes estimating all of the model parameters quite tricky. As a practical solution, we recommend the Bayesian approach. Fitting hierarchical models using Bayesian statistics is more convenient because of the many statistical algorithms that are available for jointly estimating the group and participant level parameters together. Although a detailed discussion of how to do this is outside of the scope of this chapter, the rough strategy is as follows. First, we note that the Bayesian estimate of the model parameters is derived through the *posterior distribution*, which can be written as

$$p(\theta, \delta, \Omega | N, B) \propto \mathcal{L}(\theta, \delta, \Omega | N, B) p(\theta, \delta, \Omega), \tag{3}$$

where $p(\theta, \delta, \Omega)$ denotes the prior distribution on the model parameters. The prior represents our belief about the most likely parameter values before we have observed the data N and B. As Eq. (3) shows, the posterior distribution is proportional to the likelihood function times the prior. In a Bayesian setting, we do not need to know the posterior function exactly—only proportionally, up to a constant—because the algorithms that are used to estimate the posterior will conveniently circumvent the need for evaluating the constant. Second, we need to derive the *conditional distribution* of each group of model parameters. Again, in a Bayesian setting working with conditional distributions is very natural, and only involves collecting the relevant terms for each set of parameters and disregarding the rest. For our joint model, we have the following set of conditional distributions for θ, δ, and Ω:

$$p(\theta | \delta, \Omega, N, B) \propto \mathcal{L}(\theta | B) p(\theta | \Omega), \tag{4}$$

$$p(\delta | \theta, \Omega, N, B) \propto \mathcal{L}(\delta | N) p(\delta | \Omega), \tag{5}$$

$$p(\Omega | \theta, \delta, \Omega, N, B) \propto p(\theta, \delta | \Omega) p(\Omega). \tag{6}$$

This follows from the fact that in a joint model, the likelihood function is split into two parts, one for the behavioral submodel and one for the neural submodel, such that

$$\mathcal{L}(\theta, \delta, \Omega | N, B) = \mathcal{L}(\theta | B)\mathcal{L}(\delta | N),$$

and that the priors can be specified in a hierarchical model such that the group level parameters can be separated from the participant-level parameters:

$$p(\theta, \delta, \Omega) = p(\theta, \delta | \Omega)p(\Omega).$$

With each of the conditional distributions worked out, we can use an algorithm known as Gibbs sampling to iteratively adjust each group of parameters to the right location in the parameter space. The algorithm roughly works like this:

1. Initialize all of the parameters to some starting values.
2. Evaluate the value of the first conditional distribution (4), given the current location of θ, δ, and Ω.
3. Generate a new proposal for θ, call it θ^*.
4. Evaluate the value of the first conditional distribution (4) under this new proposal θ^*.
5. If the value of the conditional is larger when using θ^* than it is when using the current position of θ, then move θ to θ^*.
6. If the value of the conditional is smaller when using θ^* than it is when using the current position of θ, then move θ to θ^* with a Metropolis-Hastings probability.
7. Repeat Steps 2–6 using the parameters δ and the second conditional distribution (5), given the current location of θ, δ, and Ω.
8. Repeat Steps 2–6 using the parameters Ω and the third conditional distribution (6), given the current location of θ, δ, and Ω.

This procedure can be repeated hundreds or thousands of times to eventually collect many samples of the full joint posterior distribution of all the model parameters.

Although one can manually program up functions that can perform Gibbs sampling, there are now many statistical software packages that can perform the estimations for you, such as in JAGS and Stan. Using minimal specification, one can define algorithmic specifications such as the number of samples to draw, number of chains to use and so on, and these programs will handle the rest.

Evaluate the fit and generalizability, and generate conclusions

Once the model has been fit to data, we need to inspect that the model is fitting the data appropriately. There are two aspects to evaluating fit. First, the model must generate predictions that are in close alignment with the data. Second, the model must not be overfitting the data. Having ensured both of these properties of a good fit, we can then draw conclusions about the associations that are inferred from the model parameters.

Evaluate agreement

Evaluating the agreement of the model with the data involves simply evaluating whether or not the model is making predictions that are close to the data that were fitted. The most straightforward way to do this is to use the best-fitting parameter values that were obtained during the data-fitting process and generate predictions for the manifest variables such as response times, choice, or neural time series for various ROIs. Once these are generated, one can compare the predictions to the observed data to ensure that they are in agreement.

Fig. 3A illustrates how such an evaluation may take place. In this example, the data are response times, shown on the log scale as the black dots for 10 trials. After having fit the model to data, one can take the best fitting parameter values, simulate data from the model, and then plot those simulated trials against the raw data. Fig. 3A shows these simulated trials as gray dots. To evaluate the accuracy of the model, one could calculate the distance between the model's predictions and the data, and aggregate these differences over trials.

A more formal way, called the posterior predictive distribution (PPD), to evaluate the agreement when using a Bayesian framework is to (1) take each posterior sample and generate predictions from the model, then (2) pool these predictions over all posterior samples. For example, when estimating a parameter δ_j we will obtain many samples from our sampling algorithm for δ_j. Because these are posterior samples, all of these values are more or less likely to be correct because in a Bayesian framework there is a natural uncertainty about every model parameter. We can propagate this uncertainty to the level of the manifest variables to see how certain our model is about the range of data that should be expected on that particular trial. To do so, we generate predictions from the model for the data point on Trial j by plugging in each of the samples. This process will create a distribution for the data on Trial j, and we can then assess whether or not the actual data we observed on Trial j matches the distribution from the model.

In some sense, if the parameters were estimated properly, there should be a good deal of overlap between the model predictions and the data. For example, if a sampled value of δ_j produced a prediction for the data on Trial j that was very inconsistent with the actual data on Trial j, then this sampled value of δ_j would not have been retained in our sampling algorithm and would not be part of our posterior estimate for δ_j. By this logic, the PPD should pretty much always be in line with the data. What is more important to assess is whether or not there is certainty in the prediction. If the model's prediction for the data on Trial j is very narrow, this indicates a great amount of certainty which can be very helpful when predicting data (e.g., behavior or brain response) to different experiments, different stimuli, or different people. This brings us to our next evaluation metric: generalizability.

Evaluate generalizability

If the model is in close agreement with the data, the next thing we must ensure is that this agreement is not too close. A problem when fitting very complex models to data is that the model might be more complex than what is warranted by the data. If this is

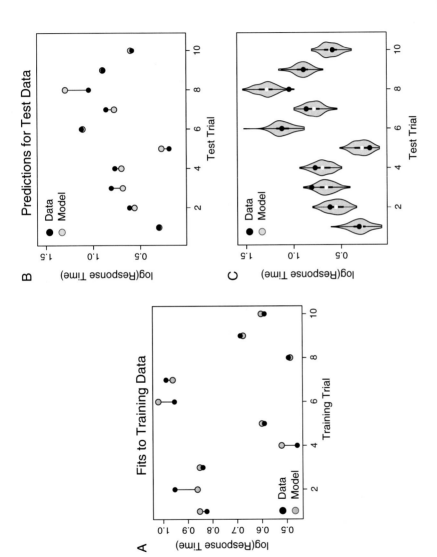

FIG. 3

Procedure for evaluating fit and generalizability. (A) This panel shows a way to evaluate agreement between the model fits (*gray dots*) and the data (*black dots*) by calculating the distance between the two points (*lines*). After aggregating over trials, we could summarize the degree to which the model is in agreement with the raw data. (B) This panel shows how we can test for generalization by using the parameters estimated from fitting one-half of the data (i.e., the training data, see Panel A) and using those parameters to make predictions for the other half (i.e., the test data). Here, the held-out data are shown in *black* and the model predictions are shown in *blue*. (C) A better test of generalization would involve many simulations of the model to construct a posterior predictive density (*blue violin plots*). Here, by integrating over the posterior distribution, we get a sense of how the uncertainty in the parameter estimates creates additional uncertainty in the model predictions. The density of the model's posterior at the location of the held-out data provides a way to evaluate the model's generalization performance.

the case, then the model fits are relatively meaningless because the complex model probably could have fit any data. This can create problems because it is likely the case that many different parameter values could have equally well fit the data, or at least, the data are not providing a sufficient amount of constraint on the model parameters such that we can reliably detect meaningful differences in the parameter values from one trial to the next, or from one participant to the next. This is a problem in joint modeling where the cognitive model parameters are meant to help us interpret changes in the brain signal through the lens of the cognitive model.

To guard against this possibility, the standard approach is to perform some type of cross-validation analysis. There are many ways to do this, but a simple example is data partitioning, where you split the data into a "training" set and a "test" set. You then fit the model to the training set and use the parameters to generate predictions for the test set. The degree of accuracy with which the model can predict the test set can be quantified and evaluated relative to other models under consideration. Fig. 3A illustrates how the model is first fit to the training data, and then the parameter estimates can be used to generate predictions for the test data in Fig. 3B, shown in blue. The distance between these model predictions (blue dots) and the held-out data (black dots) can be used to quantify the model's generalization performance (e.g., by calculating the root mean squared error). Although this approach neglects the uncertainty of the estimates of the model parameters, the aggregation of error over trials should still be useful for a quick comparison among models.

Another way to quantify the accuracy when fitting the model in Bayesian paradigm is to generate the posterior predictive density (PPD) for each data point in the test set and evaluate the density of the observed data point under the PPD (i.e., higher densities are associated with higher accuracy). Fig. 3C illustrates the PPD (blue violin plots) for each trial in the test set, shown against the test data (black dots). Here, we get a better sense of how accurate the model predictions will be in the limit, and we can use the location of each test trial's data within the PPD to evaluate performance. For example, although the model makes a very accurate prediction on Trial 9 (i.e., the black dot is perfectly centered within the PPD), the prediction on Trial 8 is much worse (i.e., the black dot is in the tail of the PPD, likely reflecting a low density value).

Generate conclusions

After having assured ourselves that the model fits are both accurate and generalizable, we can now inspect the estimated parameters and draw conclusions. As an example, when using the multivariate linking function, we can inspect the individual elements of the variance-covariance matrix Σ. Each of these off-diagonal elements in the matrix contain an estimate for the covariance parameter, which can be converted into a correlation parameter. This correlation will tell us the degree of association between, say, a cognitive model parameter and an ROI, with the usual interpretation of correlation: values different from zero are considered strong associations. In addition, one can examine the degree of correlation between ROIs to get a better sense of functional coactivation that may exist during the task.

Pitfalls of joint modeling

Although joint models are fairly robust as an analysis tool, joint models are not completely free from the risk of failure. In this section, we discuss three potential pitfalls that the would-be user of joint modeling should be aware of before considering it as a new tool for analyzing the links between brain and behavior.

Pitfall 1: Computation

Perhaps the biggest limitation to using a joint model is the amount of technical expertise that is usually needed for a custom-tailored joint model for your data. On the cognitive model side, one will need to understand cognitive theory, be able to understand how that theory is instantiated in a computational model which usually involves a complex set of equations, and then be able to implement that model on a computer which usually involves programming equations in the user's preferred language or statistical software. On the neural data side, one will need to understand how to either collect data or collaborate with someone who has the ability to collect good neural data, preprocess those data, and extract meaningful neural signatures of the cognitive process of interest (e.g., an ROI analysis with fMRI data, defining an epoch of time with EEG data). Finally, to connect the two sides, one will have to understand what a hierarchical model is, how to specify the linking function in either mathematical or statistical terms, and then write code that will perform the estimation of all the parameters of the full hierarchical model. Although we recommend Bayesian estimation when fitting the joint model to data, we also acknowledge that performing Bayesian estimation typically requires complex algorithms to estimate the model parameters accurately. For example, one standard method for estimating the joint posterior distribution is Markov chain Monte Carlo (MCMC). For the typical user, learning about MCMC requires some knowledge of statistics, and then implementing MCMC for your particular problem would require some knowledge of how to effectively program up functions for likelihoods, priors, and perturbation kernels.

Fortunately, there has been growing commitment in the field toward accessibility, leading most researchers to make available code for implementing their cognitive models, code, and pipelines for performing preprocessing, and freely available software for performing sophisticated Bayesian analyses including hierarchical modeling such as JAGS (Plummer, 2003) and Stan (Carpenter et al., 2016). Using these programs, one can specify even a complex hierarchical model with only a few lines of code, and quickly be fitting their first joint model to data. In addition, many researchers are committed to disseminating tutorials for performing analyses and doing computational modeling in the form of research articles, blogs, and even videos. For implementing joint models, we have written a tutorial that shows the reader how to link fMRI data to the parameters of the DDM (Ratcliff, 1978) in Palestro et al. (2018), with code available on GitHub (https://github.com/MbCN-lab/joint-modeling-tutorial). We have also written a book on joint modeling

(Turner et al., 2018a) in a tutorial style that has code directly inset into the book to show the reader how to map equations into code and explains what each line is intended to do. Although these two resources use specific examples and scenarios to teach the reader how to use joint models, they unfortunately may not transfer perfectly to each user's situation. In these cases, the reader may need to read materials like this chapter or reach out to joint modeling practitioners for help and guidance in developing a custom joint model for a unique situation.

Pitfall 2: Scalability

Another potential pitfall to the would-be user of a joint model is scalability. When forming a joint model, one must make a decision about how to relate the parameters of the cognitive model to the brain data. Although brain data are technically finite, the data usually consist of very large data sets. For example, a typical brain consists of about 100,000 voxels, and each voxel has a time series that can be 30–60 minutes in length, with a data point recorded every 1.5–2 seconds. EEG data are similarly large, not due to its spatial resolution (i.e., number of voxels) but due to its temporal resolution, often taking up to 500 measurements per second for 32, 64, or 128 channels, depending on the configuration of the electrode cap. Ideally, one would want to examine and test the links between the cognitive model parameters and all of these data points, but such an analysis is clearly at odds with practical implementation and the theoretical underpinnings of null hypothesis testing. For example, even if it were possible to test a link between a model parameter and every voxel in the brain, the possibility of obtaining a significant correlation by chance would be nearly guaranteed (i.e., the Type I error would be very high) if considerable precautions were not taken. Hence, some type of dimensionality reduction on the brain data is not only necessary but could dramatically benefit the power of the inference that joint modeling is intended to make.

How much reduction is necessary for the brain data to be applicable in a joint modeling context? The answer to this question centers on the linking function. For example, although the multivariate linking function provides a number of practical benefits (e.g., the mean parameter vector is the central tendency of the model parameters, the covariance matrix quantifies the correlation between a model parameter and a brain region), as one includes more brain areas in the analysis, the complexity of the covariance matrix increases quadratically, which places undue burden on the statistical methods for estimation. To reduce the rate of complexity, Turner et al. (2017b) developed a factor analytic linking function, where the goal of the joint model is to extract a set of "cognitive factors" which should be related to the theoretical ideas that motivate the cognitive model. In this approach, including more brain areas into the analysis causes only a linear increase in the complexity of the covariance matrix, which has been shown to be dramatically more efficient and robust to misspecification. For example, when data are truly generated by a multivariate normal linking function, the factor analysis linking function can still recover an accurate covariance matrix despite the reduction in the number of freely estimated covariance parameters.

In addition to the factor analytic linking function, more recent work has successfully incorporated regularization techniques such as the LASSO and Ridge methods (Kang et al., 2022b). Embracing the Bayesian perspective, in this approach a prior imposes a type of regularization that exploits the idea of shrinkage (Polson and Scott, 2011, 2012; Park and Casella, 2008) to reduce the complexity of the linking function. Essentially, the shrinkage prior enforces that neural features exhibiting zero or small correlations (or factor loadings) with the behavioral model should be pushed downward to have a zero-loading structure, whereas features with high or moderate loadings are not penalized downward as much. Regularizing the correlation parameters in this way facilitates a parsimonious structure of the brain-behavior link to emerge naturally, which facilitates clearer explanations of how cognitive mechanisms play out in mental computations. Furthermore, as shown in simulation studies (Kang et al., 2022b), allowing features with low correlations to be "zeroed out" reduces some estimation biases from occurring in some model parameters.

Regularization techniques mainly facilitate an understanding of complex brain-behavior relationships. That is, there is nothing about the regularization itself that reduces the sheer number of parameters, or number of brain-behavior links to test. However, complexity metrics such as the effective number of model parameters do reduce, which can accelerate parameter estimation conducted by posterior sampling algorithms such as MCMC. In addition, one can envision an algorithm that would discontinue posterior sampling for parameters deemed to have zero correlation. Although future work will need to verify that such procedures do not adversely affect the estimation of other parameters in the model, such a procedure would dramatically improve our ability to scale joint models to more complex data structures.

In addition to regularization methods, new methods are being developed to accelerate the efficiency of parameter estimation, especially in the Bayesian context. For example, we have found that variational inference (Galdo et al., 2020) holds great promise for establishing accurate posterior estimates for models used in cognitive science, and this accuracy can be achieved in a small fraction of the time that it would take to use even the most efficient Bayesian sampling algorithms such as differential evolution MCMC (Turner et al., 2013b). Even in the context of hierarchical models, variational inference may prove instrumental in fitting complex joint models to data.

Pitfall 3: Multicollinearity

A final pitfall is not specific to joint modeling per se but is common to any statistical method involving multiple variables. As the goal of joint modeling is to test links between cognitive model parameters and brain data, it is possible that many brain areas correlate with a model parameter. This can be a potential problem because these correlations could manifest not because the brain regions are related to the parameter of the cognitive model, but because they are related to one another. Because the brain operates off of highly dynamic and interactive communications from one region to the next, the temporal structure can be highly correlated depending on a number of factors such as structural connectivity and proximity.

Unfortunately, correlated brain measures are unavoidable given the brain's nature and an increasing pressure on better temporal and/or spatial resolution (e.g., smaller voxels will tend to correlate more than larger voxels). One could be strategic in the selection of brain measures, such as defining a small set of key regions of interest to reduce redundancy in the brain measurements, or defining key temporal windows and spatial frequencies in EEG analyses. A more statistical approach would be to focus on developing better statistical models of the brain data, such as building in the natural temporal and spatial correlations that already exist in the brain by using methods such as topographic latent source analysis (TLSA; Gershman et al., 2011) or topographic factor analysis (TFA; Manning et al., 2014). A more theoretical approach would be to build in the structural dependencies of the brain as a means to create a more informed model of the brain data, such as the efforts of dynamic causal modeling (Friston et al., 2003) or nonparametric Bayesian approaches (Kang et al., 2022a). In the end, because the brain is very complex, putting substantial effort into characterizing complex dynamics should pay dividends in terms of reducing potential pitfalls such as multicollinearity.

Open challenges and future directions

We close this chapter with a brief discussion of open challenges and future directions, which we have divided into three sections: increasing accessibility, building better models of neural dynamics, and creating more computational models of cognition that are inspired by the brain. In addition to these three specific challenges, there will always be the general challenge of making the computations more efficient in terms of maximizing accuracy and minimizing time. Although there have been considerable developments in performing estimation of complex, hierarchical models, we think there will always be room for improvement as technology continues to become more powerful.

Increasing accessibility

One of the most important challenges for joint modeling is making the methodology more accessible. As we mentioned, there are many learning curves to mastering computational modeling and processing neuroimaging data. All of these curves take time and are difficult by themselves, and can of course be daunting when considered in combination. For these reasons, joint modeling practitioners will have to create better resources for teaching users all the ins and outs of linking model parameters to neuroimaging data. Although we have made progress in creating these resources (Palestro et al., 2018; Turner et al., 2018a), a continued effort will be necessary to ensure successful dissemination of the latest, cutting-edge developments (Bahg et al., 2020; Galdo et al., 2020; Kang et al., 2022a, b).

Better models of neural dynamics

Another important direction, common to all neuroscientists, is the improvement of models of neural dynamics. Although the brain is highly complex and consists of a myriad number of interactions that take place over long ranges of the cortex in a fraction of a second, building better models of these interactions will greatly improve our ability to detect a region's involvement in a cognitive process through mainstream statistical techniques, such as the general linear model. As an example, the parametric form of the HRF is often chosen to be a combination of two gamma functions, which well describe the BOLD response in many regions of the brain (Mumford et al., 2012; Poldrack et al., 2011). However, as a standard approach, most researchers often set the parameters of this function to specific values that were estimated based on the activation of the visual cortex. This practical choice is very convenient because it avoids having to estimate the many parameters of the HRF and allows inference to be on the degree to which the activation of an ROI corresponds to the task, but it has the downfall of possible misspecification. For example, it is known that this rigid HRF function is not a great choice for some regions of the brain (Friston et al., 2000).

There are of course many other ways of improving models of neural dynamics, and many researchers have made inspiring progress in this endeavor. For example, the Balloon model considers many structural and anatomical details of the vascular system, and has found its place at the core of many applications of dynamic causal modeling (Friston et al., 2003). Of course, improving the description of the BOLD response often comes at the cost of a more complicated model, meaning more effort spent learning about the model and computation time when applying it to data, but we hope these efforts ultimately pay off as proper models of brain function are a critical next step in the area of model-based cognitive neuroscience.

Of course, fMRI is not the only type of data that could benefit from better computational models—when dealing with EEG measures, researchers often use nonparametric functions to characterize ERPs. These functions nicely approximate the ERPs, but they do not allow us to connect the EEG data to a more general theoretical description of the brain. Unfortunately, because of the inverse problem, EEG measures cannot be reliably mapped to spatial locations within the brain. It is our hope that methods of simultaneous EEG-fMRI data collection will be more available, allowing for the best of both worlds: temporal and spatial resolution. With such data, the field can build better statistical models of brain function, generating predictions for both BOLD and event-related responses (e.g., using the lead field matrix; David et al., 2005; Bojak et al., 2010).

Brain-inspired theories of cognition

Much of the utility of joint modeling relies on the assumption that the theory of cognition is appropriate for the cognition that took place during the task. A litmus test for most cognitive theories is whether or not they can generate predictions that closely match data. However, as is well established by many mathematical psychologists (Roberts and Pashler, 2000; Myung, 2000; Myung et al., 2000; Myung and Pitt, 1997), good fits to data are necessary, but not sufficient. It is possible that many of these cognitive theories that transition from explaining only behavioral data will need refinements to make close connections with neural data as well. To this end, a growing challenge for cognitive theorists will be to create computational models of cognition that carefully consider how the brain is involved in the decisions humans make.

To highlight one example of how the brain can inspire new theoretical developments in computational models, consider the leaky, competing accumulator (LCA; Usher and McClelland, 2001) model. The LCA model was proposed as a neurally plausible model of choice response time in multialternative tasks. The LCA extended other diffusion-like models at the time (e.g., Ratcliff, 1978) by including new, brain-inspired components such as information leakage and competition by way of lateral inhibition, which were shown to play a role in many different neuron dynamics. Using these new dynamics, Usher and McClelland (2001) and Tsetsos et al. (2011) have shown that human participants do not process information in a purely lossless way as suggested by other diffusion-type models: human cognition seems to be subject to both primacy and recency effects. For example, when deciding which of four alternatives has the brightest light, if one alternative is given the strongest brightness in the first half of the trial (e.g., the first 2 seconds), and another alternative is given the same brightness signal in the second half of the trial (e.g., the last 2 seconds), participants often report consistent differences in which of the two alternatives had the greatest brightness for the trial: whereas some participants report the first option (i.e., a "primacy" bias), other participants report the second option (i.e., a "recency" bias). The LCA model (Usher and McClelland, 2001; Tsetsos et al., 2011) has been shown to reliably produce both patterns of responding based on the balance of leakage and competition: primacy bias is explained by having high competition and low leakage, whereas recency bias is explained by having low competition and high leakage.

Although these differences between leakage and competition have been shown to be essential in explaining how humans integrate information over time, how have these differences materialized in our understanding of the brain? In one study, Turner et al. (2018b) showed that the balance of leakage and competition can also explain individual choice behavior in tasks involving cognitive control, such as the intertemporal choice task. In this task, participants are asked to decide between a reward that will occur sooner (a "shorter sooner" option), and another reward that is larger but will occur later (a "larger later" option). In a purely economical sense, participants should choose the larger reward, but they often succumb to the temptation that immediate rewards present. Turner et al. (2018b) derived single-trial estimates of the amount of competition (relative to leakage) by fitting the LCA to behavioral data, and then correlated these values with the BOLD response in an intertemporal choice task using FMRI data. They found that when participants chose the larger later option, a greater amount of competition was used by the LCA model, and this correlated positively with activation of the dlPFC (an area often implicated in being involved in self-control). Although the analysis did not use a complete joint model, these correlations suggest a possible link between the lateral inhibition component of the LCA model and the concept of cognitive control carried out by control networks in the brain.

Although the LCA is just one example, it is our hope that more modern cognitive theorists will use important brain nonstationarities as an inspiration for cognitive nonstationarities, such as those discussed in Bahg et al. (2020). Such developments should promote a more common set of cognitive components that will be central to explaining both behavioral and neural data, and as such, will make applications of joint models far more direct and successful.

Take-home points

- Joint models provide a principled framework for estimating the link (i.e., the strength of association) between brain and behavior through theoretically motivated cognitive mechanisms.
- Joint models can be constructed to investigate links at different levels of analysis, such as across subjects, across trials, or across moments in time.
- Joint models can be used to test for links between brain and behavior in either a confirmatory or exploratory fashion.
- Recent work has focused on making high-dimensional linking functions a reality through factor analytic techniques and dimensionality reduction through regularization.
- Although joint models can be mathematically complicated, there are now many powerful open source software programs that can fit joint models to data with minimum specification from the user, which significantly decreases the initial cost of learning how to do joint modeling.

References

Bahg, G., Evans, D., Galdo, M., Turner, B.M., 2020. Gaussian process linking functions for mind, brain, and behavior. Proc. Natl Acad. Sci. 117, 29398–29406.

Bojak, I., Oostendorp, T.F., Reid, A.T., Kötter, R., 2010. Connecting mean field models of neural activity to EEG and fMRI data. Brain Topogr. 23 (2), 139–149.

Brindley, G.S., 1970. Physiology of Retina and Visual Pathways, second ed. Williams & Wilkins, Baltimore, MD.

Brown, S., Heathcote, A., 2008. The simplest complete model of choice reaction time: linear ballistic accumulation. Cogn. Psychol. 57, 153–178.

Carpenter, B., Gelman, A., Hoffman, M.D., Lee, D., Goodrich, B., Betancourt, M., Brubaker, M., Guo, J., Li, P., Riddell, A., 2016. Stan: a probabilistic programming language. J. Stat. Softw. 76, 1–37.

David, O., Harrison, L., Friston, K.J., 2005. Modelling event-related responses in the brain. NeuroImage 25 (3), 756–770.

Donkin, C., Brown, S., Heathcote, A., 2011a. Drawing conclusions from choice response time models: a tutorial. J. Math. Psychol. 55, 140–151.

Donkin, C., Brown, S., Heathcote, A., Wagenmakers, E.J., 2011b. Diffusion versus linear ballistic accumulation: different models for response time, same conclusions about psychological mechanisms? Psychon. Bull. Rev. 18, 61–69.

Eichele, T., Debener, S., Calhoun, V.D., Specht, K., Engel, A.K., Hugdahl, K., von Cramon, D.Y., Ullsperger, M., 2008. Prediction of human errors by maladaptive changes in event-related brain networks. Proc. Natl Acad. Sci. USA 16, 6173–6178.

Forstmann, B.U., Dutilh, G., Brown, S., Neumann, J., von Cramon, D.Y., Ridderinkhof, K.R., Wagenmakers, E.J., 2008. Striatum and pre-SMA facilitate decision-making under time pressure. Proc. Natl Acad. Sci. 105, 17538–17542.

Forstmann, B.U., Anwander, A., Schäfer, A., Neumann, J., Brown, S., Wagenmakers, E.J., Bogacz, R., Turner, R., 2010. Cortico-striatal connections predict control over speed and accuracy in perceptual decision making. Proc. Natl Acad. Sci. 107, 15916–15920.

Friston, K.J., Mechelli, A., Turner, R., Price, C.J., 2000. Nonlinear responses in fMRI: the balloon model, Volterra kernels, and other hemodynamics. NeuroImage 12 (4), 466–477.

Friston, K., Harisson, L., Penny, W., 2003. Dynamic causal modeling. NeuroImage 19, 1273–1302.

Galdo, M., Bahg, G., Turner, B.M., 2020. Variational Bayesian methods for cognitive science. Psychol. Methods 25, 535–559.

Gershman, S.J., Blei, D.M., Pereira, F., Norman, K.A., 2011. A topographic latent source model for fMRI data. NeuroImage 57, 89–100.

Kang, I., Galdo, M., Turner, B.M., 2022a. Constraining functional coactivation with a cluster-based structural connectivity network. Netw. Neurosci. 6, 1032–1065.

Kang, I., Yi, W., Turner, B.M., 2022b. A regularization method for linking brain and behavior. Psychol. Methods 27, 400–425.

Ly, A., Boehm, U., Heathcote, A., Turner, B.M., Forstmann, B., Marsman, M., Matzke, D., 2017. A flexible and efficient hierarchical Bayesian approach to the exploration of individual differences in cognitive-model-based neuroscience. In: Moustafa, A.A. (Ed.), Computational Models of Brain and Behavior. vol. 1. Wiley Blackwell, pp. 467–480.

Manning, J.R., Ranganath, R., Norman, K.A., Blei, D.M., 2014. Topographic factor analysis: a Bayesian model for inferring brain networks from neural data. PLoS ONE 9, e94914.

Mumford, J.A., Turner, B.O., Ashby, F.G., Poldrack, R.A., 2012. Deconvolving BOLD activation in event-related designs for multivoxel pattern classification analyses. NeuroImage 59, 2636–2643.

Myung, I.J., 2000. The importance of complexity in model selection. J. Math. Psychol. 44, 190–204.

Myung, I.J., Pitt, M.A., 1997. Applying Occam's razor in modeling cognition: a Bayesian approach. Psychon. Bull. Rev. 4, 79–95.

Myung, I.J., Forster, M., Browne, M.W., 2000. Special issue on model selection. J. Math. Psychol. 44, 1–2.

Palestro, J.J., Bahg, G., Sederberg, P.B., Lu, Z.L., Steyvers, M., Turner, B.M., 2018. A tutorial on joint models of neural and behavioral measures of cognition. J. Math. Psychol. 84, 20–48.

Park, T., Casella, G., 2008. The Bayesian Lasso. J. Am. Stat. Assoc. 103 (482), 681–686. https://doi.org/10.1198/016214508000000337.

Pitt, M.A., Myung, I.J., 2002. When a good fit can be bad. Trends Cogn. Sci. 6, 421–425.

Plummer, M., 2003. JAGS: a program for analysis of Bayesian graphical models using Gibbs sampling. In: Proceedings of the 3rd International Workshop on Distributed Statistical Computing.

Poldrack, R.A., Mumford, J.A., Nichols, T.E., 2011. Handbook of Functional MRI Data Analysis. Cambridge University Press, New York, NY.

Polson, N.G., Scott, J.G., 2011. Shrink globally, act locally: sparse Bayesian regularization and prediction. In: Bayesian Statistics 9, Oxford University Press, https://doi.org/10.1093/acprof:oso/9780199694587.003.0017.

Polson, N.G., Scott, J.G., 2012. On the half-Cauchy prior for a global scale parameter. Bayesian Anal. 7 (4), 887–902.

Ratcliff, R., 1978. A theory of memory retrieval. Psychol. Rev. 85, 59–108.

Roberts, S., Pashler, H., 2000. How persuasive is a good fit? Psychol. Rev. 107, 358–367.

Schall, J.D., 2004. On building a bridge between brain and behavior. Ann. Rev. Psychol. 55, 23–50.

Shiffrin, R.M., Nobel, P.A., 1997. The art of model development and testing. Behav. Res. Meth. Instrum. Comput. 29, 6–14.

Teller, D.Y., 1984. Linking propositions. Vis. Res. 24, 1233–1246.

Tsetsos, K., Usher, M., McClelland, J.L., 2011. Testing multi-alternative decision models with non-stationary evidence. Front. Neurosci. 5, 1–18.

Turner, B.M., 2015. Constraining cognitive abstractions through Bayesian modeling. In: Forstmann, B.U., Wagenmakers, E.J. (Eds.), An Introduction to Model-Based Cognitive Neuroscience. Springer, New York, NY, pp. 199–220.

Turner, B.M., 2019. Toward a common representational framework for adaptation. Psychol. Rev. 126, 660–692.

Turner, B.M., Forstmann, B.U., Wagenmakers, E.J., Brown, S.D., Sederberg, P.B., Steyvers, M., 2013a. A Bayesian framework for simultaneously modeling neural and behavioral data. NeuroImage 72, 193–206.

Turner, B.M., Sederberg, P.B., Brown, S., Steyvers, M., 2013b. A method for efficiently sampling from distributions with correlated dimensions. Psychol. Methods 18, 368–384.

Turner, B.M., Van Maanen, L., Forstmann, B.U., 2015. Combining cognitive abstractions with neurophysiology: the neural drift diffusion model. Psychol. Rev. 122, 312–336.

Turner, B.M., Rodriguez, C.A., Norcia, T., Steyvers, M., McClure, S.M., 2016. Why more is better: a method for simultaneously modeling EEG, fMRI, and behavior. NeuroImage 128, 96–115.

Turner, B.M., Forstmann, B.U., Love, B.C., Palmeri, T.J., Van Maanen, L., 2017a. Approaches to analysis in model-based cognitive neuroscience. J. Math. Psychol. 76, 65–79.

Turner, B.M., Wang, T., Merkel, E., 2017b. Factor analysis linking functions for simultaneously modeling neural and behavioral data. NeuroImage 153, 28–48.

Turner, B.M., Forstmann, B.U., Steyvers, M., 2018a. Computational approaches to cognition and perception. In: Criss, A.H. (Ed.), Simultaneous Modeling of Neural and Behavioral Data. Springer International Publishing, Switzerland.

Turner, B.M., Rodriguez, C.A., Liu, Q., Molloy, M.F., Hoogendijk, M., McClure, S.M., 2018b. On the neural and mechanistic bases of self-control. Cereb. Cortex 29, 1–19.

Turner, B.M., Palestro, J.J., Miletic, S., Forstmann, B.U., 2019. Advances in techniques for imposing reciprocity in brain-behavior relations. Neurosci. Biobehav. Rev. 102, 327–336.

Usher, M., McClelland, J.L., 2001. On the time course of perceptual choice: the leaky competing accumulator model. Psychol. Rev. 108, 550–592.

Network modeling: The explanatory power of activity flow models of brain function

Michael W. Cole

The Center for Molecular and Behavioral Neuroscience, Rutgers University,
Newark, NJ, United States

Introduction to activity flow modeling

Activity flow is defined as the movement of activity between neural populations (Cole et al., 2016). The concept of brain activity flow is everywhere and nowhere in the neuroscience literature. It is everywhere in the sense that the standard model of neural transmission—wherein action potentials flow along axons to influence downstream neurons via neurotransmitter release impacting their dendrites—involves the flow of "activity" (electrochemical signals). Yet, activity flow is nowhere in neuroscience in the sense that neuroscientific inferences are normally made using *either* activity patterns or connectivity. Seven years of studies focusing on combining activity patterns and connectivity (typically using functional/effective connectivity (FC)) to build activity flow models—simulations of the generation of neurocognitive functions via activity flow processes—has demonstrated the broad utility of this approach above and beyond standard activity *or* connectivity approaches alone. Modeling integration between task-evoked activity and connectivity to make strong inferences about brain function will likely be essential for developing rich causal explanations of the neural basis of cognitive functions, for the purpose of fundamental understanding and developing treatments for brain disorders.

To better illustrate the relevance of the activity flow modeling approach, let us consider a hypothetical scenario wherein an alien technology lands on Earth: the optimal brain imager (OBI). After some fiddling, human scientists discover that the OBI noninvasively reads out every aspect of the entire human brain at the atomic level (including electromagnetic fields) at microsecond resolution. Immediately, full brain scans of human brain anatomy and human brains performing all manner of tasks are collected, and the data are rapidly analyzed by the accompanying alien computer (capable of handling the massive datasets produced by the OBI). These analyses map the human connectome at full molecular resolution, along with cellular-level whole brain maps of neural activity patterns accompanying every

Computational and Network Modeling of Neuroimaging Data. https://doi.org/10.1016/B978-0-443-13480-7.00004-1

stimulus and task variable. Neuroscientists predict that all of the mysteries of the brain will soon be solved. Certainly, some mysteries are rapidly solved, and neuroscientists rejoice.

However, it eventually becomes clear that the data and analyses derived from the OBI present a major barrier to fundamental understanding of how brain activity generates cognition and behavior. Specifically, even after full mapping of brain connectivity and function-related brain activity, it remains unclear how these "parts" work together to generate neurocognitive functions. Such *generative* understanding of function is akin to understanding how the parts of a car work together (causally interact) to generate the emergent properties of rapid and controlled movement (Ito et al., 2020). For example, understanding how pressing a car's accelerator generates movement requires knowledge of the causal relationships between the system's components (e.g., the pedal, the transmission, the engine, and the wheels). Without such an understanding, cars would appear to be mysterious to us, with the practical problem that no one would be able to fix a car if it were to break down. This is largely the situation with the human brain—many of its functions remain mysterious to us, and we have only a very limited ability to fix it when brain disorders arise.

Activity flow modeling is one possible solution to the problem of generative understanding. Rather than treating mapping of brain connectivity and brain activity patterns as goals in and of themselves, these become starting points in the quest for generative understanding of brain function. Indeed, activity flow models are directly built on the combination of brain connectivity and brain activity patterns: First, empirical connectivity between all nodes (neural populations) of interest are identified, followed by input of empirical task-evoked activity into those nodes. A subset of nodes are "held-out"—meaning their empirical task-evoked activities are not used as input—such that the activity of those nodes can instead be generated by the model. The basic generative process is the same as most neural network simulations: a node's activity is determined by the connectivity-weighted sum of the activity inputs into that node (Fig. 1A). Model accuracy is assessed by comparing the generated task-evoked activity with the empirical task-evoked activity (Fig. 1B). To the extent that the generated task-evoked activity is accurate for a given neural population, the contributing activity, connectivity, and activity flow (i.e., activity-connectivity interaction) processes can be analyzed to infer details of the generative processes driving task-evoked activity in that neural population.

Going back to the hypothetical OBI, systematic application of activity flow modeling to test generative hypotheses could be an effective strategy for achieving a generative understanding of brain function. For example, generative understanding of how an individual is able to read aloud could involve OBI-based recording of their brain activity during vocalized reading of a long passage of text. This individual's detailed connectome would be used in a network simulation, with activity recorded from the individual's primary auditory cortex during verbal instructions (to read the text) input into the model. The model would then simulate the resulting activity flows moving throughout the brain as the individual prepares for the reading task. The text stimuli would then be input into the simulation via visual activity input (e.g., empirical activity patterns from primary visual cortex during stimulus

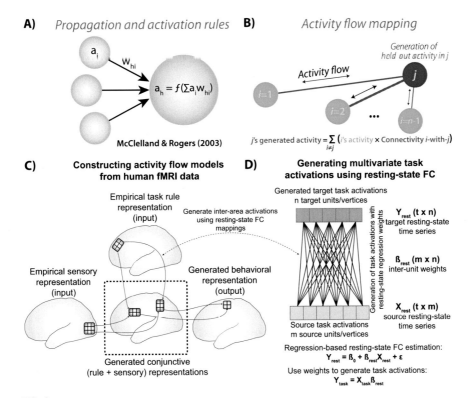

FIG. 1

Overview of activity flow modeling in terms of theoretical origin, methodology, and practical steps for model building. (A) Theoretical origin of activity flow modeling. The same propagation and activation rules used in connectionist artificial neural networks are used in activity flow models, but the parameters are set from empirical data. (B) Basic methodological schematic for activity flow modeling. The standard activity flow mapping approach (Cole et al., 2016) involves generating task-evoked activity in each region one-at-a-time based on activity in all other regions and the connectivity with the to-be-generated region. The connectivity with the target region, along with the equation (transfer function) at the bottom of the panel, constitutes that target region's activity flow model. (C) An example activity flow model based on Ito et al. (2022), going from sensory inputs and task rule representations to motor outputs in a context-dependent decision-making task. This example goes beyond the one-at-a-time activity flow mapping approach in panel (B), linking sensory inputs to cognitive transformations and behavior (motor responses). (D) How to build an activity flow model. Most activity flow models use standard multiple regression to estimate functional/effective connectivity (FC) (model weights) between brain regions. Resting-state brain data are typically used to estimate functional connections, though task data can also be used. These FC patterns are then used (along with the previous layers' task-evoked activity) to generate task-evoked activity in the next layer in the multistep activity flow procedure.

(A) Adapted from McClelland, J.L., Rogers, T.T., 2003. The parallel distributed processing approach to semantic cognition. Nat. Rev. Neurosci. 4 (4), 310–22. (B) Adapted from Cole, M.W., Ito, T., Bassett, D.S., Schultz, D.H., 2016. Activity flow over resting-state networks shapes cognitive task activations. Nat. Neurosci. https://doi.org/10.1038/nn.4406. (C and D) Adapted from Ito, T., Robert Yang, G., Laurent, P., Schultz, D.H., Cole, M.W., 2022. Constructing neural network models from brain data reveals representational transformations linked to adaptive behavior. Nat. Commun. 13 (673). https://doi.org/10.1038/s41467-022-28323-7.

presentation), with the resulting activity flows throughout the brain mixing with the activity previously initiated during instructions. To the extent that activity patterns following the initial inputs are accurately generated—especially the motor activity patterns directly driving the behavior of interest (reading aloud)—the various simulated activity flows driving those processes can be analyzed for insights into the generation of reading aloud behavior. Furthermore, changes to those activity flows can be made within the model to better understand their impact, resulting in yet deeper understanding of the generative processes involved. Ultimately, activity flow modeling would go beyond the comprehensive neural activity and connectivity information obtained by the OBI to provide insights into how activity and connectivity interact to produce cognition and behavior.

Core principles of activity flow modeling

There are four principles that are central to activity flow modeling (Table 1). The first of these principles was described in the previous section: *generativity*, wherein functions of interest (e.g., face selectivity in the fusiform face area, or behavioral responses via primary motor cortex) are produced in the act of modeling brain function. Generating the function of interest allows quantification of the success of the proposed mechanism for producing that function (Ito et al., 2020). Thus, generativity is key to the utility of activity flow modeling in providing evidence regarding the likely brain processes underlying functions of interest. This contrasts with most data analysis approaches (such as estimating connectivity or classifying task-evoked activation patterns), which are descriptive rather than generative.

Table 1 Core principles underlying activity flow modeling.

Activity flow modeling principle	Description
Generative	Use activity and connectivity to generate held-out (independent) activity in one or more neural population, allowing assessment of the causal sufficiency of the used activity and connectivity features in generating functions of interest
Simplicity/abstraction	Increase interpretability and identification of essential model features through model simplicity and abstraction (e.g., focusing on activity of entire brain regions rather than within-region activity patterns), increasing complexity only as necessary based on empirical evidence and the to-be-generated function(s) for a given study
Mechanistic/causal	Add accurate causal constraints (e.g., to connectivity estimates) whenever possible to increase the likelihood that modeled processes match the causal mechanisms used in the brain
Empirically constrained (data-driven)	Model features should be directly estimated by empirical brain data when possible, grounding the model in reality and reducing the number of modeling assumptions

Generativity follows from the central equation in activity flow modeling, which is also used in most artificial neural networks: $a_h = f(\Sigma a_i w_{hi})$, where a_h is the to-be-generated neural population's activity, w_{hi} is the connectivity to region a_h, and f is a transfer function, such as a sigmoid, linear, or rectified linear (threshold) transformation (Fig. 1A). Thus, activity flow modeling can be considered an approach to generate empirically estimated neural network simulations, aiding interpretation of neural data by testing its ability to generate neural data in independent/held-out neural populations (Ito et al., 2020).

Another core principle in the development of activity flow models has been *simplicity/abstraction*—keeping the models as simple as possible until there is an explicit need to make them more complex (Table 1). This includes the concept of abstraction, wherein features at lower levels of organization are aggregated over in the interest of explanatory simplicity. For example, we abstract over the quantum and atomic scale when describing the behavior of individual neurons, and so it may be that the most effective generative explanations of neurocognitive functions are at yet larger scales (Saxena and Cunningham, 2019). While simplicity and abstraction typically result in activity flow models lacking many details included in more biophysically detailed models, this simplicity principle has several important advantages. First, simplicity reduces the chance of the models containing a large number of unjustified assumptions that turn out to be incorrect. Instead, we start out simple with new assumptions treated as hypotheses to be supported by data and/or simulations. Second, simplicity can reduce overfitting of model features to particular problems, tasks, or data types, increasing the generality of findings (Li and Spratling, 2023; Hansen, 2020). Third, simplicity is obtained in part by using the methods that most researchers are actively using (e.g., using general linear models (GLMs) for fMRI activity estimates and Pearson correlation for fMRI connectivity estimates), making activity flow modeling results easier for most researchers to understand, as well as maximally relating results to previous studies. Thus, simplicity has been advantageous for activity flow models, even as they become more complex (and thus able to account for more phenomena) with each study.

To briefly illustrate this simple-to-complex transition, let us start with the original activity flow modeling study (Cole et al., 2016), which used standard Pearson-correlation FC and GLMs to maximally relate to the existing fMRI literature. That same study then demonstrated (in both simulations and empirical fMRI data) the increased causal validity and generative performance of using multiple regression as an FC measure. Thus, results were maximally applicable to the existing literature, while advancing the literature using a more complex and causally principled FC approach (multiple-regression FC). This simple-to-complex trajectory has led most recently to a multistep activity flow approach (Ito et al., 2022) (Fig. 1C and D), using multiple-regression FC across "layers" (sets of brain regions) from (1) visual and auditory inputs to (2) intermediate "conjunction regions" implementing cognitive information transformations to (3) motor outputs in primary motor cortex. These model-generated motor activations are then decoded (based on a decoding model trained using empirical motor activations), such that the model generates task-performing behavior. I will be unpacking the details of these models next, but I briefly described them here to illustrate the scale of the simple-to-complex transition so far (across 17 studies published between 2016 and 2023).

Another core principle of activity flow modeling is to focus on *mechanistic* explanations, by way of identifying causal constraints whenever possible (Table 1). The importance of this principle derives from the observation that an explanation provides less generative understanding if the generating/predicting process is unrelated to causal mechanisms within the system of interest. For example, pure (nonmechanistic) prediction of a motor response from a statistical model (e.g., based on the previous history of motor responses) rather than one grounded in causality would provide less insight into the processes generating that motor response than alternate predictions based on causally grounded generative processes.

The mechanistic principle is perhaps the most challenging principle to enact, given the difficulty of causal inference in general and in the context of brain interactions in particular (Reid et al., 2019; Mill et al., 2017a,b). The challenges of causal inferences are numerous and complex (see Pearl, 2009), but an example is confounding. To illustrate, consider that pain killer use (such as acetaminophen) is strongly correlated with mortality (Lipworth et al., 2003). This may lead one to avoid pain killer use, yet this is a noncausal relationship due to confounding. Specifically, many diseases (e.g., cancer) cause pain (and so pain killer use), as well as mortality. This scenario can be represented as a simple causal graph ($A \leftarrow C \rightarrow B$), with C being disease, A being pain killer use, and B being mortality. Critically, a causal inference method (e.g., simple correlation) that does not take into account confounding would suggest the wrong conclusion that pain killer use causes death. The same basic problem arises frequently in neuroscience (Reid et al., 2019).

Despite these challenges, causal/mechanistic explanation is a core principle as causality is central to scientific understanding and application of that understanding, such as causal interventions (i.e., treatments) to cure brain diseases. As an example of this principle in action, consider the use of multiple-regression FC in the original activity flow modeling study (Cole et al., 2016), which was based on the improved causal inferences possible using multiple-regression FC (due to accounting for causal confounds and causal chains) relative to the more standard Pearson-correlation FC approach (Fig. 2A and B). Furthermore, subsequent progress was made, however, with a recent study, demonstrating that multiple-regression FC (and partial-correlation FC) is less accurate in capturing ground truth FC than Pearson-correlation FC in the case of causal colliders (Fig. 2C) (Sanchez-Romero and Cole, 2021). That same study demonstrated a method—termed combinedFC—that combines the advantages of both multiple-regression FC (or partial-correlation FC) and Pearson-correlation FC, increasing the causal validity of FC beyond either method alone. Furthermore, that study used empirical fMRI data to find massive reductions in the number of estimated connections with partial-correlation FC relative to regular Pearson-correlation FC, suggesting confounders and chains (affecting correlation FC) are a much bigger problem in practice than colliders (affecting multiple regression/partial correlation). Accordingly, if a method

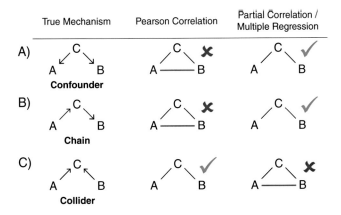

FIG. 2

Relationship between underlying causal mechanisms and common FC measures. Three common causal graph motifs are shown to illustrate fundamental principles of relating causal mechanisms to FC measures. CombinedFC (Sanchez-Romero and Cole, 2021) is an FC method that combines the best of both bivariate Pearson correlation and partial-correlation/multiple-regression FC. Other methods that achieve similar (and sometimes better) results also exist (Sanchez-Romero et al., 2023). (A) A causal confounder (also termed a *fork*) is when a neural population causes activity in two or more others, causing correlations in the downstream populations' time series. This can lead to spurious associations/causal inferences. Common FC measures, such as Pearson correlation and coherence, are susceptible to false positives from confounders (Reid et al., 2019). (B) A causal chain is when a neural population influences another via a third neural population. Not accounting for the third neural population (i.e., a mediator) can lead to false positive connections (though these could be considered "indirect" connections). (C) A causal collider occurs when two or more neural populations influence another neural population. In this case common methods such as Pearson correlation or coherence make proper inferences, but methods that control for confounding (e.g., partial correlation) create false connections. Note that, at least with fMRI data, it was found that the confounding problem (panel A) was much more problematic in the human brain than the collider problem, though both are present (Sanchez-Romero and Cole, 2021).

Figure adapted from Sanchez-Romero, R., Cole, M.W., 2021. Combining multiple functional connectivity methods to improve causal inferences. J. Cogn. Neurosci. 33 (2), 180–94.

like combinedFC (or the Peter-Clark algorithm (Sanchez-Romero et al., 2023)) is not used then multiple-regression FC (or partial-correlation FC) is preferable to Pearson-correlation FC.

Also building on the mechanistic/causal principle, a recent expansion of activity flow modeling to source-localized high-density electroencephalography (EEG) datasets (Mill et al., 2022b) improves causal inferences of activity flow modeling further by taking into account the temporal order of causal events (Fig. 3). These approaches have also been combined with simulated lesions and other innovations to improve

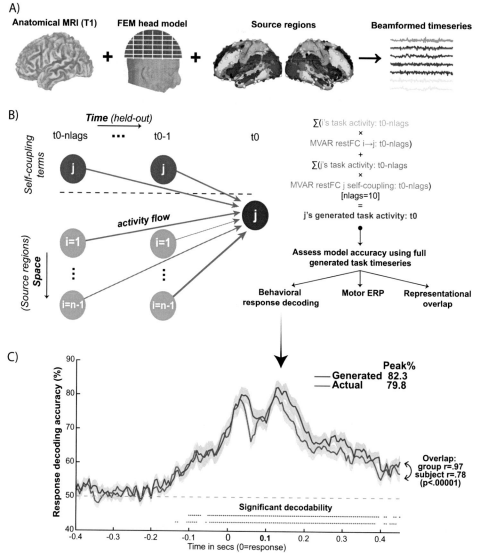

FIG. 3

Improving causal mechanistic inferences via extension of activity flow modeling/mapping to high-temporal-resolution neural recordings (here EEG). Approach and results from Mill et al. (2022b). (A) High-density EEG data in human participants were used along with individual-subject structural MRI to implement region-level source localization. (B) The dynamic activity flow mapping framework. Instead of just generating activity in spatially held-out regions, this approach also generates *temporally* held-out (i.e., future) activity. This improves the causal inference given the centrality of the direction of time in causal interactions. (C) The decoded motor response information time course based on generated cortical motor network activity (blue) from an activity flow model (including all nonmotor regions) during a simple sensory-motor mapping task. The decoded motor response information time course from the actual cortical motor network activity (pink) is shown for comparison.

Figure adapted from Mill, R., Flinker, A., Cole, M.W., 2022a. Invasive human neural recording links resting-state connectivity to generation of task activity; Mill, R.D., Hamilton, J.L., Winfield, E.C., Lalta, N., Chen, R.H., Cole, M.W., 2022b. Network modeling of dynamic brain interactions predicts emergence of neural information that supports human cognitive behavior. PLoS Biol. 20 (8), e3001686.

causal inferences once activity flow models are built (Hearne et al., 2021; Mill et al., 2022b; Ito et al., 2022). Despite these advances in FC methodology and the consequent improvement in mechanistic inferences with activity flow modeling, numerous opportunities to increase the causal validity of activity flow models remain.

Finally, perhaps the most fundamental principle in activity flow modeling is the centrality of *empirical constraints* (Table 1). Indeed, activity flow modeling begins and ends with empirical data, such that it can be considered both a data analysis and computational modeling framework. The approach begins with empirical data in the sense that model features—connections and input activity—are directly estimated from empirical brain data. The approach ends with empirical data in the sense that the model-generated activity is compared to actual empirical brain activity to test the validity of the model for generating the brain function(s) of interest. For example, a recent study (Cocuzza et al., 2022) used a causal FC approach (combinedFC; see Fig. 2) with fMRI to estimate the empirical connectivity between brain regions in the visual system, then used those empirical connections along with empirical activations in V1 as input into the model. This model was then tested for its ability to generate well-known category-selectivity activity in higher-level visual regions, such as face selectivity in the fusiform face area (Fig. 4). The success of the model in generating such category-selective activity demonstrated the sufficiency of fMRI connectivity and activity—when combined in an activity flow model—for providing an (important but partial) explanation of how category-selective visual brain activity is generated via brain network interactions.

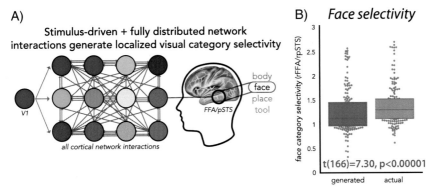

FIG. 4

Activity flow modeling of visual processing generates canonical visual category selectivity via distributed processes. Approach and results from Cocuzza et al. (2022). (A) A multistep activity flow model was used, with the only empirical task-evoked activations coming from V1 activity patterns. All other task-evoked activations were generated via activity flow over resting-state FC (with combinedFC). Three activity flow steps were simulated to generate task-evoked activity in visual-category-selective region sets (panel B). (B) Generated and actual category selectivity for an example visual-category-selective region set (fusiform face complex: fusiform face area + other face-selective regions) are plotted. Statistical significance >1 (no category selectivity) for each set of generated activations are shown (blue text).

Certainly, there are many details left out of such a model, yet the reliance on empirical data grounds the model such that there is a clear path toward adding those details (e.g., fine-grained temporal dynamics) in future work. As a counterexample to illustrate the utility of the empirical constraint principle of activity flow modeling, consider the immense flexibility of modeling category-selectivity visual responses without such a constraint. As discussed further next, there are an enormous number of ways to generate a category-selective pattern from exemplars (e.g., any of a variety of machine learning models or artificial neural networks (Cybenko, 1989)), such that we could spend decades demonstrating this variety of possibilities without narrowing down to a small set of empirically plausible options. The empirical constraint principle can help more rapidly narrow down the possibilities, reducing the scientific search space to those options consistent with both empirical connectivity and empirical activity patterns. Ideally, all relevant empirical constraints (e.g., the activity and connectivity of every neuron in the human brain) would be available to inform activity flow model-based explanations, but the work so far has demonstrated that we can still make tremendous progress without every relevant constraint. Indeed, the simplicity principle pushes against such a complex model, such that even if we had access to every neural event in the brain we would want to use simulated lesioning (Ito et al., 2022; Mill et al., 2022b) or related dimensionality reduction methods to simplify the model and identify the core set of activity flows underlying functions of interest.

Examples of successful activity flow models

We originally developed activity flow modeling to make two specific kinds of inference regarding neurocognitive phenomena (Cole et al., 2016). First, we wanted to determine where the human brain is on the continuum from primary localized (within-region) versus primarily distributed (inter-region) processing across a wide variety of tasks. Simulations were used to validate this inference—activity flow mapping only worked to the extent that underlying processes were distributed. The finding that activity flow mapping also performed well with empirical fMRI data suggested that the human brain's task-evoked activations are mostly determined by highly distributed processes. The second inference was to determine the functional relevance of resting-state FC to task-related processing. Prior studies had already shown a statistical link between resting-state FC and task-evoked activations (Smith et al., 2009), yet activity flow mapping attributed this to a common cause—that the activity flow pathways detected by resting-state FC were also those involved in generating task-evoked activations. Furthermore, we developed an FC method that used multiple regression to control for (observed) causal confounds, likely resulting in a more causally valid FC method relative to field-standard Pearson-correlation FC (see Sanchez-Romero and Cole, 2021). These (types 1 and 6) and other categories of activity-flow-based explanations are listed in Table 2, with the explanation types linked to individual studies in Table 3.

Table 2 Major forms of activity-flow-based explanations of neurocognitive phenomena (so far).

Type	What is being explained?	What is explanation based on?
1	Task-evoked activations	Distributed processes (activity flows)
2	Task-evoked activations	Specific connections and/or (other) activations
3	Dysfunctions/differences (in activations)	Specific connections and/or activations
4	Behavior (via motor activations)	Specific connections and/or activations
5	Task-related information (MVPA)	Specific connections and/or activations
6	Functional relevance of connection/activation	Contribution of connection/activation to another activation/dysfunction/behavior

Table 3 List of studies that use activity flow modeling, what kinds of explanation they include, what kind of connectivity they use, and what kind of activity they use.

Study	Explanation type	Imaging method	Connectivity type	Activity type
Cole et al. (2016)	1,6	fMRI	RSFC with correlation, Multreg, PCR	7 diverse tasks, whole cortex; simulations
Ito et al. (2017)	5	fMRI	RSFC with PCR	12 task rules, whole cortex; simulations
Mill et al. (2020)	3,6	fMRI	RSFC with PCR	Aging-related dysfunctional task-evoked activations
Ito et al. (2020)	1	fMRI	RSFC with Multreg	Activity in transmodal (vs sensory-motor) cortex is better predicted by distributed processes
Cole et al. (2021)	2,6	fMRI	Task-state FC with correlation, Multreg, PCR	24 diverse task conditions, whole cortex
Keane et al. (2021)	2	fMRI	RSFC with Multreg	Visual shape completion task activations
Hearne et al. (2021)	3, 2	fMRI	Cross-task average FC with Multreg	Working-memory-related activation dysfunction in schizophrenia

Continued

Table 3 List of studies that use activity flow modeling, what kinds of explanation they include, what kind of connectivity they use, and what kind of activity they use—cont'd

Study	Explanation type	Imaging method	Connectivity type	Activity type
Yan et al. (2021)	6	DWI, fMRI	DWI structural connectivity	2 tasks, whole cortex
Schultz et al. (2022)	1,2,5	fMRI	RSFC with correlation	12 task rules, multiple demand regions
McCormick et al. (2022)	6	fMRI	Latent FC (rest & task) with correlation	24 diverse task conditions, whole cortex
Ito et al. (2022)	4,2,6	fMRI	RSFC with Multreg	64 context-dependent tasks, from sensory & rule activations to motor responses (in M1)
Mill et al. (2022b)	4,2,5,6	EEG	RSFC with MVAR and PCR	Decoded motor information from source-localized motor cortex EEG activity
Hwang et al. (2022)	2,3	fMRI	RSFC with correlation	100+ task conditions, thalamus as source and cortex as target
Zhu et al. (2023)	6	fMRI	Probabilistic correlation RSFC	7 diverse tasks, whole cortex
Keane et al. (2023)	3,6	fMRI	RSFC with Multreg	Dysfunctional visual shape completion task activation in schizophrenia
Sanchez-Romero et al. (2023)	6,2	fMRI	RSFC with correlation, Multreg, combinedFC, and PC algorithm	24 task conditions, whole cortex; prefrontal working memory activation; simulations
Cocuzza et al. (2022)	1,2	fMRI	RSFC with combinedFC	4 visual category activations, visual-category-selective regions

DWI = diffusion weighted imaging; EEG = electroencephalography; RSFC = resting-state functional connectivity; Multreg = multiple regression; PCR = principal components regression; MVAR = multivariate autoregression (related to Granger causality); PC algorithm = Peter-Clark algorithm; CombinedFC = combined functional connectivity.
From Sanchez-Romero, R., Cole, M.W., 2021. Combining multiple functional connectivity methods to improve causal inferences. J. Cogn. Neurosci. 33 (2), 180–94.

Explanation type 2 (Table 2) is a more targeted inference, wherein a specific connection or activation (or a set of connections or activations) helps explain the generation of a task-evoked activation. For example, Hearne et al. (2021) calculated individual activity flow estimates—the [activations * connectivity] values (prior to summing to produce the target activation). These values were then used to infer the source activations and connections that drove working-memory-related activations. Furthermore, this approach was extended to group differences (explanation type 3), to identify the likely sources of dysfunctional working memory activations in schizophrenia patients. Other explanation type 2 efforts have involved simulated lesions (Mill et al., 2022b) or network subset analyses (Keane et al., 2023) to isolate the sources of task-evoked activations of interest. This approach has also been extended to characterize the flow of task-related information via multivariate pattern analysis (explanation type 5), revealing specific information flows between brain regions (Ito et al., 2017) and the role of resting-state FC in determining the multi-functionality of cognitive control regions (Schultz et al., 2022).

Behavior holds a privileged place in the demonstration of neurocognitive functionality (Krakauer et al., 2017). We have therefore sought ways to link activity flow processes to behavior (explanation type 4). Based on our mechanistic/causal principle (Table 1), we determined that proper modeling of behavior would involve identifying the activity flows that generate motor responses via primary motor cortex. In Ito et al. (2022), this involved linking sensory inputs (visual and auditory) to task-rule-related activations via resting-state FC, ultimately resulting in generated activations in primary motor cortex (M1) (Fig. 1C). These M1 activations were then decoded to identify which button would have been pressed given each M1 activation pattern. This resulted in above-chance performance of a complex context-dependent cognitive task. The activity flow model that generated this behavior could then be analyzed for insights into the likely mechanisms that led to this nontrivial cognitive task performance. Motor behavior generation was also the goal in Mill et al. (2022b) (Fig. 2C), with the advantage of high temporal resolution to improve the directionality of causal inferences (via generating future brain activity).

Assumptions of activity flow models

The core assumption of activity flow modeling is that activity flows—the movement/propagation of activity between neural populations—support neurocognitive computation. This assumption has so far not been problematic, however, given that the activity flow framework builds in a way to test this assumption via prediction of (typically already available) empirical activations. As emphasized in the previous discussion of explanation type 1 previous (Table 2), this contrasts with the alternative possibility that processes within neural populations (recurrent processes) are essential for neurocognitive computation (and cannot be accounted for with simple shifts in connectivity weights or blurring of time). Note that the recently developed high-temporal-resolution version of activity flow mapping can incorporate local recurrent processing, as well as cross-region activity flows (Mill et al., 2022b), potentially allowing direct comparisons between distributed and local processes.

From a graph theoretical perspective, activity flow mapping has so far assumed that activity propagates in the brain via a flow-based or diffusion-based routing protocol (Avena-Koenigsberger et al., 2017). Flow-based routing involves each signal (here task-evoked activations) propagating along all available direct connections rather than (for example) the most efficient/shortest path to a target node. This graph theoretical characterization suggests that—relative to shortest paths—activity flow models have assumed brain network communication optimizes for high parallel processing and low information costs (connectivity weights alone determine routing rather than a central router), but with high metabolic costs (since unnecessary nodes are activated). Note that these high metabolic costs may be offset by the sparsity of activations (due to local inhibition) (Rozell et al., 2008) and sparse connectivity (Sacramento et al., 2015). It will be important for future work to test this assumption in activity flow models, while also testing the optimization trade-offs predicted by graph theory.

When using FC, activity flow models assume that observed correlations (or other statistical associations) between neural populations are driven by similar activity flows as occur between task-evoked activations. For example, with resting-state FC, this assumes that spontaneous activity flows at rest use the same (or similar) pathways as task-evoked activity flows. For task-state FC (with task-evoked activity confounds removed (Cole et al., 2019)), this assumes that spontaneous activity flows (or activity flow variation driven by trial-by-trial variability in stimuli) during a given task use the same pathways as task-evoked activity flows. Finally, when using nondirectional FC, this assumes that all connections are bidirectional and equally weighted. Note that weighted and directed connectivity in macaque monkeys suggests this tends to be the case (Markov et al., 2014). However, Sanchez-Romero et al. (2023) demonstrated the utility of directed functional connectivity (sometimes termed effective connectivity) approaches for making directional activity flow inferences with fMRI data, while Mill et al. (2022b) did so with EEG data. Overall, the assumption that the same pathways are involved in generating the data used for estimating FC and the data used for estimating activations are not a problematic assumption, since activity flow predictions are unlikely to be accurate if this were not true.

Structural connectivity-based activity flow models (Yan et al., 2021) assume that activity flows occur over structural connections, which is almost certainly true. Such models also assume that all structural connections are bidirectional and symmetrically weighted, while the aggregate effect of synaptic weights (and perhaps other functional details captured by FC estimates) is negligible on activity flows. This is not a problematic assumption in the sense that activity flow-based predictions are unlikely to be accurate if this is not true.

An important assumption regarding the data used in activity flow models is the statistical independence of the connectivity and activity estimates. As covered in the "Pitfalls of activity flow modeling" section, there are several ways that such statistical circularity can bias activity-flow-based inferences. For instance, if task fMRI data are used for both FC and activation estimation, it is likely that activity-flow-based predictions would be biased toward overly accurate predictions of "held-out" activations. This is due to the same variance (e.g., large task-evoked activations in two regions) driving both increased FC estimates and increase task-evoked activation estimates. This particular case is due to causal confounding (Fig. 2) from external stimuli, such that regression of stimulus timing can remove this issue in some cases (Cole et al., 2019), making it possible to use task-state FC with activity flow modeling (Cole et al., 2021).

Most activity flow models have so far assumed stationarity of both FC and task-evoked activations, given their focus on time-averaged effects. Thus, most activity flow models will not capture changes to activity flows based on FC changes over time or task-evoked activations varying from trial-to-trial. Note, however, that some activity flow dynamics driven by transient task-induced activity are captured in the EEG activity flow modeling approach (Fig. 3). It will be important for future studies to go beyond time-averaged effects to better characterize activity flow dynamics.

Building, testing, interpreting activity flow models

From a broad perspective, activity flow mapping involves building a connectionist artificial neural network from empirical connectivity estimates. In principle, substantial biological detail can be added to these models, but the immense functional expressiveness of connectionist-level modeling (Rogers and McClelland, 2014) suggests that more can be learned by abstracting away from much of that detail when possible. This reveals a tension between the mechanistic and simplicity principles (Table 1), yet the optimal outcome from this tension should be the minimum mechanistic details necessary to explain/generate a given function of interest. Once this connectionist model is derived from empirical data, task-evoked activations are added, followed by the propagation and activation rules from standard connectionist modeling (Ito et al., 2020). This "animates" the model, simulating functions of interest and allowing inferences regarding the neural mechanisms underlying those functions to the extent that the model's constraints are empirically valid, the function is accurately generated by the model, and subsequent analyses isolate key factors within the model producing the function of interest.

More practically, one can build an activity flow model whenever one has both connectivity and activity estimates for a set of neural nodes (neurons or neural populations). Building and evaluating an activity flow model involves six steps. For visual overviews of how to build and test activity flow models, see Fig. 1 for slow imaging methods (e.g., fMRI) and Fig. 3 for fast imaging/recording methods (e.g., EEG). Step 1 is estimating connectivity among all nodes, such as using multiple regression (Cole et al., 2016). Step 2 is to estimate activity in all nodes, such as using an fMRI general linear model to estimate average task-evoked activations across trials of each condition. Alternatively, it is possible to use activity time series with activity flow modeling (see Fig. 3), rather than averaging activity over time. Step 3 is to decide, based on the goals of the study, which nodes are *source nodes* and which are *target nodes*. Source nodes' activity levels are set to match empirical levels of activity in those nodes, while target nodes' activity levels are generated by the model. Step 4 involves running the model to generate activity in the target nodes across all task conditions of interest. Step 5 is to evaluate model accuracy by comparing generated to empirical activity in the target nodes. This can be done by a variety of similarity measures, such as correlation, R^2, or mean squared error. Assuming the generated activity matches empirical reality, step 6 involves gaining additional insights into the generative activity flow processes by interpreting the model features and intermediate activity (and activity flows) generated by the model. This can involve description of model features or interventions on model features (e.g., connections or activity levels) to observe the impact of those interventions on the generated activity.

For a detailed overview of how to implement activity flow modeling using the open source Brain Activity Flow Toolbox please, see Cocuzza et al. (2022) and the toolbox website: https://colelab.github.io/ActflowToolbox/. Activity flow mapping was originally developed to relate standard task-evoked activations and standard FC as estimated using fMRI (Cole et al., 2016). Thus, these standard measures can be used to make activity-flow-based inferences (see Table 2). When optimizing for the theoretical principles underlying activity flow modeling (Table 1), however, other activity and connectivity estimates may be desired. The Brain Activity Flow Toolbox is highly flexible, allowing use of a wide variety of possible activity and connectivity types. Note, however, that activity flow mapping has only been adapted to high-temporal-resolution data (in this case source-localized EEG) recently. Until this is incorporated into the full toolbox, this adaptation of the approach can be found here: https://github.com/ColeLab/DynamicSensoryMotorEGI_release.

As a further guide to conducting studies using activity flow modeling, a list of steps are included next based on a set of recent activity flow modeling studies from my lab (Hearne et al., 2021; Mill et al., 2022b; Ito et al., 2022; Sanchez-Romero et al., 2023), which focus on identifying connectivity-based explanations of specific neurocognitive phenomena:

1) Identify a brain function of interest, and obtain a reliable empirical measurement of that function. Examples: cortical hemispheric lateralization during language tasks; face selectivity in FFA (Fig. 4); motor selection and responding in M1 (Fig. 3).

2) Work backward from the function of interest to hypothesized source regions/ times, excluding the to-be-generated function from the source set. Note that it is possible to remove the function via artificial means, such as averaging across hemispheres for a lateralization-generation study. For example, averaging the source task-evoked activations across hemispheres would allow the activity flow model to demonstrate that the model's network architecture was sufficient for generating lateralization, rather than simply spreading pre-existing lateralization to held-out brain regions. As a best practice, it is usually important to run an

analysis verifying that the function of interest is not present in the source set, such that a connectivity-based transformation is required to generate that function in the model.

3) Use activity flow modeling to generate the function of interest based on the source activity flowing over connectivity, testing for above-chance predicted-to-actual similarity.

4) Identify the key connectivity (and/or activity) properties that allowed for the generation of the function of interest. Some example approaches include simulated lesions (see Mill et al., 2022b; Ito et al., 2022), plotting/analyzing the activity flow graph (see Hearne et al., 2021), scrambling/permuting connectivity patterns to show dependence on those specific patterns, and graph theory to identify connectivity properties (ideally comparing to a model with those properties removed, or identifying correlations between the strength of those properties and variation in the function of interest).

Pitfalls of activity flow models

Perhaps the most commonly encountered pitfall in activity flow modeling is the possibility of analysis circularity (Kriegeskorte et al., 2009) in the comparison between generated and actual task-evoked activations. This typically arises from mixing source and target node activity during activity flow simulation. This leads to some portion of the to-be-predicted (target) data being erroneously added to the model-produced activity, rather than having the model generate that activity independently. For example, the spatial smoothness inherent in fMRI data creates circularity in activity flow inferences between nearby voxels. Cole et al. (2016) dealt with this issue by using a brain region atlas with regions 10 mm apart and, when performing voxelwise analyses, excluding all voxels within 10 mm of each target voxel from the set of source flows. This has become common practice in activity flow mapping with fMRI data. Note, however, that tests of the impact of circularity with brain region atlases (using cross-voxel averaged time series) that share borders have revealed that there are typically only minimal impacts of circularity in these cases (see Activity Flow Toolbox demo: https://colelab.github.io/ActflowToolbox/HCP_example. html). Nonetheless, it is a best practice to avoid source-target pairs that are within 5 mm of each other (and likely more with voxels larger than 2.5 mm) with fMRI data.

The situation would appear to be much worse with EEG or MEG data, as volume conduction can cause a source signal to traverse the entire brain, causing analysis circularity for every possible source-target pair. Mill et al. (2022b) used a rigorous multipronged approach to eliminate this issue, however. Most consequentially, simulated activity flows generated *future* activity to completely avoid the possibility that volume conduction (which occurs with zero lag) could result in analysis circularity. Additional steps increased the precision of activity flow inferences. First, high-density EEG data and structural MRIs were used to localize sources with beamforming. Second, sources were localized to the same set of regions with 10 mm gaps used by Cole et al. (2016), reducing the chance of substantially mis-localized sources. Third, rather than using standard temporal filters we used causal filters, which prevent temporal circularity by avoiding leakage of future activity back in time. Finally, the target region's zero-lag time series was regressed out of all source time series, while also being fit in an autoregressive manner, further improving isolation of the target time series. Together, these steps both eliminate the chance of spatial analysis circularity while substantially improving inferential precision.

Another source of potential circularity is specific to using task-state FC (rather than resting-state FC or structural connectivity): Simultaneous sensory inputs leading to inflation of FC and inflating activity flow prediction accuracy (Cole et al., 2019). For example, watching a video of a car exploding will simultaneously stimulate your primary auditory and visual cortices (A1 and V1), creating strong correlations between those regions despite there being no direct connectivity between them. This situation would create a false functional connection between A1 and V1, which could then be used with an activity flow model to predict V1 activity based on A1 activity (and vice versa). The resulting misleading inference would be due to a causal confound, wherein an external common cause has created the appearance of direct causal influence for both the FC estimation algorithm and the activity flow model. Cole et al. (2019) identified this confound and demonstrated that subtraction (or regressing out) of cross-trial mean task-evoked activity is a way to remove the impact of the common cause and correct for this confound. This is therefore the best practice when using activity flow modeling with task-state FC (Cole et al., 2021).

A general pitfall of activity flow modeling is shared with any connectivity-based study: the chance that inaccurate connectivity estimates will lead to an inaccurate activity flow-based inference. Multiple studies have demonstrated that improving FC estimation also improves activity flow-based task-evoked activation prediction accuracies (Cole et al., 2016; Sanchez-Romero et al., 2023). We define FC improvement as improved estimation of cross-neural-population causal interactions (Reid et al., 2019), such as reducing the number of causal confounds. While prediction accuracy typically increases with improved causal FC (since the true causal model is among the best possible predictive models), it remains possible for prediction accuracy to be improved by causal confounds (Reid et al., 2019; Sanchez-Romero et al., 2023). The example previous with task-state FC confounds is a case in point—the causal confound of correlated inputs can enhance prediction accuracy of A1 activity from V1 (and vice versa). Therefore, it is essential to rely on bedrock causal principles—rather than activity flow prediction accuracy alone—when developing connectivity methods for parameterizing activity flow models (Sanchez-Romero and Cole, 2021; Sanchez-Romero et al., 2023).

A nonobvious case of analysis circularity can occur when a causal inference is in the wrong direction. For example, it is known that primary motor cortex sends efferent copies of its output back to brain regions that are inputs to primary motor cortex (Khan and Hofer, 2018). This is less of a problem with high-temporal-resolution methods like EEG, since the feedforward and feedback processes can be easily separated using dynamic activity flow mapping (Mill et al., 2022b). Separation of feedforward and feedback processes is much more challenging with a low-temporal-resolution method like fMRI, however, since the feedforward and feedback processes (which likely occur on the order of 100–200 ms) are mixed into single fMRI task-evoked activation estimates. There are three basic strategies (so far) to deal with this issue:

First, simply make a weaker inference, wherein source activations caused target activations *and/or* target activations caused source activations. Cole et al. (2016) ran simulations to show that even this weaker inference is useful, as it can reveal the extent to which distributed processes support the generation of task-evoked activations (explanation type 1 in Table 2).

Second, use multistep activity flow modeling starting from sensory inputs, wherein only activation patterns in primary sensory regions (e.g., V1) are empirical and all subsequent activity flows are based on that input (Cocuzza et al., 2022). This isolates variance to feedforward processes from the inputs, eliminating the chance for causal circularity after that initial step. There is some chance of feedback processes impacting the initial input node, however, resulting in causal circularity in the initial activity flow step. Cocuzza et al. (2022) dealt with this issue via signal normalization to ensure that there was no selectivity for each visual category of interest in V1 (the input node). In other words, the input node did not contain the function of interest, ensuring that the function of interest was ultimately generated via transformations on the input node's activity.

Third, use of careful counterbalancing and averaging to remove the to-be-generated effects of interest from the source task-evoked activations, such that feedback processing cannot explain observed effects. For example, Ito et al. (2022) use of a counterbalanced factorial design, wherein all stimuli were paired with each task rule and each motor response across trials. They then averaged across all task rules and motor responses when estimating each stimulus input activation. The resulting counterbalanced-and-averaged task-evoked activation estimates guaranteed that motor responses generated via mixing of stimulus and task rule activity flows would occur in a noncircular manner.

Together, these strategies reveal the flexibility of the activity flow modeling framework. Even in a situation where causal circularity seemed inevitable, clever use of activity flow algorithm development (e.g., a multistep approach) and experimental design (e.g., counterbalancing) led to inferential improvements. I expect future innovations to overcome current and future challenges to the activity flow modeling framework.

How activity flow modeling relates to other approaches

Activity flow modeling is highly related to a variety of other approaches to brain data analysis and neural network modeling. Despite being similar to these approaches, activity flow modeling does not appear to be redundant with any of them and can add something unique to each of them. Indeed, it appears that any possible connectivity method—and any possible activity estimation method—can be incorporated into activity flow models, with unique inferences added in each case. Furthermore, activity flow modeling can add unique inferences to standard task-evoked neural activation studies, as well as theoretical computational models. These points are illustrated next. Fig. 5 provides a general map along mechanistic and empirical axes, illustrating the relationship to activity flow modeling to some other approaches.

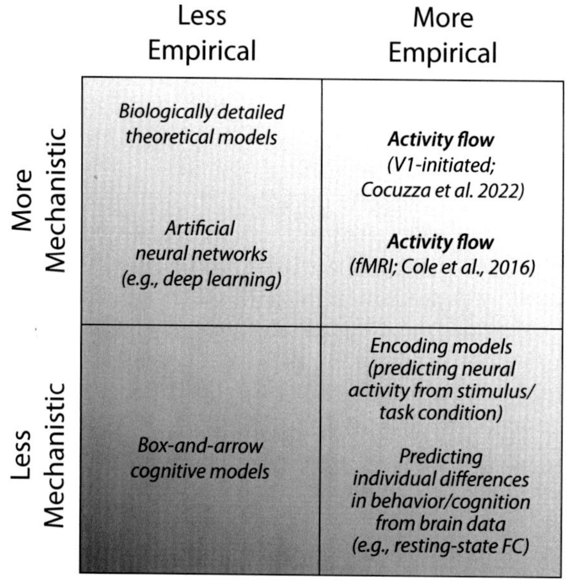

	Less Empirical	More Empirical
More Mechanistic	*Biologically detailed theoretical models* *Artificial neural networks (e.g., deep learning)*	*Activity flow (V1-initiated; Cocuzza et al. 2022)* *Activity flow (fMRI; Cole et al., 2016)*
Less Mechanistic	*Box-and-arrow cognitive models*	*Encoding models (predicting neural activity from stimulus/task condition)* *Predicting individual differences in behavior/cognition from brain data (e.g., resting-state FC)*

FIG. 5

Relationship between activity flow modeling and some other approaches to modeling neurocognitive phenomena (behavior and task-evoked activations). A subset of neurocognitive modeling approaches are shown along two axes: mechanistic and empirical. The placement of the text describing each approach indicates approximately (even within each quadrant) its relative level of mechanistic vs empirical properties. *Mechanistic* is defined here as the level of detail regarding the generative causal processes hypothesized to underly the phenomena of interest. For example, box-and-arrow cognitive models (e.g., Baddeley, 2000) include fewer details regarding the causal events hypothesized to generate cognitive processes than connectionist artificial neural networks, and even fewer than biologically detailed theoretical models (e.g., Babadi and Abbott, 2010). Activity flow modeling was developed to be more empirically constrained and empirically validated than artificial neural network models, and more mechanistic than models that predict behavior (or task-evoked activations) based on brain data (e.g., Smith et al., 2009, 2015). Two activity flow models are shown, with the Cocuzza et al. (2022) V1-initiated model (Fig. 4B) being more mechanistic than the original activity flow modeling approach (Cole et al. (2016)). This is because of the additional details included regarding the generative causal processes underlying the generation of the processes of interest (in this case visual category selectivity in human visual cortex). Note that so far no activity flow model has been as mechanistically detailed as biologically detailed theoretical models of neural function (but see Lee et al., 2022).

Computational models (deep neural networks, recurrent neural networks, etc.). Activity flow models can be considered to be computational models that—unlike standard computational models—are derived from empirical brain data. Indeed, activity flow models were developed based on connectionist artificial neural networks (ANNs), which consist of nodes and connections over which activity flows according to a standard "propagation rule" (Rogers and McClelland, 2014; Ito et al., 2020). Importantly, hundreds of studies over decades have revealed the seemingly limitless potential of ANNs to model human cognition (Rumelhart et al., 1986a,b; McClelland and Rogers, 2003) and can even exceed human cognition in "deep neural network" versions of ANNs with many layers (Bengio et al., 2021; Vaswani et al., 2017; Silver et al., 2018). This flexibility of ANNs is both a blessing and a curse, as ANNs can be used to model anything (Hornik et al., 1989), yet they often fit the training data differently than the human brain does. Indeed, this divergence from the human brain is highly likely, given the immense architectural differences between ANNs and the human brain. A major motivation for developing the activity flow framework was to add extensive empirical constraints in order to improve ANN-based inferences regarding the human brain and human behavior. Note that some prior studies already used empirical data as inputs to ANNs (Hanson and Hanson, 1996), illustrating the utility of empirical data constraints (as opposed to broad empirical constraints, such as use of activity propagation among units to model brain processing) on neural network-based inferences. Furthermore, given that humans have computational/cognitive abilities that ANNs do not, there is potential to develop computational models with state-of-the-art abilities based on brain data-derived modeling. In principle, using activity flow modeling to create an ANN architecture directly from empirical brain connections should (1) provide empirical neural data analyses with the theoretical insights typical of computational modeling and (2) provide computational modeling with the scientific conclusiveness and grounding in reality of empirical neural data analyses.

Some recent deep ANNs have taken inspiration from the network architecture of the primate visual system, roughly matching the number of layers/steps used by primates to process visual stimuli (Yamins et al., 2014; see Yamins and DiCarlo, 2016, for review). These studies have shown that—despite primarily being shaped by learning an object recognition task—each ANN layer's representations are similar to the kinds of representations present in empirical multiunit recording data from nonhuman primate brains. Together, these results demonstrate that even rough approximations of the correct network architecture can (when augmented by connectivity adjustments from task learning objectives) result in important insights into neural computations. Activity flow modeling goes beyond these models by using network architectures *directly specified by brain data* (rather than task learning objectives), incorporating many more brain network architectural features and thus providing results that are likely to be even more informative regarding the network-based cognitive computations carried out by the brain.

Resting-state FC (with fMRI, magnetoencephalography (MEG)/EEG, intracranial EEG, etc.). Activity flow modeling has so far been primarily used to add insight into the computational and cognitive contributions of resting-state FC

(Cole et al., 2016; Ito et al., 2020). This complements other approaches that associate resting-state FC with cognition, such as correlating resting-state FC with activity patterns (Smith et al., 2009) or individual differences in behavior (Cole et al., 2012; Smith et al., 2015). This is accomplished by parameterizing activity flow models using resting-state FC estimates and a subset of task-evoked activations, then testing for the resulting model's ability to generate held-out task-evoked activations linked to cognitive phenomena. High generated-to-actual similarity (compared, e.g., to randomly permuted models) provides evidence for the validity of the activity flow model, and thus supports the associated model-based explanation for the generation of task-evoked activations and associated cognitive phenomena. Notably, the mechanistic principle described previously further increases confidence in the validity of the model-generated explanation, since it provides additional constraints on the model beyond parameterization with empirical connectivity and activity (and testing of the model via comparison with empirical data). As a case in point, we typically use multiple regression (rather than field-standard Pearson correlation) to estimate FC for fMRI activity flow predictions (Cole et al., 2016, 2021), given the improved mechanistic/causal validity of multiple-regression FC due to reduced FC confounds (Fig. 2).

A major reason to use resting-state FC (as opposed to FC during any given task state) is the desire to identify FC architectures that generalize beyond specific states. This idea of identifying a general network architecture has resulted in the development of latent FC (McCormick et al., 2022), wherein factor analysis is used to identify the FC weight underlying a wide variety of task (and rest) brain states. Even without using latent FC (which, in its standard form, requires each subject to perform a task battery), this concept can be used to strengthen the generalizability of activity flow inferences by simply estimating FC with an independent brain state (e.g., rest, or a different task) from the task state(s) of interest. For example, Hearne et al. (2021) averaged FC across rest and a variety of tasks, excluding the task of interest from FC estimation to reduce the chance of analysis circularity and help ensure the results were based on a state-general brain network architecture.

Individual differences resting-state FC. As briefly described previously, a variety of approaches use resting-state FC to predict individual differences in cognition and behavior (Cole et al., 2012; Varoquaux and Poldrack, 2018; Smith et al., 2015; Shen et al., 2017). Furthermore, some recent approaches use resting-state FC and/or structural connectivity to predict individual differences in task-evoked activations (Bernstein-Eliav and Tavor, 2022; Tavor et al., 2016; Osher et al., 2016). Like these approaches, the activity flow modeling approach is predictive. However, rather than predicting data from held-out individuals, the activity flow approach predicts held-out activations *within individuals*. While prediction can be implemented just as well between individuals, causal brain mechanisms ultimately occur within individual brains and are thus better served via within-individual models using causally grounded connectivity estimates. As an illustration of this point, consider how challenging it would be to use an individual differences approach to infer the functions of the human heart and circulatory system, given how little the relevant mechanisms differ between individuals (resulting in very small individual difference correlations) and how the relevant biological mechanisms all interact in complex ways within individual bodies. Thus, activity flow modeling differs from these other methods in its primarily mechanistic and causal explanatory (rather than primarily predictive) goal (see Table 1).

Task-state FC (with fMRI, MEG/EEG, intracranial EEG, etc.). Task-state FC (also termed task-related FC)—FC estimated from brain data collected during task performance—can be used nearly as easily as resting-state FC to make inferences with activity flow modeling. This was recently demonstrated by Cole et al. (2021), wherein task-state FC was shown to consistently improve activity flow predictions beyond those obtained using resting-state FC. This suggests that task-state FC estimated from the same task as the to-be-predicted task activations contains additional features that determine the flow of activity above and beyond those present in other states such as rest. This makes an entirely new line of research possible, wherein task-state FC changes specific to a given task condition are identified and their functional relevance (for generating task-evoked activations) determined via activity flow modeling. Note, however, the need to account for potential analysis circularity when using task-state FC (covered next).

Structural connectivity (with diffusion weighted imaging (DWI), tract tracing, etc.). A recent study demonstrated that DWI-based structural connectivity can be used effectively in place of resting-state FC for activity flow modeling (Yan et al., 2021). The theoretical inferences possible with structural connectivity are very similar to those of resting-state FC. Indeed, one can think of resting-state FC as structural connectivity with additional information regarding the aggregate effects of synaptic weights. However, relative to correlation-based FC, structural connections are much more likely to reflect direct connections between neural populations (Damoiseaux and Greicius, 2009). (Note that other FC methods—such as regularized partial correlation—can also be used to estimate direct connectivity.) Furthermore, unlike FC estimates (and some forms of tract tracing), DWI-based structural connectivity has a strong distance-based bias toward false-negative connectivity (Donahue et al., 2016). Thus, there are some advantages and some disadvantages to using structural connectivity relative to FC for activity flow modeling.

Task-evoked activation-based approaches (GLMs, MEG/EEG event-related potentials, multivariate pattern analysis (MVPA), etc.). Task-evoked activations—changes in neural activity amplitude—are both the inputs and the to-be-generated outputs of activity flow models. This reflects the principle of empirical grounding (see Table 1), wherein we maximize contact with empirical data by using empirical activations as input into each model. Importantly, our explanatory inferences also focus on activations, given the mechanistic centrality of activations in standard neural theory. Specifically, the construct of a neural action potential entails both activation (i.e., an increase in neural activity amplitude) and the flow of activity to other neurons. Neural activity amplitudes are thus ultimately based in action potentials/ spike rates, which cause local field potentials, large-scale electromagnetic fields, synaptic activity, hemodynamic responses (Lee et al., 2010), and other measures of neural activity. Thus, activity flow modeling leverages standard, well-established neural theory to interpret task-evoked activations and their flow/movement to other neural populations via connectivity patterns. These inferences are then verified empirically by comparing the generated task-evoked activations to the actual task-evoked activations in each neural population for each task condition (Fig. 1B). This theoretical background reveals that the ideal activation estimation would use multiunit spike recording. In the absence of such ideal data in humans, we use task GLMs with fMRI (Figs. 1 and 4), event-related potentials with MEG/ EEG (Fig. 3), or any of a number of possible activation-derived functional/cognitive signatures.

Activity flow mapping has also been applied to gain insights into the flow of neural information content via modeling multivariate patterns of task-evoked activations (Ito et al., 2017). To illustrate one possible approach, consider using MVPA to decode information content in each brain region, then using the activity flow approach to predict (based on other regions' information content) downstream information content based on the connectivity strength between the regions. In the interest of maintaining the mechanistic principle (see Table 1) to improve our inferences, however, we used a more nuanced approach (but see Schultz et al., 2022 for a simpler approach). This involved first applying standard activity flow mapping between individual vertices (the smallest unit of spatial measurement with cortical surface fMRI data). All vertex-to-vertex resting-state functional connections between each pair of cortical regions were estimated. Empirical task-evoked activity for each vertex in the source region was used as input, then the activity flow calculation was applied to predict task-evoked activation patterns in the target region. MVPA was then applied to this predicted activity pattern, based on a decoder trained to distinguish the actual activity patterns across task conditions. This procedure allowed us to infer the degree to which brain regions shared task-related information via fine-grained resting-state FC topology. More generally, this study demonstrated that activity flow mapping can be readily applied to make inferences about multivariate patterns of activity (and connectivity), suggesting the ability to extend activity flow mapping to a variety of research questions relating to neural information representation. For instance, this study was the basis of several recent studies that incorporated inferences regarding the flow of task-evoked information-carrying activity patterns (Ito et al., 2022; Schultz et al., 2022; Mill et al., 2022b; Hwang et al., 2022).

Encoding models. As discussed by Ito et al. (2020), activity flow models can be considered to be connectivity-based encoding models. This contrasts with standard encoding models (Huth et al., 2012; Naselaris et al., 2011) (Fig. 6A), which predict task-evoked activations directly from stimuli (or task conditions) mapped (via a regression weight) to a given neural population (e.g., an fMRI voxel). Note that standard task fMRI GLMs can be considered encoding models, though they are not typically tested as predictive models using properly held-out data. Activity flow models also predict task-evoked activations but, rather than using a direct mapping from stimulus features (or task conditions) to task-evoked activations, brain connectivity is used to map task-evoked activations in one or more neural populations to

A) **Function-structure mapping**

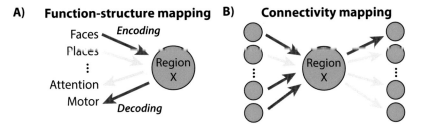

B) **Connectivity mapping**

C) **Activity flow modeling (connectivity + function-structure mapping)**

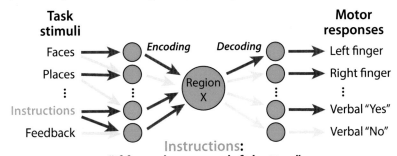

Instructions:
"If face, then press left button"

D)

Input
Task rule representations
Stimulus representations

Decoding (reading out) rules + stimuli
Encoding conjunctive representations

Decoding (reading out) conjunctions
Encoding motor representations

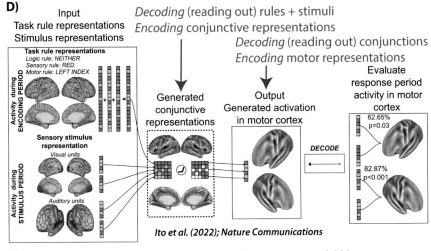

Ito et al. (2022); Nature Communications

E) $Y = f (X_{rule} W_{rule \to hidden} + X_{stimulus} W_{stimulus \to hidden}) W_{hidden \to output}$

Y : *predicted motor activation pattern*
W : *Estimated connectivity weights from A ~ B*
f : *Rectified linear function (threshold negative values)*
X : *Input activation patterns (either rule activations or sensory stimulus activations)*

FIG. 6

From encoding and decoding to activity flow modeling. (A) Function–structure mapping (Henson, 2005), such as standard encoding models (e.g., fMRI GLMs) and decoding models (Naselaris et al., 2011). (B) Connectivity mapping, such as using diffusion MRI to estimate structural connectivity or using fMRI to estimate functional connectivity. (C) Activity flow modeling combines connectivity mapping with function–structure mapping, inferring how

(Continued)

FIG. 6—CONT'D

functions are generated via activity flows over connectivity patterns. Rather than the experimenter doing the encoding/decoding, activity flow modeling allows us to infer how neural populations decode each other (de-Wit et al., 2016) to encode information when implementing neurocognitive processes. (D) Results from Ito et al. (2022)—an example of a full activity flow model mapping representational transformations all the way from stimulus inputs to motor outputs in a context-dependent cognitive task. Each intermediate step can be interpreted as both a decoding model (decoding input activity patterns) and an encoding model (encoding output activity patterns). Decoding is used as the final step to determine what behavior (i.e., motor actions) is generated by the model. The model generated above-chance task performance. (E) Equation specifying motor (output layer) activity patterns in panel (D), based on activity flowing between model layers. Note that the "hidden" layer is labeled as "Generated conjunctive representations" in panel (D).

Panels A–C adapted from Ito, T., Hearne, L., Mill, R., Cocuzza, C., Cole, M.W., 2020. Discovering the computational relevance of brain network organization. Trends Cogn. Sci. https://doi.org/10.1016/j.tics.2019. 10.005. Panels D and E adapted from Ito, T., Yang, G.R., Laurent, P., Schultz, D.H., Cole, M.W., 2022. Constructing neural network models from brain data reveals representational transformations linked to adaptive behavior. Nat. Commun. 13 (673). https://doi.org/10.1038/s41467-022-28323-7.

task-evoked activations in another neural population (Fig. 6C and D). Critically, this connectivity-based approach is more mechanistic than standard encoding models (Fig. 5), since connectivity is thought to be a primary mechanism by which task-evoked activations are generated in the brain (Cole et al., 2016; Ito et al., 2020; Mars et al., 2018; Passingham et al., 2002). Thus, even if an activity flow model makes less accurate predictions of task-evoked activations than a standard encoding model, that activity flow model will still provide insight due its additional connectivity-based mechanistic details. The Ito et al. (2022) study is the most complete example to date of the full stimulus-to-response activity flow modeling approach illustrated in Fig. 6C. That study (Figs. 1C and 6D) started with standard encoding models (visual, auditory, and task set activations) and combined them with empirical connectivity patterns to generate cognitive representations. Those cognitive representations then generated motor representations implementing context-dependent cognitive task performance (as verified by a decoding model).

Open challenges and future directions

The success of activity flow modeling depends on the quality of the methods used to estimate the task-evoked activations and connectivity that parameterize activity flow models. Thus, improvements to connectivity and task-evoked activation estimation are inevitably also improvements to activity flow mapping. For example, we recently made the case for improving FC approaches by shifting the goal of FC research from estimating associations to estimating causal relationships (Reid et al., 2019). Several of our recent studies have supported improved causal inferences with improved FC measures (Cole et al., 2016; Sanchez-Romero and Cole, 2021), resulting in improved activity flow inferences (Sanchez-Romero et al., 2023).

Switching to a causal goal for FC also opens up new means to validate (and improve) FC methods using causal stimulation approaches. For example, combining a stimulation approach like transcranial magnetic stimulation (TMS) and a neuroimaging approach like EEG can yield causally grounded connectivity maps (Esposito et al., 2020). This is based on evidence that TMS causes localized action potentials (Mueller et al., 2014), along with the assumption that this results in stimulation-evoked activity in brain regions downstream of the stimulated site. Importantly, this assumption is based on the activity flow construct—movement of experimenter-evoked activity between neural populations. Thus, activity flows are theoretically central to this general approach, with activity flow modeling empirical testing this via assessment of whether actual causal interventions (stimulations) applied to brain regions aligns with activity flow-generated effects of such interventions. Some alternative combinations of causal stimulation and brain recording that can support this FC validation approach include transcranial electric stimulation and fMRI (Kar et al., 2020), intracranial stimulation and fMRI (Thompson et al., 2020), and intracranial stimulation and recording (Sheth et al., 2022). Given the centrality of activity flows to these simultaneous stimulation-recording approaches, these approaches have the potential to both benefit and benefit from the activity flow framework. Indeed, activity flow modeling has strong potential for providing individualized predictions of stimulation interventions to reduce symptoms (i.e., treatments) across a wide variety of brain disorders (Sanchez-Romero et al., 2023).

Activity flows are directly tied to action potentials, given that action potentials are the only known mechanism for long-distance neural signal propagation in the brain. This theoretical link supports the likely centrality of activity flows in neural processing. There remains an important open challenge to confirm this link, however, given the implication that measures like fMRI task-evoked activations and FC could be more directly tied to action potentials (spike rates) with a simple mathematical transformation (i.e., estimated activity flows). As Carl Sagan said, extraordinary claims require extraordinary evidence. Such evidence in this case could come from simultaneous blood oxygen level dependent (BOLD) response measurements and multiunit recording in brain region pairs, ideally with varying levels of connectivity among region pairs. The prediction that activity flow estimation brings fMRI/BOLD signals closer to spike rates would be supported if task-evoked activity flows better reflect spike rates than the BOLD task-evoked activations and FC estimates that went into those activity flow estimates.

An open challenge for the recently developed high-temporal-resolution activity flow mapping approach (Mill et al., 2022b) is the possibility of subtle analysis circularity driven by temporal autocorrelation. Specifically, it will be important to use activity flow to disambiguate between truly generating the onset of task-related information (information transformation) vs modeling its spread (information transfer) across the brain after that initial generation. For example, if a temporally distributed task-evoked activation is the function of interest, the start of this activation may not be predicted, yet once the start of the activation becomes the input to the model it can accurately predict the rest of the activation's spatiotemporal pattern. In the end such a model would be just pattern completing the temporal profile of the activation rather than generating it. One potential solution is to define the entire activation as the function of interest, then ensure the inputs to the activity flow model do not include any portion of the activation itself, based on either spatial or temporal (or both) restriction of the inputs. This issue reveals how activity flow modeling touches on the deep philosophical question of emergence (Mediano et al., 2022). It will be important to clarify best practices surrounding the identification of neurocognitive phenomena of interest and what features can interact within activity flow models to properly explain their emergence.

Another open challenge for activity flow modeling is to expand it to more invasive methods, potentially improving causal inferences by measuring more direct neural signals relative to the noninvasive methods used so far. However, most invasive methods have a much smaller field of view than fMRI and MEG/EEG, reducing the chance of detecting and controlling for causal confounds when estimating FC (Reid et al., 2019; Pearl and Mackenzie, 2018). Thus, it is not necessarily the case that invasive methods will improve activity flow inferences. Nonetheless, the fundamental role of action potentials in neural processing, and the clear link between action potentials and activity flows, suggests that activity flow modeling can add insights into all manner of neuroscience data. Some of this potential has been realized in the form of a multiregion multiunit recording study in nonhuman primates (Ito, 2021; Chapter 4). This study demonstrated the potential for activity flow mapping to link intrinsic FC and task-evoked activations even when using invasive spike recordings. Similarly, an intracranial EEG study was conducted with human participants with intractable epilepsy, revealing that high-temporal-resolution activity flow mapping is successful in modeling the generation of spectrally resolved language processes based on resting-state FC (Mill et al., 2022a). Together, these studies point the way toward utilizing activity flow modeling across the wide variety of available neuroscience methods, extracting new insights from those methods and ultimately driving discovery of distributed mechanisms underlying brain functions.

Take-home points

- Activity flow is the movement of activity between neural populations (also termed: activity propagation, information flow, activity spread, or activity diffusion)
- Activity flow modeling is a flexible approach that allows for building (based on empirical brain data) and empirically testing network models of brain function
- Task-evoked brain activity is used as input to simulate the flow of activity over empirically estimated brain connections, which is then tested for the ability to generate a neural or cognitive phenomenon of interest

- Examples of generated neurocognitive phenomena from previous activity flow mapping studies: face selectivity in the fusiform face area, task rule representations in the frontoparietal control network, and motor responses decoded from M1 activity (i.e., behavior) during a context-dependent decision-making task
- Four general principles guide activity flow modeling, together optimizing for clear and empirically grounded explanations of neurocognitive phenomena: (1) Simplicity/abstraction, (2) Generative, (3) Mechanistic/causal, and (4) Empirically constrained (data-driven)
- Any connectivity method (and any activity estimation method) can be incorporated into activity flow models
- Activity flow modeling can add unique inferences to standard task-evoked neural activation studies, as well as theoretical computational models.

Acknowledgments

I would like to thank Ravi Mill, Ruben Sanchez-Romero, Carrisa Cocuzza, Kirsten Peterson, Alexandros Tzalavras, and Lakshman Chakravarthy for their feedback on an earlier version of this chapter. This work was supported by US National Science Foundation grant 2219323.

Appendix: How activity flow modeling relates to additional other approaches

Dynamic causal modeling (DCM) (fMRI, MEG/EEG). DCM is a functional connectivity approach (often termed effective connectivity) that estimates causal interactions from neural time series by fitting highly parameterized models using a variational Bayesian estimation approach (Friston et al., 2003). In principle, DCM could be used with activity flow modeling just like any other FC measure. Indeed, DCM is highly compatible with the activity flow framework, given the emphasis on causal mechanisms and connectivity in both approaches. However, activity flow modeling goes further than DCM in characterizing the generative mechanism of neurocognitive processes of interest. This is due to the additional building of generative models—which could be based on DCM estimates in this case—focused on predicting each neurocognitive process of interest in independent data. This added value of activity flow modeling reflects the fact that the DCM framework does not demonstrate an ability to predict/generate neural effects in independent data. Additional inferential value can be added using recent activity flow modeling innovations, such as simulated lesioning to infer the importance of each model feature in generating the neurocognitive process of interest (Ito et al., 2022; Mill et al., 2022b).

Other complex neural modeling approaches. A wide variety of complex neural (and cognitive) modeling approaches exist. In principle, activity flow modeling could be integrated with and complement any of them. For example, the "joint modeling of neural and behavior" approach (Palestro et al., 2018) identifies neural variance linked to behavioral variance, which could also be linked to activity flow processes as mediators between sensory input and motor output. A similar modeling framework (in that it encompasses both neural and behavioral data) is the dynamic neural field theory modeling framework (Wijeakumar et al., 2017). These models appear to be initially hand-built (based on researchers' assumptions/hypotheses), then researcher-specified free parameters are fit to neural and behavioral data. Activity flow modeling could be used to empirically specify/constrain the network architecture within each neural field theory model. This integration has the potential to improve inferences regarding how empirical brain connectivity specifies cognitive functionality in the human brain.

As another example, The Virtual Brain (Ritter et al., 2013) may initially appear to encompass activity flow modeling given that it generates activity time series based on (typically structural) connectivity. However, it appears the inferential algorithm underlying activity flow mapping—using empirical activity and connectivity to generate held-out task-related neurocognitive functions—is not a standard part of The Virtual Brain. Instead, The Virtual Brain uses complex biophysical simulation equations to model neural mass model dynamics, typically using randomly generated spontaneous activity as a starting point (rather than empirical activity). Thus, activity flow modeling could be readily incorporated into The Virtual Brain to expand the inferential power of the framework. One promising avenue for this incorporation of activity flow modeling is the ability to input task stimulus timing (or brain stimulation) time series into The Virtual Brain simulations. It appears this feature has not been widely utilized in published The Virtual Brain studies, as the primary utilization of this modeling framework has been to simulate resting-state brain dynamics via simulated dynamics over structural connectomes (Martí-Juan et al., 2023). More generally, it is worth noting that the simplicity/abstraction principle underlying the activity flow modeling framework (see Table 1) suggests the need to test The Virtual Brain's assumption that complex neural mass modeling equations are necessary to accurately generate neurocognitive functionality.

One important exception to the focus on resting state (rather than task state) processes with The Virtual Brain has been a series of studies incorporating a biologically realistic task-performing model (Tagamets and Horwitz, 1998) into The Virtual Brain modeling software (Ulloa and Horwitz, 2016; Liu et al., 2022; Ulloa and Horwitz, 2018). The primary advantages of this integration appears to be making more specific predictions regarding the anatomical locations corresponding to each of the original model's nodes, as well as more realistic generation of noise via The Virtual Brain's standard structural connectome and dynamical equations. Critically, however, the model's task-performing elements remain completely specified by the original version of the model, which was created via a mix of the researchers' domain knowledge and hand-coded connectivity (Tagamets and Horwitz, 1998). Thus, unlike activity flow modeling (Ito et al., 2022), it appears that task-related functionality has not been generated by lower-level activity and connectivity constraints within The Virtual Brain. This represents a major opportunity to increase the task-related cognitive relevance of The Virtual Brain simulations via incorporation of activity-flow-like modeling concepts.

Feedforward predictive modeling of the visual system. Activity flow modeling has recently been applied specifically to the visual system (Cocuzza et al., 2022), leveraging rich knowledge regarding visual processes in the brain to enhance activity flow models (and gain further insight into the visual system). Several other approaches have been developed based on similar insights regarding the role of activity flows in generating activity throughout the visual system. First, Haak et al. (2013) developed connective field modeling, which fits Gaussian spatial models that map visual receptive fields between cortical areas. This approach is directly related to the concept of activity flow over connectivity, but is more specifically related to topographic organization of representations within the visual system. It appears that the activity flow modeling approach could extend this approach to gain insight into the activity flow relationships among nontopographic features in the visual system. By the same token, connective field modeling could refine activity flow modeling to better reveal when activity flow projections are topographic in organization. A second approach, voxel-to-voxel predictive modeling (Mell et al., 2021), is again focused exclusively on the visual system, but is more directly related to activity flow modeling. Like activity flow modeling, relationships between neural populations (weights) are estimated, then source activity multiplied by the weights are used to predict target activity. However, rather than using connectivity per se, the voxel-to-voxel weights appear to be based on the task-evoked activities themselves. This results in a distinct set of inferences, wherein the role of connectivity in the predicted relationships is less clear. It appears that the voxel-to-voxel predictive modeling approach could be enhanced by the use of functional or structural connectivity to gain a more mechanistic understanding of the relationships between visual areas. Activity flow modeling could also be enhanced by taking insights from voxel-to-voxel predictive modeling, especially with regard to inferring receptive field properties, the relationship to stimulus encoding models, and comparing results to deep neural network models of the visual system.

References

Avena-Koenigsberger, A., Misic, B., Sporns, O., 2017. Communication dynamics in complex brain networks. Nat. Rev. Neurosci. 19 (1), 17–33.

Babadi, B., Abbott, L.F., 2010. Intrinsic stability of temporally shifted spike-timing dependent plasticity. PLoS Comput. Biol. 6 (11), e1000961.

Baddeley, A., 2000. The episodic buffer: a new component of working memory? Trends Cogn. Sci. 4 (11), 417–423.

Bengio, Y., Lecun, Y., Hinton, G., 2021. Deep learning for AI. Commun. ACM 64 (7), 58–65.

Bernstein-Eliav, M., Tavor, I., 2022. The prediction of brain activity from connectivity: advances and applications. Neuroscientist. https://doi.org/10.1177/10738584221130974.

Cocuzza, C.V., Ruben, S.-R., Ito, T., Mill, R.D., Keane, B.P., Cole, M.W., 2022. Distributed network flows generate localized category selectivity in human visual cortex. BioRxiv. https://doi.org/10.1101/2022.02.19.481103.

Cole, M.W., Yarkoni, T., Repovs, G., Anticevic, A., Braver, T.S., 2012. Global connectivity of prefrontal cortex predicts cognitive control and intelligence. J. Neurosci. Off. J. Soc. Neurosci. 32 (26), 8988–8999.

Cole, M.W., Ito, T., Bassett, D.S., Schultz, D.H., 2016. Activity flow over resting-state networks shapes cognitive task activations. Nat. Neurosci. 19 (12), 1718–1726.

Cole, M.W., Ito, T., Schultz, D., Mill, R., Chen, R., Cocuzza, C., 2019. Task activations produce spurious but systematic inflation of task functional connectivity estimates. Neuroimage 189 (April), 1–18.

Cole, M.W., Ito, T., Cocuzza, C., Sanchez-Romero, R., 2021. The functional relevance of task-state functional connectivity. J. Neurosci. Off. J. Soc. Neurosci. 41 (12), 2684–2702.

Cybenko, G., 1989. Approximation by superpositions of a sigmoidal function. Math. Control Signals Syst. 2 (4), 303–314.

Damoiseaux, J.S., Greicius, M.D., 2009. Greater than the sum of its parts: a review of studies combining structural connectivity and resting-state functional connectivity. Brain Struct. Funct. 213 (6), 525–533.

de-Wit, L., Alexander, D., Ekroll, V., Wagemans, J., 2016. Is neuroimaging measuring information in the brain? Psychon. Bull. Rev. 23 (5), 1415–1428.

Donahue, C.J., Sotiropoulos, S.N., Jbabdi, S., Hernandez-Fernandez, M., Behrens, T.E., Dyrby, T.B., Coalson, T., et al., 2016. Using diffusion tractography to predict cortical connection strength and distance: a quantitative comparison with tracers in the monkey. J. Neurosci. Off. J. Soc. Neurosci. 36 (25), 6758–6770.

Esposito, R., Bortoletto, M., Miniussi, C., 2020. Integrating TMS, EEG, and MRI as an approach for studying brain connectivity. Neuroscientist 26 (5–6), 471–486.

Friston, K.J., Harrison, L., Penny, W., 2003. Dynamic causal modelling. Neuroimage 19 (4), 1273–1302.

Haak, K.V., Winawer, J., Harvey, B.M., Renken, R., Dumoulin, S.O., Wandell, B.A., Cornelissen, F.W., 2013. Connective field modeling. Neuroimage 66 (C), 376–384.

Hansen, K.B., 2020. The virtue of simplicity: on machine learning models in algorithmic trading. Big Data Soc. 7 (1). https://doi.org/10.1177/2053951720926558.

Hanson, C., Hanson, S.J., 1996. Development of schemata during event parsing: Neisser's perceptual cycle as a recurrent connectionist network. J. Cogn. Neurosci. 8 (2), 119–134.

Hearne, L.J., Mill, R.D., Keane, B.P., Repovš, G., Anticevic, A., Cole, M.W., 2021. Activity flow underlying abnormalities in brain activations and cognition in Schizophrenia. Sci. Adv. 7 (29). 2020.12.16.423109.

Henson, R., 2005. What can functional neuroimaging tell the experimental psychologist? The Q. J. Exp. Psychol. A 58 (2), 193–233.

Hornik, K., Stinchcombe, M., White, H., 1989. Multilayer feedforward networks are universal approximators. Neural Netw. 2 (5), 359–366.

Huth, A.G., Nishimoto, S., Vu, A.T., Gallant, J.L., 2012. A continuous semantic space describes the representation of thousands of object and action categories across the human brain. Neuron 76 (6), 1210–1224.

Hwang, K., Shine, J.M., Cole, M.W., Sorenson, E., 2022. Thalamocortical contributions to cognitive task activity. Elife 11 (December). https://doi.org/10.7554/eLife.81282.

Ito, T., 2021. In: Cole, M.W. (Ed.), Cognitive Information Transformation in Functional Brain Networks. Rutgers University-Newark.

Ito, T., Kulkarni, K.R., Schultz, D.H., Mill, R.D., Chen, R.H., Solomyak, L.I., Cole, M.W., 2017. Cognitive task information is transferred between brain regions via resting-state network topology. Nat. Commun. 8 (1), 1027.

Ito, T., Hearne, L., Mill, R., Cocuzza, C., Cole, M.W., 2020. Discovering the computational relevance of brain network organization. Trends Cogn. Sci. https://doi.org/10.1016/j.tics.2019.10.005.

Ito, T., Yang, G.R., Laurent, P., Schultz, D.H., Cole, M.W., 2022. Constructing neural network models from brain data reveals representational transformations linked to adaptive behavior. Nat. Commun. 13 (673). https://doi.org/10.1038/s41467-022-28323-7.

Kar, K., Ito, T., Cole, M.W., Krekelberg, B., 2020. Transcranial alternating current stimulation attenuates BOLD adaptation and increases functional connectivity. J. Neurophysiol. 123 (1), 428–438.

Keane, B., Barch, D., Mill, R., Silverstein, S., Krekelberg, B., Cole, M., 2021. Brain network mechanisms of visual shape completion. NeuroImage 236, 118069. https://doi.org/10.1016/j.neuroimage.2021.118069.

Keane, B.P., Krekelberg, B., Mill, R.D., Silverstein, S.M., Thompson, J.L., Serody, M.R., Barch, D.M., Cole, M.W., 2023. Dorsal attention network activity during perceptual organization is distinct in schizophrenia and predictive of cognitive disorganization. Eur. J. Neurosci. 57 (3), 458–478.

Khan, A.G., Hofer, S.B., 2018. Contextual signals in visual cortex. Curr. Opin. Neurobiol. 52 (October), 131–138.

Krakauer, J.W., Ghazanfar, A.A., Gomez-Marin, A., MacIver, M.A., Poeppel, D., 2017. Neuroscience needs behavior: correcting a reductionist Bias. Neuron 93 (3), 480–490.

Kriegeskorte, N., Kyle Simmons, W., Bellgowan, P.S.F., Baker, C.I., 2009. Circular analysis in systems neuroscience: the dangers of double dipping. Nat. Neurosci. 12 (5), 535–540.

Lee, J.H., Durand, R., Gradinaru, V., Zhang, F., Goshen, I., Kim, D.-S., Fenno, L.E., Ramakrishnan, C., Deisseroth, K., 2010. Global and local FMRI signals driven by neurons defined optogenetically by type and wiring. Nature 465, 788–792.

Lee, J.H., Liu, Q., Dadgar-Kiani, E., 2022. Solving brain circuit function and dysfunction with computational modeling and optogenetic FMRI. Science 378 (6619), 493–499.

Li, L., Spratling, M., 2023. Understanding and combating robust overfitting via input loss landscape analysis and regularization. Pattern Recogn. 136 (109229), 109229.

Lipworth, L., Friis, S., Mellemkjær, L., Signorello, L.B., Johnsen, S.P., Nielsen, G.L., McLaughlin, J.K., Blot, W.J., Olsen, J.H., 2003. A population-based cohort study of mortality among adults prescribed paracetamol in Denmark. J. Clin. Epidemiol. 56 (8), 796–801.

Liu, Q., Ulloa, A., Horwitz, B., 2022. The spatiotemporal neural dynamics of Intersensory attention capture of salient stimuli: a large-scale auditory-visual modeling study. Front. Comput. Neurosci. 16 (May), 876652.

Markov, N.T., Ercsey-Ravasz, M.M., Ribeiro Gomes, A.R., Lamy, C., Magrou, L., Vezoli, J., Misery, P., et al., 2014. A weighted and directed Interareal connectivity matrix for macaque cerebral cortex. Cereb. Cortex 24 (1), 17–36.

Mars, R.B., Passingham, R.E., Jbabdi, S., 2018. Connectivity fingerprints: from areal descriptions to abstract spaces. Trends Cogn. Sci. 22 (11), 1026–1037.

Martí-Juan, G., Sastre-Garriga, J., Martinez-Heras, E., Vidal-Jordana, A., Llufriu, S., Groppa, S., Gonzalez-Escamilla, G., et al., 2023. Using The Virtual Brain to study the relationship between structural and functional connectivity in patients with multiple sclerosis: a multicenter study. Cerebral Cortex. https://doi.org/10.1093/cercor/bhad041.

McClelland, J.L., Rogers, T.T., 2003. The parallel distributed processing approach to semantic cognition. Nat. Rev. Neurosci. 4 (4), 310–322.

McCormick, E.M., Arnemann, K.L., Ito, T., Hanson, S.J., Cole, M.W., 2022. Latent functional connectivity underlying multiple brain states. Netw. Neurosci. (Camb. Mass) 6 (2), 570–590.

Mediano, P.A.M., Rosas, F.E., Luppi, A.I., Jensen, H.J., Seth, A.K., Barrett, A.B., Carhart-Harris, R.L., Bor, D., 2022. Greater than the parts: a review of the information decomposition approach to causal emergence. Philos. Transact. A Math. Phys. Eng. Sci. 380 (2227). 20210246.

Mell, M.M., St-Yves, G., Naselaris, T., 2021. Voxel-to-voxel predictive models reveal unexpected structure in unexplained variance. Neuroimage 238 (September), 118266.

Mill, R.D., Bagic, A., Bostan, A., Schneider, W., Cole, M.W., 2017a. Empirical validation of directed functional connectivity. Neuroimage 146 (February), 275–287.

Mill, R.D., Ito, T., Cole, M.W., 2017b. From connectome to cognition: the search for mechanism in human functional brain networks. Neuroimage 160 (October), 124–139.

Mill, R., Gordon, B., Balota, D., Cole, M., 2020. Predicting dysfunctional age-related task activations from resting-state network alterations. NeuroImage 221, 117167. https://doi.org/10.1016/j.neuroimage.2020.117167.

Mill, R., Flinker, A., Cole, M.W., 2022a. Invasive Human Neural Recording Links Resting-State Connectivity to Generation of Task Activity.

Mill, R.D., Hamilton, J.L., Winfield, E.C., Lalta, N., Chen, R.H., Cole, M.W., 2022b. Network modeling of dynamic brain interactions predicts emergence of neural information that supports human cognitive behavior. PLoS Biol. 20 (8), e3001686.

Mueller, J.K., Grigsby, E.M., Prevosto, V., Petraglia 3rd, F.W., Rao, H., Deng, Z.-D., Peterchev, A.V., et al., 2014. Simultaneous transcranial magnetic stimulation and single-neuron recording in alert non-human primates. Nat. Neurosci. 17 (8), 1130–1136.

Naselaris, T., Kay, K.N., Nishimoto, S., Gallant, J.L., 2011. Encoding and decoding in FMRI. Neuroimage 56 (2), 400–410.

Osher, D.E., Saxe, R.R., Koldewyn, K., Gabrieli, J.D.E., Kanwisher, N., Saygin, Z.M., 2016. Structural connectivity fingerprints predict cortical selectivity for multiple visual categories across cortex. Cerebral Cortex 26 (4), 1668–1683.

Palestro, J.J., Bahg, G., Sederberg, P.B., Zhong-Lin, L., Steyvers, M., Turner, B.M., 2018. A tutorial on joint models of neural and behavioral measures of cognition. J. Math. Psychol. 84 (June), 20–48.

Passingham, R.E., Stephan, K.E., Kötter, R., 2002. The anatomical basis of functional localization in the cortex. Nat. Rev. Neurosci. 3 (8), 606–616.

Pearl, J., 2009. Causal inference in statistics: an overview. Stat. Surv. 3 (none), 96–146.

Pearl, J., Mackenzie, D., 2018. The Book of Why: *The New Science of Cause and Effect*. Basic Books.

Reid, A.T., Headley, D.B., Mill, R.D., Sanchez-Romero, R., Uddin, L.Q., Marinazzo, D., Lurie, D.J., et al., 2019. Advancing functional connectivity research from association to causation. Nat. Neurosci. https://doi.org/10.1038/s41593-019-0510-4.

Ritter, P., Schirner, M., McIntosh, A.R., Jirsa, V.K., 2013. The virtual brain integrates computational modeling and multimodal neuroimaging. Brain Connect. 3 (2), 121–145.

Rogers, T.T., McClelland, J.L., 2014. Parallel distributed processing at 25: further explorations in the microstructure of cognition. Cognit. Sci. 38 (6), 1024–1077.

Rozell, C.J., Johnson, D.H., Baraniuk, R.G., Olshausen, B.A., 2008. Sparse coding via thresholding and local competition in neural circuits. Neural Comput. 20 (10), 2526–2563.

Rumelhart, D.E., Hinton, G.E., McClelland, J.L., Others., 1986a. A general framework for parallel distributed processing. In: Parallel Distributed Processing: Explorations in the Microstructure of Cognition. vol. 1, pp. 45–76.

Rumelhart, D.E., Hinton, G.E., Williams, R.J., 1986b. Learning representations by back-propagating errors. Nature. https://doi.org/10.1038/323533a0.

Sacramento, J., Wichert, A., van Rossum, M.C.W., 2015. Energy efficient sparse connectivity from imbalanced synaptic plasticity rules. PLoS Comput. Biol. 11 (6), e1004265.

Sanchez-Romero, R., Cole, M.W., 2021. Combining multiple functional connectivity methods to improve causal inferences. J. Cogn. Neurosci. 33 (2), 180–194.

Sanchez-Romero, R., Ito, T., Mill, R.D., Hanson, S.J., Cole, M.W., 2023. Causally informed activity flow models provide mechanistic insight into network-generated cognitive activations. Neuroimage 278 (120300), 120300.

Saxena, S., Cunningham, J.P., 2019. Towards the neural population doctrine. Curr. Opin. Neurobiol. 55 (April), 103–111.

Schultz, D.H., Ito, T., Cole, M.W., 2022. Global connectivity fingerprints predict the domain generality of multiple-demand regions. Cereb. Cortex 32 (20), 4464–4479.

Shen, X., Finn, E.S., Scheinost, D., Rosenberg, M.D., Chun, M.M., Papademetris, X., Todd Constable, R., 2017. Using connectome-based predictive modeling to predict individual behavior from brain connectivity. Nat. Protoc. 12 (3), 506–518.

Sheth, S.A., Bijanki, K.R., Metzger, B., Allawala, A., Pirtle, V., Adkinson, J.A., Myers, J., et al., 2022. Deep brain stimulation for depression informed by intracranial recordings. Biol. Psychiatry 92 (3), 246–251.

Silver, D., Hubert, T., Schrittwieser, J., Antonoglou, I., Lai, M., Guez, A., Lanctot, M., et al., 2018. A general reinforcement learning algorithm that masters chess, shogi, and Go through self-play. Science 362 (6419), 1140–1144.

Smith, S.M., Fox, P.T., Miller, K.L., Glahn, D.C., Mickle Fox, P., Mackay, C.E., Filippini, N., et al., 2009. Correspondence of the Brain's functional architecture during activation and rest. Proc. Natl. Acad. Sci. U. S. A 106 (31), 13040–13045.

Smith, S.M., Nichols, T.E., Vidaurre, D., Winkler, A.M., Behrens, T.E.J., Glasser, M.F., Ugurbil, K., Barch, D.M., Van Essen, D.C., Miller, K.L., 2015. A positive-negative mode of population covariation links brain connectivity, demographics and behavior. Nat. Neurosci. 18 (11), 1565–1567.

Tagamets, M.A., Horwitz, B., 1998. Integrating electrophysiological and anatomical experimental data to create a large-scale model that simulates a delayed match-to-sample human brain imaging study. Cerebral Cortex 8 (4), 310–320.

Tavor, I., Parker Jones, O., Mars, R.B., Smith, S.M., Behrens, T.E., Jbabdi, S., 2016. Task-free MRI predicts individual differences in brain activity during task performance. Science 352 (6282), 216–220.

Thompson, W.H., Nair, R., Oya, H., Esteban, O., Shine, J.M., Petkov, C.I., Poldrack, R.A., Howard, M., Adolphs, R., 2020. A data resource from concurrent intracranial stimulation and functional MRI of the human brain. Scientific Data 7 (1), 258.

Ulloa, A., Horwitz, B., 2016. Embedding task-based neural models into a connectome-based model of the cerebral cortex. Front. Neuroinform. 10 (August), 32.

Ulloa, A., Horwitz, B., 2018. Quantifying differences between passive and task-evoked intrinsic functional connectivity in a large-scale brain simulation. Brain Connect. 8 (10), 637–652.

Varoquaux, G., Poldrack, R.A., 2018. Predictive models avoid excessive reductionism in cognitive neuroimaging. Curr. Opin. Neurobiol. 55 (December), 1–6.

Vaswani, A., Shazeer, N., Parmar, N., Uszkoreit, J., Jones, L., Gomez, A.N., Kaiser, L., Polosukhin, I., 2017. Attention is All you Need. ArXiv. http://arxiv.org/abs/1706.03762.

Wijeakumar, S., Ambrose, J.P., Spencer, J.P., Curtu, R., 2017. Model-based functional neuroimaging using dynamic neural fields: an integrative cognitive neuroscience approach. J. Math. Psychol. 76 (Pt B), 212–235.

Yamins, D.L.K., DiCarlo, J.J., 2016. Using goal-driven deep learning models to understand sensory cortex. Nat. Neurosci. 19 (3), 356–365.

Yamins, D.L.K., Hong, H., Cadieu, C.F., Solomon, E.A., Seibert, D., DiCarlo, J.J., 2014. Performance-optimized hierarchical models predict neural responses in higher visual cortex. Proc. Natl. Acad. Sci. U. S. A. 111 (23), 8619–8624.

Yan, T., Liu, T., Ai, J., Shi, Z., Zhang, J., Pei, G., Jinglong, W., 2021. Task-induced activation transmitted by structural connectivity is associated with behavioral performance. *Brain Struct. Funct.* https://doi.org/10.1007/s00429-021-02249-0.

Zhu, H., Huang, Z., Yang, Y., Su, K., Fan, M., Zou, Y., Li, T., Yin, D., 2023. Activity flow mapping over probabilistic functional connectivity. Hum. Brain Mapp. 44 (2), 341–361. https://doi.org/10.1002/hbm.26044.

Glossary

Activity flow The movement of activity between neural populations.

Activity flow models Empirically constrained simulations of the generation of neurocognitive functions via activity flow processes.

Functional/effective connectivity (FC) FC methods estimate the statistical association between neural time series. The typical theoretical target of FC methods is the causal relationship between the neural populations contributing those time series. However, FC methods vary substantially in their degree of causal validity, with no current FC method thought to yield perfect causal inferences.

Generative understanding A way of comprehending a phenomenon and the system that generates it such that the necessary and sufficient conditions of the system are known for what generates that phenomenon. This is especially useful for predicting the impact of novel configurations of the system on the phenomenon of interest, such as a system break down (e.g., brain disorder) or intervention (e.g., brain treatment or brain enhancement).

Held-out activity Brain activity that is independent from the activity used to build a model, such that a prediction of that activity from the model would be logically noncircular. In the context of activity flow modeling, this refers to independence between the target activity that is to be generated/predicted by a model and the data used to build that model.

Neurocognitive function A neural phenomenon associated with a cognitive or behavioral process, which can be a target of an explanation. Note that function does not imply that the to-be-explained phenomenon is the ultimate purpose of the involved neural populations. Examples: face selectivity in the fusiform face area, behavior being driven by primary motor cortex activity.

Prediction (versus causality) The generation of data that can be compared to a phenomenon of interest, with a better prediction resulting from a better match between the generated and actual phenomenon. Some noncausal (e.g., statistical) models can produce accurate predictions, such that the true causal model/process is only one of a variety of good predictive models. However, most possible (e.g., random) models are not good predictors, making good prediction a necessary but not sufficient condition for a model being the true causal model.

Resting-state functional connectivity FC estimated using functional brain data (such as fMRI) while a participant rests.

Task-evoked activity Activity elicited by experimenter-presented stimuli, such as sensory input and task instructions.

Biophysical modeling: An approach for understanding the physiological fingerprint of the BOLD fMRI signal

Mario Gilberto Báez-Yáñez and Natalia Petridou

Translational Neuroimaging Group, Center for Image Sciences, University Medical Center Utrecht, Utrecht, The Netherlands

Introduction to biophysical modeling
What is the BOLD fMRI signal?

Blood oxygenation level-dependent functional magnetic resonance imaging (BOLD fMRI) is a powerful noninvasive technique that utilizes susceptibility-sensitive imaging to map brain functioning in vivo. The BOLD fMRI signal, although indirect, provides insights into neuronal activity. It is influenced by hemodynamic changes—such as cerebral blood flow (CBF), cerebral blood volume, and metabolic oxygen consumption—resulting from neurovascular coupling, the vascular architecture within the imaged volume, and the biophysical interactions between strong magnetic fields, deoxygenated blood, and tissue (Ogawa et al., 1993; Bandettini et al., 1994; Belliveau et al., 1990).

The sequence of events from neuronal activity to the BOLD fMRI signal is illustrated in Fig. 1. When neurons are activated during cognitive tasks, sensory stimuli, or resting-state fluctuations, they consume oxygen and glucose, triggering an increase in CBF to meet their metabolic demands through neurovascular coupling (Iadecola, 2004; Attwell and Iadecola, 2002; Harris et al., 2011; Schaeffer and Iadecola, 2021; Vazquez et al., 2014; Iadecola, 2017). This increased blood flow delivers more oxygenated hemoglobin to the active brain regions. As neurons consume oxygen, the ratio of oxyhemoglobin to deoxyhemoglobin changes locally, leading to alterations in localized blood oxygenation in venous vessels that drain blood from active sites. Because deoxygenated hemoglobin exhibits paramagnetic effects (Grgac et al., 2017, 2013; Stanisz et al., 1998), alterations in blood oxygenation modify the blood magnetic susceptibility (Fisel et al., 1991; Fujita, 2001; Kiselev and Posse, 1998, 1999; Kiselev and Novikov, 2002; Ogawa et al., 1993).

Computational and Network Modeling of Neuroimaging Data. https://doi.org/10.1016/B978-0-443-13480-7.00008-9

This alteration can be detected using different MRI sequences, such as gradient-echo, spin-echo, and balanced and non-balanced steady-state free-precession (SSFP) fMRI techniques (Sukstanskii and Yablonskiy, 2001, 2002, 2003; Kiselev and Posse, 1999; Dickson et al., 2011; Bieri and Scheffler, 2007; Scheffler et al., 2001; Sun et al., 2015; Uludag et al., 2009). As a result, while the BOLD signal originates from intravascular changes (local vessel oxygenation), it encompasses contributions to the MRI signal both within and outside blood vessels. It is crucial to highlight that the paramagnetic intravascular susceptibility changes lead to proton dephasing within and around blood vessels (Boxerman et al., 1995a,b; Kennan et al., 1994; Kiselev, 2001; Haacke et al., 2009; Weisskoff et al., 1994) (Fig. 1).

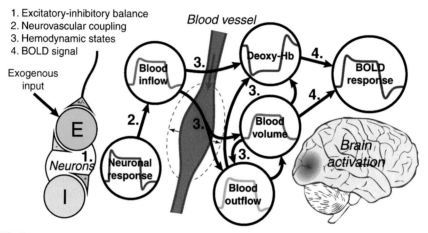

FIG. 1

A schematic illustration of the causal sequence of events between the neuronal and the BOLD response. The neuronal model (1) reflects the excitatory–inhibitory balance, which evokes hemodynamic changes via feedforward neurovascular coupling (2) and causes changes in blood inflow. Changes in blood inflow are accompanied by changes in blood outflow, blood volume, and deoxyhemoglobin content (3). The rate of blood outflow is influenced by the change in the blood volume. Changes in blood volume and deoxyhemoglobin content are then reflected in the BOLD response (4).

From Havlicek, M., et al., 2015. Physiologically informed dynamic causal modeling of fMRI data, Neuroimage, 122, 355–372.

The hemodynamic response to neuronal activation occurs via an intricate but highly organized vascular tree of several groups of vessels arranged according to their diameter (~5–100 μm) and the cortical depth they serve (Duvernoy et al., 1981; Cassot et al., 2006; Kim & Ogawa 2012) (see Fig. 2). Upon neuronal activation, oxygenated blood flows from arteries at the pial surface toward the capillary bed via penetrating arteries and arterioles, and subsequently

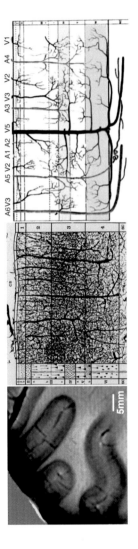

FIG. 2

Cortical human vasculature organization. Left: The image shows a zoomed view of the parietal cortex acquired with T2* weighted imaging reconstructed as phase (shown) and magnitude. Arrows point to visible intracortical veins (dark lines). Center: Illustration of the intracortical vasculature. Vascular layers are denoted on the right; neuronal layers are denoted on the left of the schematic. Right: Illustration of the intracortical vascular organization according to penetration depth and diameter, from the pial surface (top) to white matter (bottom). A denotes arteries and V denotes veins; the numbers indicate the vessel group.

Left: Adapted from Petridou, N., et al., 2010. Investigating the effect of blood susceptibility on phase contrast in the human brain. Neuroimage, 50 (2), 491–8. Center. Reproduced from Duvernoy HM et al. Cortical blood vessels of the human brain. Brain Res. Bull. (1981) 7:519–579. Right: Taken from Kim, S.G., Ogawa, S., 2012. Biophysical and physiological origins of blood oxygenation level-dependent fMRI signals. J. Cereb. Blood Flow Metab. 32 (7), 1188–2.

deoxygenated blood flows back toward the pial veins via venules and ascending veins (Tsai et al., 2003, 2009; Weber et al., 2008; Blinder et al., 2013; Reichold et al., 2009). The capillary bed is a complex interconnected network, showing the smallest element of the vascular angioarchitecture (diameters $\sim$5 μm per vessel) within deeper cortical layers with the highest metabolic demand; it is the bridge between the arterial and venous side, and it is where oxygen exchange takes place to maintain the metabolic needs of the surrounding neuronal tissue (Lauwers et al., 2008; Lorthois et al., 2011; Schmid et al., 2017, 2019; Fang et al., 2008) (Fig. 2).

Moreover, neuronal activity gives rise to alterations in oxidative metabolism, commonly referred to as the cerebral metabolic rate of oxygen (CMRO2). This leads to an elevation in the quantity of deoxyhemoglobin, causing a reduction in blood oxygenation and an increase in magnetic susceptibility (Duyn, 2013; Duyn and Schenck, 2017; Koch et al., 2006; Lee et al., 2010; Li and Leigh, 2004; Liu, 2010; Marques and Bowtell, 2004, 2008; Petridou et al., 2010; Pathak et al., 2008; Reichenbach, 2012). However, the increased deoxyhemoglobin levels during activation are counterbalanced by the inflow of oxygenated blood from arteries, due to the increase in CBF and cerebral blood volume (CBV). This compensatory mechanism offsets the deoxyhemoglobin increase. Furthermore, adjustments in CBV within blood vessels containing deoxyhemoglobin also contribute to the decrease in deoxyhemoglobin levels (Krieger et al., 2012; Mandeville et al., 1998; Payne and El-Bouri, 2018; Krishnamurthy et al., 2014). Consequently, the combined effect of CMRO2, CBV, and CBF culminates in an overall reduction in deoxyhemoglobin levels and a concurrent elevation in blood oxygenation, as reflected by the BOLD signal (Buxton, 2012, 2013; Van der Zwaag et al., 2009; Uludag and Blinder, 2018; Devor et al., 2012) (see Fig. 1).

Thus, during stimuli or tasks, the BOLD signal exhibits distinct temporal deviations from its baseline state, commonly known as temporal features of the hemodynamic response function (HRF) (Bandettini et al., 1997; de Zwart et al., 2005; Buckner, 1998). These temporal features are characterized by a gradual increase in signal intensity in relation to neuronal activity, often displaying a peak after some seconds (Logothetis et al., 2001). In some cases, an initial decrease in the BOLD signal from its baseline, termed the "initial dip," is noticeable. Following the cessation of stimulation, there is often an observed poststimulus undershoot. The BOLD HRF elicited by the stimulus can exhibit variations in both magnitude and temporal profile among subjects and cerebral regions (Siero et al., 2011; Aguirre et al., 1998) (see Fig. 1(4)—the BOLD signal response).

It might appear that the nature of the BOLD signal is quite well defined; however, the BOLD signal is not fully understood due to several reasons:

(1) The multifaceted nature of hemodynamic response: Understanding how CBF, CBV, and CMRO2 interact and contribute to the BOLD signal is challenging.

(2) Spatial and temporal resolution: The BOLD signal typically acquired has limitations in spatial and temporal resolution. It reflects changes in activity averaged over a relatively large region of the brain—limited by the voxel size—and is slower than changes in neuronal activity. This makes it difficult to capture rapid and precise neural dynamics.

(3) Biophysical underpinnings: the BOLD signal is influenced by a range of physiological factors, including oxygen extraction fraction (OEF), vascular architecture, and tissue composition. These factors can vary across individuals and brain regions, making it challenging to establish a universal model of the BOLD response.

(4) Indirect measure: The BOLD signal is an indirect measure of neural activity, and the relationship between neural firing rates and the BOLD response is complex. Neural activity influences hemodynamics, but the exact translation from neural firing to the BOLD signal is not fully understood.

(5) Neurovascular coupling: The coupling between neural activity and changes in blood flow and oxygenation is not uniform across brain regions and can vary based on factors such as age, health, and disease. This variability adds to the complexity of interpreting BOLD signals, and finally

(6) Data contaminants: BOLD data can be affected by various artifacts and confounding features, such as motion artifacts, physiological noise, and susceptibility-induced distortions, which can make it difficult to isolate the true underlying neural activity.

What is a biophysical model of the BOLD fMRI signal?

Biophysical models of the BOLD signal can help address several broad questions related to brain function, neurovascular coupling, and the mechanisms underlying the observed fMRI response. A biophysical model of the BOLD signal attempts to explain the relationship between neural activity, localized hemodynamic changes, and the observed changes in the MRI signal (Zheng et al., 2002, 2005; Stephan et al., 2007, Stephan and Roebroeck, 2012; Davis et al., 1998). Such a model integrates various physiological processes to simulate this observed signal (Puckett et al., 2016; Markuerkiaga et al., 2016). It typically considers the physiological hemodynamic and metabolic changes such as oxygen metabolism, CBF, cerebral blood volume (Huppert et al., 2007; Angleys et al., 2018; Havlicek and Uludağ, 2020; Hartung et al., 2021), and the magnetic properties of blood, and, thus, it aims to capture the complex interplay between these factors and neuronal activity—dependent on the complexity of the proposed model (see section "*Examples of successful biophysical models of the BOLD fMRI signal*").

A biophysical model of the BOLD signal involves mathematical equations either solved in an analytical way, or by means of numerical approximations using computer simulations, to describe the dynamics of oxygenation, blood flow, and other relevant physiological variables. Biophysical models take into account factors like the rate of oxygen consumption by neurons, the dilation of blood vessels in response to neural activity (neurovascular coupling), and the interaction between oxygenated vessels and the main magnetic field (Weisskoff, 2012; El-Bouri et al., 2015; Buschle et al., 2015; Ziener et al., 2006, 2007a,b; Buxton, 2012).

Different biophysical models have been proposed over the years, each with its own assumptions and level of complexity (Weisskoff et al., 1994; Yablonskiy and Haacke, 1994, Yablonskiy and Sukstanskii, 2010; He and Yablonskiy, 2009). These models have helped to interpret fMRI data, to estimate physiological parameters from the observed signal, and to gain insights into the underlying neuronal processes. It is important to note that while biophysical models provide valuable insights, they are simplifications of the complex biological processes involved (Yablonskiy et al., 2012, Yablonskiy and Sukstanskii, 2015). The models are continually refined and updated as our understanding of neurovascular coupling and the BOLD signal improves through experimental studies that directly manipulate neural activity and measure the corresponding BOLD response (Yu et al., 2012, 2014).

Finally, in order to broaden our understanding on the nature of the BOLD signal, biophysical models are often integrated with other imaging modalities and techniques, such as arterial spin labeling (ASL) for measuring CBF (Wells et al., 2013) or VASO (Huber et al., 2019) for measuring cerebral blood volume; measuring cerebrovascular reactivity (CVR) or simultaneous EEG-fMRI recordings for combining neural and hemodynamic data (Zaldivar et al., 2018; Akbari et al., 2023). This multimodal integration helps in refining and validating the models further.

The development of biophysical models of the BOLD signal has its roots in the early days of functional MRI research and our growing understanding of the underlying physiological processes. A brief historical background and theoretical context for these models involve the *discovery of the BOLD effect* by Seiji Ogawa and colleagues in the early 1990s (Ogawa et al., 1990). Initially, researchers interpreted the BOLD signal as a direct measure of neural activity (Belliveau et al., 1990; Bandettini et al., 1997). However, it soon became clear that the relationship between neuronal activity and the BOLD signal is complex and indirect. As already described earlier, the BOLD signal reflects changes in blood flow, blood volume, and oxygenation, which are influenced by neurovascular coupling and metabolic processes (functional hyperemia). As we sought to understand the mechanisms underlying the BOLD signal, we began developing more complex biophysical models that could explain the observed phenomena aiming to integrate physiological processes and provide a theoretical framework for the generation of the BOLD response.

How can biophysical models help in understanding and interpreting the BOLD fMRI signal?

Some examples of the questions that biophysical models can help to answer are given as follows:

How does neural activity translate into the observed fMRI signal? Biophysical models can elucidate the relationship between neuronal activity and the resulting hemodynamic response captured by the BOLD signal. They can help explain how changes in synaptic activity, action potentials, and neurotransmitter release give rise to alterations in CBF, oxygenation, and the magnetic properties of blood. For example, this question was addressed in the work of Havlicek et al. (Havlicek and Uludağ, 2020; Havlicek et al., 2015), which is based on a biophysical model that comprises a nonlinear system reduced to a coupled system of differential equations showing the impact of neuronal activity and the corresponding BOLD signal response (Havlicek and Uludağ, 2020). Another example is the computational framework proposed by Hermes et al. (Hermes et al., 2017) that models BOLD signals as the sum of the metabolic demand of a population of active neurons, expressed as the sum of the power of activity-induced current. Electrophysiological signals, however, for the same neuronal population were better predicted as the power of the summed activity-induced current per neuron.

What are the physiological parameters associated with brain function? Biophysical models allow estimation and quantification of physiological parameters relevant to brain function. These parameters include CBF, oxygen consumption, cerebral metabolic rate, OEF, and others. By fitting the model to fMRI data, we can infer these parameters and gain insights into how neural activity leads to changes in CBF, oxygenation, and the resulting BOLD signal, as well as their variations during different cognitive tasks or neurological conditions (Buxton, 2012). For example, a category of calibrated BOLD fMRI methods was used by Blockley et al. (2015) for the quantitative assessment of the baseline OEF. These techniques rely on two respiratory challenges and a mathematical model to estimate OEF based on the resulting changes in the BOLD fMRI signal. However, this mathematical model does not encompass all contributing factors to the BOLD signal, relies on several physiological assumptions, and may be influenced by intersubject physiological variations (Davis et al., 1998). That study aimed to explore the sources of systematic error and their impact on OEF estimation, achieved through simulations employing a detailed BOLD signal model. Despite these variations, the uncertainty in the relationship between the measured BOLD signals and OEF remained relatively low. Investigations into the underlying physiological assumptions of the mathematical model revealed that OEF measurements may be prone to overestimation in cases of oxygen metabolism changes during hypercapnia or CBF alterations during hyperoxia (Schellekens et al., 2023; Bhogal et al. 2014).

How do alterations in neurovascular coupling impact the BOLD signal? Neurovascular coupling refers to the relationship between neural activity and the local changes in blood flow and oxygenation (Logothetis et al., 2001). Biophysical models can investigate the effects of impaired or altered neurovascular coupling on the BOLD signal. They can shed light on how physiological changes such as changes in vessel dilation, oxygen extraction, or vascular reactivity that are associated with neural activity affect the observed fMRI response (e.g., Boas et al., 2008). This knowledge has been pivotal in interpreting fMRI signals and gaining insights into how alterations in neurovascular coupling and/or cerebrovascular dysfunction may contribute to various brain disorders such as stroke, neurodegenerative diseases, or psychiatric disorders (El-Bouri et al., 2021).

What are the spatial and temporal characteristics of the BOLD response? Biophysical models can provide insights into the spatial and temporal dynamics of the BOLD signal. For example, by incorporating parameters on the vascular topography (Lorthois et al., 2011), the timing and distribution of blood flow in the vascular tree, and the resulting changes in local susceptibility and magnetic fields. Thus, they can help answer questions about the time course of the hemodynamic response following neural activation, the spatial specificity of the response, and the factors influencing the shape and amplitude of the BOLD signal (e.g., Gagnon et al., 2015; Báez-Yánez et al., 2017, 2023; Hartung et al., 2021). Understanding these characteristics is crucial for accurate interpretation and analysis of fMRI data (Siero et al., 2011; Fracasso et al., 2018, Viessmann et al. 2019; Petridou and Siero, 2017; Polimeni et al., 2010).

How can we optimize experimental design and data analysis in fMRI studies? Biophysical models can guide experimental design by predicting the expected fMRI response under different conditions. By simulating the expected fMRI response under different conditions, these models enable to generate hypotheses and design experiments to test specific theories about brain function. This guidance has been valuable in optimizing stimulus paradigms, determining appropriate scan durations, or identifying the most informative imaging parameters (Scheffler et al. 2001, 2018; Shmuel et al., 2007). For instance, in the work done by Havlicek et al. (2017), they observed that the BOLD response exhibits nonlinear dependence on echo time (TE), and this nonlinearity varies throughout the entire time course (Pfaffenrot et al., 2021). When fitting a linear model to this nonlinear relationship, they found a positive intercept at TE = 0 ms. The time course of this intercept displays both rapid and slow modulations, which are distinct from both the BOLD response and CBF. To better understand the TE dependence of the BOLD signal and the time course of the intercept, they conducted simulations using a nonlinear two-compartmental BOLD signal model in conjunction with the dynamic balloon model (Buxton et al., 1998). The modeling suggests that the time course of the intercept reflects a weighted combination of changes in deoxyhemoglobin concentration and venous CBV signal.

Additionally, these models can aid in developing data analysis approaches that account for the underlying physiological processes and improve the detection and characterization of brain activity. By comparing the simulated fMRI response with empirical data, the evaluation of the performance and accuracy of novel algorithms, imaging sequences, and preprocessing methods can be made, leading to the development of more robust and accurate fMRI analysis tools—such as the open-source *SPM software* developed by Friston et al. (Stephan et al., 2010, Stephan and Roebroeck, 2012; Penny et al., 2006).

Can we improve the diagnosis or monitoring of brain disorders? By incorporating disease-specific alterations in neurovascular coupling, vascular functioning, or metabolism and generating predictions of resulting hemodynamics and expected BOLD fMRI data, biophysical models can contribute to the development of diagnostic tools or biomarkers for brain disorders. They can help identify abnormalities in the BOLD response that are associated with specific conditions and potentially assist in the early detection, monitoring, or assessment of treatment efficacy.

Examples of successful biophysical models

There have been several classic examples of successful biophysical models of the BOLD signal that have contributed significantly to our understanding of the relationship between neuronal activity and hemodynamic responses and the formation of the BOLD fMRI signal.

One of the first developed models that include the complex interaction of hemodynamics is the *Balloon Model*: Buxton et al. in 1998 (Buxton et al., 1998) along with Mandeville and colleagues in 1999 (Mandeville et al., 1999) have pioneered the development of biophysical models relating changes in CBF and deoxyhemoglobin content to the BOLD signal. The model describes the neurovascular coupling between neural activity, CBF, and the BOLD response.

The balloon model is based on the idea that neural activity leads to an increased demand for oxygen, which triggers a cascade of physiological responses in the brain vasculature. The model incorporates three main components: the balloon component, the flow component, and the volume component. The balloon component represents the hemodynamic response to neural activity. It assumes that neuronal firing causes an increase in the concentration of oxygen in the blood vessels, leading to local vasodilation and increased blood volume. This vasodilation is modeled as a balloon-like expansion. The rate of change of the balloon volume represents the rate of change of the blood volume in response to neural activity. The flow component describes the dynamics of CBF in response to the change in blood volume. It assumes that the increase in blood volume leads to an increase in CBF due to the vasodilation of the blood vessels. The rate of change of CBF is proportional to the rate of change of the balloon volume. Finally, the volume component represents the dynamics of the venous blood volume. It accounts for the fact that the increased blood inflow and

CBF cause an increase in venous blood volume. The rate of change of the venous blood volume is proportional to the CBF. By integrating these components, the balloon model simulates the temporal dynamics of the BOLD response following neural activity. The model parameters, such as the time constants and coupling constants, can be estimated from experimental data to fit the observed BOLD response. The intrinsic idea behind the balloon model lies in the next two equations.

$$\frac{dq}{dt} = \frac{1}{\tau_0}\left[f_{in}(t)\frac{E(t)}{E_0} - f_{out}(v)\frac{q(t)}{v(t)}\right]$$

$$\frac{dv}{dt} = \frac{1}{\tau_0}[f_{in}(t) - f_{out}(v)]$$

where q is the deoxyhemoglobin content, v is the volume, f is the flow, $\tau_0 = v/f$ is the mean transit time through the venous compartment at rest, and E is the resting net extraction of O2, by the capillary bed. For explicit details on the balloon model, see Buxton et al. (1998)

Thus, the balloon model provides a simplified representation of the complex neurovascular coupling mechanisms. It assumes a linear relationship between neural activity and the BOLD response, neglecting certain nonlinearities and complexities in the underlying physiology. Nonetheless, the model has been successful in capturing the essential features of the BOLD response and it has been widely used in fMRI studies to interpret and analyze the observed signal.

Another successful model developed early on is the Davis model. The Davis model, developed by Thomas L. Davis and colleagues in 1998, focuses on the dynamics of oxygen metabolism and blood flow in response to neural activity. It incorporates a detailed mathematical description of oxygen delivery, consumption, and extraction in brain tissue. The Davis model has been influential in understanding the relationships between neuronal firing rates, CBF, and oxygenation changes that underlie the BOLD signal (Davis et al., 1998). The Davis model states that,

$$\Delta R_2^*(t) \propto f_v(t)\left(\frac{CMR_{O2}(t)}{CBF(t)}\right)^\beta - f_v(0)\left(\frac{CMR_{O2}(0)}{CBF(0)}\right)^\beta$$

Thus, they postulated that the BOLD effect manifests as a change in the observed MR transverse relaxation rate, ΔR_2^*. This change is assumed to have a linear dependence on the blood volume fraction, f_v, and is influenced by the magnetic susceptibility difference between blood and tissue. The latter is modulated by CBF and CMRO2, both raised to the power of β (a free parameter contingent upon the MRI pulse sequence). For a comprehensive elucidation of the derivation of the Davis model, see Davis et al., 1998.

Over time, biophysical models became more sophisticated, incorporating additional factors and complexities. These models included elements such as the OEF, arterial and venous compartments, nonlinearities in the oxygenation response, realistic vascular networks, novel numerical methods to approximate the hemodynamic solutions, and more. Below we outline some of these models.

Examples of models that use nonrealistic vessels to simulate MRI signals

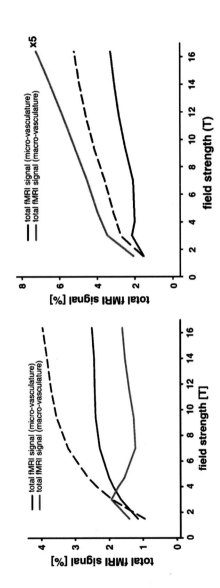

EXAMPLE 1

An integrative model of fMRI signals accounting for intra- and extravascular contributions as a function of field strength, echo time, and gradient-echo (GE) and spin-echo (SE) data acquisition techniques. The figure illustrates simulation results as a function of field strength for the following: left: SE BOLD fMRI and echo time TE = T2 tissue, and right: GE BOLD fMRI and echo time TE = T2* tissue. The graphs illustrate the total fMRI signal for microvessels (black) and macrovessels (red).

Adapted from Uludağ, K., Müller-Bierl, B., Uğurbil, K., 2009. An integrative model for neuronal activity-induced signal changes for gradient and spin echo functiona imaging. Neuroimage. 48 (1),150–65. https://doi.org/10.1016/j.neuroimage.2009.05.051. Epub 2009 May 27. PMID: 19481163.

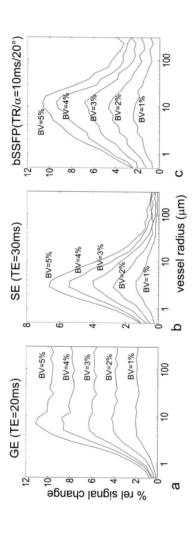

EXAMPLE 2

Model developed to examine the oxygenation-based BOLD signal changes for (A) GE, (B) SE, and (C) pass-band balanced steady-state free precession (SSFP) acquisitions. The figure shows the total BOLD signal change of GE, SE, and bSSFP as a function of vessel radius (randomly oriented cylinders) for different fractional blood volumes. Blood oxygenation of $Y = 77\%$ and $Y = 85\%$ was assumed for the resting and activated state, respectively, at a field strength of 9.4T.

From Báez-Yánez, M.G., et al., 2017. The impact of vessel size, orientation and intravascular contribution on the neurovascular fingerprint of BOLD bSSFP fMRI, NeuroImage, 163, 13–23.

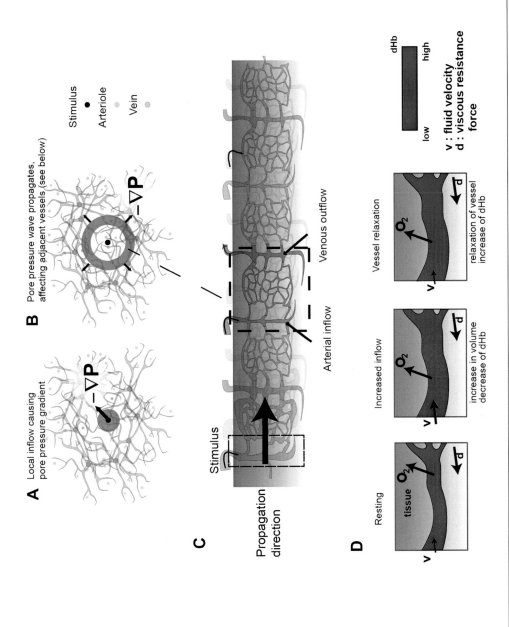

EXAMPLE 3

The spatiotemporal hemodynamic model developed to examine the spatiotemporal properties of BOLD responses given a set of physiological assumptions: (A) The hemodynamic response is driven by a localized spatiotemporal input, which represents neural activity and causes a change in arterial inflow of blood, which sets up a pressure gradient $= P$. (B) This mass inflow deforms surrounding tissue and thus exerts pressure on nearby vessels. (C) The rise in pressure causes further volume changes in adjacent vessels, which propagate outward via interactions with successive regions of tissue. Damping of the response through blood viscosity and losses via outflow. (D) The increase of local inflow increases oxygenated hemoglobin (oHb), reducing the amount of local deoxygenated hemoglobin (dHb). As oxygen is simultaneously passively extracted from blood, oHb is converted to dHb, causing a delayed rise of dHb during vessel relaxation.

Image taken from Aquino, K.M., Schira, M.M., Robinson, P.A., Drysdale, P.M., Breakspear, M., 2012. Hemodynamic traveling waves in human visual cortex. PLoS Comput. Biol. 8 (3), e1002435. https://doi.org/10.1371/journal.pcbi.1002435.

Examples of models that approximate a realistic architecture

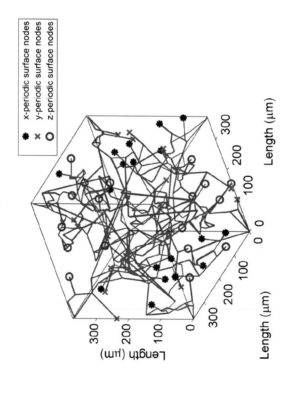

EXAMPLE 4

Typical cerebral capillary network, generated in a 375 μm cube with the additional constraint of spatial periodicity. The model was developed to examine the blood flow within the capillary bed of the human cortex, which gives insight into the hemodynamics underlying the BOLD signal.

Image taken from El-Bouri, W., Payne, S., 2015. Multi-scale homogenization of blood flow in 3-dimensional human cerebral microvascular networks. J. Theor. Biol. 380 (Sept 2015), 40–47.

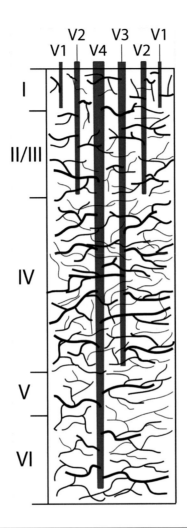

EXAMPLE 5

Schematic representation of the vascular model developed to examine the laminar specificity of the BOLD signal. Intracortical veins (in blue) run perpendicular to the cortical surface and drain blood from all layers they pass through. The microvascular density is highest in layer IV.

Image taken from Markuerkiaga, I., Barth, M., Norris, D.G., 2016. A cortical vascular model for examining the specificity of the laminar BOLD signal. Neuroimage. 132, 491–498. https://doi.org/10.1016/j.neuroimage. 2016.02.073. Epub 2016 Mar 4. PMID: 26952195.

Examples of models that use realistic vascular architectures based on mouse microscopy data

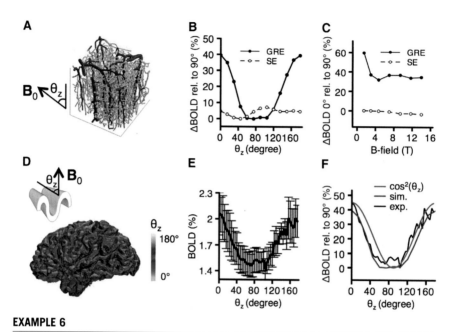

EXAMPLE 6

Model developed to give insight into the microvascular origins of the BOLD signal based on the mouse cortical vascular architecture and physiology as may be observed in a typical fMRI voxel. The figure illustrates the predicted BOLD dependence on the local folding of the cortex. (A) Convention for the angle (θz) between the vector normal to the local cortical surface and the external magnetic field (represented by the arrow). (B) Variation in the extravascular BOLD response at 3 T predicted from simulations with θz ranging from 0° to 180° normalized by the BOLD response simulated for $\theta z = 90°$. (C) Difference in the extravascular BOLD response simulated at $\theta z = 0°$ relative to $\theta z = 90°$ (in percentage relative to $\theta z = 90°$) for different B-field strengths (TE = T2, tissue*). (D) Illustration of θz values computed in the gray matter of the human brain. (E) Averaged BOLD responses measured in the gray matter of human subjects during a hypercapnic challenge as a function of θz. (F) Comparison of angular dependence predicted from our simulations with angular dependence measured during the hypercapnic challenge in humans ($n = 5$ human subjects for experimental; $n = 6$ animals for simulations).

Image taken from Gagnon, L., et al., 2015. Quantifying the microvascular origin of BOLD-fMRI from first principles with two-photon microscopy and an oxygen-sensitive nanoprobe. J. Neurosci. 35 (8), 3663–3675.

Similar computational concepts have been applied by Báez-Yáñez et al. (Báez-Yáñez et al., 2023) in a recent work aiming to simulate the human vascular architecture and associated MRI signals, given assumptions based on human histology and subsequently by mathematically converting mouse microscopy data into a statistical approximation of the human vasculature.

These models have their foundation in the model developed by Boas et al. (2008) that connects changes in vascular tone and oxygen dynamics to hemodynamics, illustrated as follows (Fig. 3):

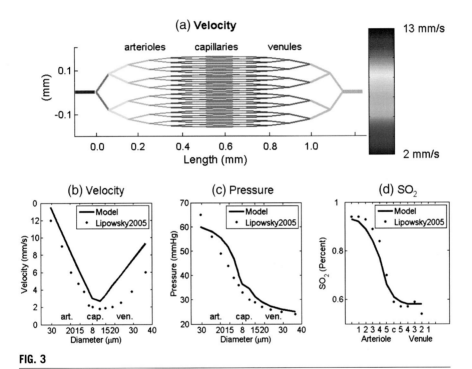

FIG. 3

The vascular anatomical network model. (A) The 2D vascular network model and the velocity in each segment calculated from the arterial–venous pressure difference and the resistance of each vascular segment. (B) Velocity, (C) pressure in different diameter segments, and (D) hemoglobin oxygen saturation in different vascular branches agree well with experimental measurements from Lipowsky (2005) and Vovenko (1999). From left to right are arterioles, capillaries, and venules.

Image taken from Boas, D.A., Jones, S.R., Devor, A., Huppert, T.J., Dale, A.M., 2008. A vascular anatomical network model of the spatio-temporal response to brain activation, NeuroImage, 40 (3), 1116–1129, ISSN 1053-8119, https://doi.org/10.1016/j.neuroimage.2007.12.061.

EXAMPLE 7

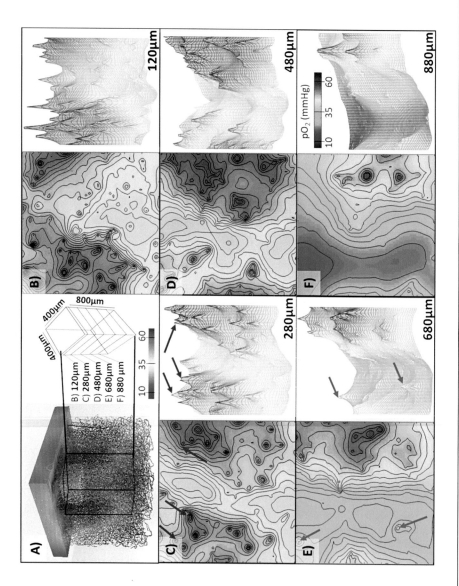

Model of oxygen dynamics and cerebral blood flow in the microvasculature aiming at improved computational efficiency. This figure illustrates the oxygen distribution throughout a microvascular network specimen (E4.1) from the vibrissa primary sensory cortex in a mouse. (A) Visualization of a portion of the extravascular space with a block cutout defined by five observation planes at different cortical depths (120, 280, 480, 680, and 880 μm below the cortical surface). (B–F) The visualization of oxygen distribution at the individual layers. (C) Steep oxygen gradients form around penetrating arterioles—marked with red arrows. (E) Oxygen tension is also elevated around veins—marked with blue arrows. In the deeper regions, (E and F) a hypoxic valley begins to form, seen as deep blue basins.

Image taken from Hartung, G., et al., 2021. Voxelized simulation of cerebral oxygen perfusion elucidates hypoxia in aged mouse cortex, PLOS Comp Biol.

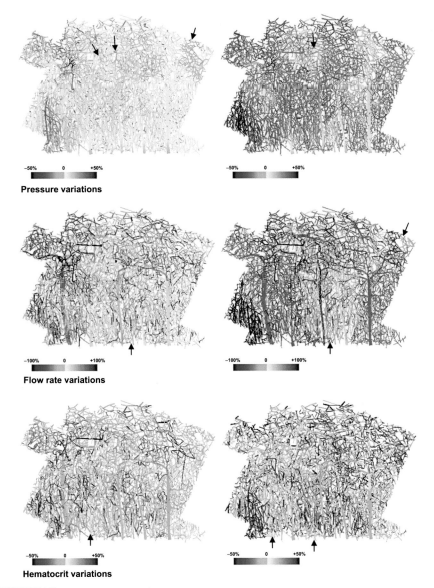

Pressure variations

Flow rate variations

Hematocrit variations

EXAMPLE 8

Model developed to investigate the hemodynamics of brain activation underlying BOLD fMRI based on fluid dynamics. The figure illustrates global vasodilations: spatial distributions of pressure variations (first row), flow rate variations (second row), and hematocrit variations (third row) calculated for the maximal value of fvaso under study (fvaso = 2) relative to baseline for zero-flow (Case 1) boundary condition (left) and assigned pressure (Case 2) boundary condition (right) and H in = 0.4. Numerical values of mean pressure variation and flow rate variation in each segment relative to baseline are color-coded as indicated in the color scale. For clarity of representation, diameters have been increased threefold compared to lengths. Hot colors (yellow to reddish) display positive variations, whereas cold colors (bluish colors) display negative variations. Green segments are segments exhibiting no variation relative to the baseline.

Image taken from Lorthois, S., Cassot, F., Lauwers, F., 2011. Simulation study of brain blood flow regulation by intra-cortical arterioles in an anatomically accurate large human vascular network. Part II: Flow variations induced by global or localized modifications of arteriolar diameters. NeuroImage, 54 (4), 2840–2853, ISSN 1053-8119, https://doi.org/10.1016/j.neuroimage.2010.10.040.

Example of a model that explores the characteristics of hemodynamic responses influencing the BOLD signal

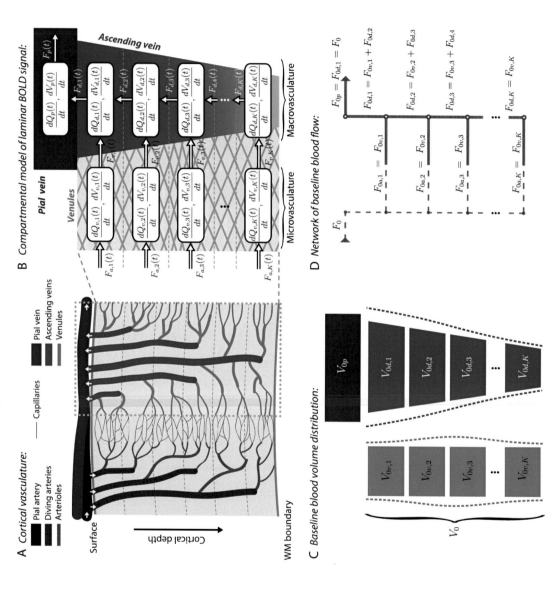

A Cortical vasculature:

B Compartmental model of laminar BOLD signal:

C Baseline blood volume distribution:

D Network of baseline blood flow:

EXAMPLE 9

See legend on next page

(Continued)

EXAMPLE 9, CONT'D

Dynamical model of the laminar BOLD response that accounts for the cortical vascular architecture and the spatiotemporal distribution of neuronal activity patterns. (A) Schematic illustration of depth-dependent distribution of vasculature—from arteries to veins—in the cortex. White arrows indicate direction of the blood flow. Dotted rectangle emphasizes the venous vasculature that mainly contributes to the GE-fMRI signal. (B) Flow diagram representing the compartmentalized version of the cortical vascular (venous) network, distinguishing microvasculature (venules) and macrovasculature (ascending and pial veins). Compartments are connected with blood flow, and each compartment models the changes in CBV and dHb content. (C) Each compartment is further described by baseline CBV. Dashed lines illustrate the possibility to specify diverse laminar distributions of baseline CBV. (D) Network of baseline CBF, representing the merging between venules and ascending vein (based on mass conservation law).

Image taken from Havlicek, M., Uludağ, K., 2020. A dynamical model of the laminar BOLD response. NeuroImage, 204, 116209.

It's worth noting that biophysical models continue to evolve and new models are continually being proposed to refine our understanding of the BOLD signal. Also worth highlighting is that we outline some examples, but several other models exist (see, for example, review articles and references therein such as Gauthier and Fan (2019) and Uludağ (2023). Biophysical models have laid the foundation for subsequent research and have significantly advanced our knowledge of the physiological processes underlying fMRI measurements.

Assumptions of biophysical models

Biophysical models of the BOLD signal are based on several fundamental assumptions that simplify the complex physiological processes underlying fMRI measurements. These assumptions serve as a starting point for modeling and provide a framework for understanding the relationship between neuronal activity and the observed hemodynamic response. Some examples of assumptions commonly made in biophysical modeling of the BOLD signal can be summarized into these six important points.

(1) Linearity and additivity: Biophysical models often assume linearity and additivity, meaning that the overall response is a linear sum of individual contributions. This assumption allows for the simplification of complex interactions between various physiological factors and enables the modeling of the overall hemodynamic response as a linear combination of the underlying neural activity.

(2) About the neurovascular coupling: Biophysical models assume that there is a close coupling between neural activity and the associated changes in blood flow, blood volume, and oxygenation. This neurovascular coupling is a key aspect of the models, as it describes the link between neuronal activity and the subsequent hemodynamic response. The assumption is that increased neuronal activity leads to increased metabolic demand, resulting in changes in blood flow and oxygenation in the activated regions.

(3) Homogeneity in vascular architecture and functionality: Biophysical models often assume homogeneity within brain regions of interest. This assumption implies that the properties and responses of the vasculature, blood flow, and oxygen metabolism are relatively uniform within a given region. While there may be variations across different brain regions, assuming homogeneity helps simplify the modeling process and allows for the estimation of average responses.

(4) Time invariance: Some biophysical models often assume time invariance, meaning that the underlying physiological processes and parameters remain constant over time. This assumption allows for the modeling of steady-state or quasi-steady-state conditions, where the physiological variables reach a relatively stable state. However, it should be noted that there are dynamic aspects to neurovascular coupling, and models accounting for time-varying processes are also available.

(5) Spatial localization: Models assume that the BOLD response is spatially localized, meaning that the observed signal primarily reflects changes in the local vasculature and tissue properties within a specific brain region. This assumption allows to make inferences about localized neural activity based on the measured fMRI signal.

(6) Homoscedasticity: Some biophysical models often assume homoscedasticity, which means that the variance of the measured fMRI signal is constant across time or experimental conditions. This assumption simplifies the modeling process and facilitates parameter estimation.

It's important to recognize that these assumptions are simplifications of the complex physiological processes occurring in the neuronal and vascular system. They provide a starting point for modeling and can be adjusted or expanded upon based on the specific questions and available data.

The accuracy of assumptions that underlie biophysical models of the BOLD signal holds significant implications for various aspects. These implications span multiple dimensions, affecting the performance and predictive power of the models, the interpretation of research results, experimental design and analysis approaches, the development of new or refined models, and even the advancement of experimental techniques themselves. First, the precision of these assumptions directly influences the performance and reliability of biophysical models, impacting their capacity to replicate observed BOLD signals and estimate physiological parameters. Any significant deviation from real-world conditions can undermine the model's ability to simulate the complex neurovascular coupling accurately, thereby limiting its utility in deciphering fMRI data and drawing meaningful insights into brain activity and function. Furthermore, the interpretational framework built upon biophysical models can be substantially influenced by inaccurate assumptions. Inaccuracies may compromise the reliability of conclusions drawn from model outputs, potentially leading to less trustworthy or even misleading interpretations of research findings.

In instances where inaccuracies are identified between simulations and data, we may be prompted to develop novel models or enhance existing ones—revisit underlying assumptions, striving to align models more closely with empirical observations. This pursuit of accuracy may culminate in the creation of advanced, context-specific models that more accurately capture the intricate details of the BOLD signal, ultimately yielding more reliable and nuanced results.

Beyond modeling, inaccuracies can catalyze progress in experimental techniques. Researchers may be driven to pioneer new methods that directly measure or validate the underlying physiological processes. Innovations may encompass novel imaging approaches, the integration of multimodal data, or the application of invasive measurements to provide a more comprehensive understanding of neurovascular coupling and the BOLD signal.

Building, testing, interpreting biophysical models

As described in the introduction, most of the biophysical models of the BOLD signal are built upon core foundations that capture the essential physiological processes and principles underlying the observed fMRI response. These core foundations form the basis for understanding the relationship between neuronal activity and hemodynamic changes. The most relevant "ingredients" or factors that need to be taken into account to simulate the BOLD signal can be specifically defined as follows:

Neurovascular coupling: the models are based on the principle of neurovascular coupling, which describes the relationship between neuronal activity and the associated changes in CBF, blood volume, and oxygenation. This coupling forms the foundation for linking neural activity to the observed hemodynamic response in fMRI. Biophysical models aim to capture the mechanisms by which changes in neuronal activity lead to alterations in the local vasculature and subsequently affect the BOLD signal (Iadecola, 2017; Attwell and Iadecola, 2002).

Oxygen metabolism: Biophysical models need to incorporate the fundamental processes of oxygen metabolism in brain tissue including factors such as oxygen consumption, delivery, and extraction, as well as the metabolic demands of neurons. These factors account for the dynamic interplay between neuronal energy consumption, oxygen utilization, and the resulting changes in oxygenation levels in the blood (Fang et al., 2008; Jain et al., 2012; Lyons et al., 2016; Secomb et al., 2004).

Vascular dynamics: Vascular dynamics have an impact on the formation of the BOLD signal. These dynamics include the effects of changes in blood flow, blood volume, and vessel dilation/constriction on the observed fMRI response (Gould et al., 2017; Pries et al., 1990; Su et al., 2012; Kjolby et al., 2006). Biophysical models often incorporate mathematical descriptions of hemodynamic processes, such as the behavior of blood vessels and the dynamics of oxygenation and deoxygenation of hemoglobin (Blockley et al., 2008).

Magnetic resonance principles: Biophysical models take into account the principles of MRI that underlie the acquisition of fMRI data. These principles include the interaction of magnetic fields with the protons in the tissues, the relaxation times of tissues, and the signal generation mechanisms such as water diffusion and susceptibility-induced processes. Biophysical models incorporate the effects of the MRI pulse sequence used to acquire the fMRI data and account for the influence of magnetic field strength, TE, and other imaging parameters on the observed signal (Asslaender et al., 2017; Audoly et al., 2003; Barzykin, 1998; Beaulieu, 2002; Bloch, 1946; Cheng et al., 2001; Durrant et al., 2003; Elst et al., 2002; Freed et al., 2001; Frohlich et al., 2005; Ganter, 2006a,b; Goa et al., 2014; Grebenkov, 2007; Hahn, 1950; Hall and Alexander, 2009; Hall, 2016; Jenkinson et al., 2004; Jensen et al., 2006, Jensen and Chandra, 2000; Khajehim et al., 2017; Kim et al., 2012; Kurz et al., 2014, 2016a,b,c, 2017; Bihan, 2012; Lee et al., 1999; Luo et al., 2010, 2014; Luz and Meiboom, 1963; Ma et al., 2013; Miller and Jezzard, 2008, Miller, 2010a,b; Mueller-Bierl et al., 2004, 2007; Murase et al., 2011; Park et al., 2011; Pflugfelder et al., 2011; Rozenman et al., 1990; Ruh et al., 2018; Sen, 2004; Semmineh et al., 2014; Seppenwoolde et al., 2005; Shmueli et al., 2011; Torrey, 1956; Vuong et al., 2011; Ye and Allen, 1995; Yeh et al., 2013; Zhong et al., 2007; Zhou et al., 2012; Zielinski and Sen, 2003; Zur et al., 1990).

Experimental data and validation: Biophysical models are built upon experimental data and are often validated using empirical measurements (Petridou et al., 2010; Barth et al., 2007,2010; Goense and Logothetis, 2006, Goense et al., 2016; Pannetier et al., 2014; Panagiotaki et al., 2012). It is well established to use task-based or resting-state fMRI data, as well as other physiological measurements, to calibrate and refine the models. Comparisons between model predictions and observed data provide a basis for evaluating the accuracy and performance of the models.

Conducting biophysical modeling of the BOLD signal demands a solid grasp of mathematical, statistical, and computational methodologies. These skills provide the groundwork for crafting intricate equations, analyzing data, and executing essential simulations inherent to these models. Key technical foundations encompass mathematical modeling, relying on differential equations to capture physiological relationships, statistical analysis for parameter estimation and model assessment, and numerical methods to find solutions to complex equations, and computational modeling involving simulations, data analysis and visualization techniques, and algorithm development for tackling modeling challenges.

The effective practice of biophysical modeling of the BOLD signal demands adherence to a comprehensive array of guidelines and best practices, serving as a compass to circumvent potential pitfalls and misconceptions. To effectively engage in biophysical modeling of the BOLD signal, it is recommended to begin by immersing oneself in the existing literature and grasping foundational concepts as outlined above, assumptions, and methodologies.

These principles offer a robust framework for conducting modeling with accuracy and precision. Among the key recommendations are the following:

- Defining clear research objectives and articulating specific aspects of the BOLD signal to be modeled.
- The transparent articulation of underlying assumptions and explicit acknowledgment of model limitations, ensuring a contextual understanding for interpreting outcomes across various ways of communication.
- Constructing a mathematical model that encapsulates key physiological processes such as neurovascular coupling and oxygen metabolism is essential, (e.g., see section *Examples of successful biophysical models of the BOLD fMRI signal*), followed by its translation into a computational framework using well-documented code.
- The consideration and integration of temporal dynamics are highlighted, advocating for models that encapsulate time delays and nonlinearities to capture the intricate evolution of the BOLD response.
- In addressing the regional heterogeneity intrinsic to the brain, a call is made to incorporate pertinent regional-specific factors, thereby enriching the model's predictive ability.
- Sensitivity analyses can be used as a tool to evaluate the model's robustness, permitting an exploration of how variations in parameter values, assumptions, or experimental conditions influence outcomes.
- Model validation is essential, urging to employ independent datasets or divergent experimental designs to ascertain the model's veracity and generalizability. Model validation against empirical data and subsequent refinement should be carried out iteratively to enhance accuracy and precision. Gathering relevant data, aligning with research goals and modeling assumptions is crucial. Rigorous data quality control measures should be employed, supporting meticulous preprocessing techniques to counteract artifacts and confounding factors and appropriate statistical analyses, coupled with transparent documentation.
- Collaboration with experts in related fields is advisable, ensuring insights from biophysics, neuroimaging, and computational modeling are connected. Such collaboration can help in fostering improvements and validating modeling strategies.
- Reproducibility takes a central stage, accentuating the sharing of resources, data, and code to encourage independent validation, transparency, and the identification of potential discrepancies.
- Prudent extrapolation for circumspect application of model predictions beyond their original scope and advocating for validation when delving into uncharted territories.

Through these steps, the technical process of biophysical modeling can be navigated skillfully, contributing to an enhanced understanding of the BOLD signal's implications for brain function.

Pitfalls in biophysical modeling

Biophysical modeling of the BOLD signal presents challenges that we should address for a robust approach. Although these challenges don't invalidate the method, they underscore important considerations. Common pitfalls include over-simplifying complex processes, relying on assumptions with acknowledged limitations, poor data quality, and availability for model validation, inaccurate parameter estimation in complex models, not validating models for generalizability, interpretation limitations for causal inference, and limited collaboration and feedback from experts. These pitfalls emphasize the importance of cautious interpretation of model outcomes. Examples include misinterpreting causality, overlooking temporal dynamics, neglecting regional differences, overrelying on simplified models, disregarding model-data discrepancies, and extrapolating models to untested conditions.

It is essential to be aware of these pitfalls and critically evaluate the assumptions, limitations, and uncertainties associated with the model. Engaging in rigorous validation, considering alternative explanations, and seeking independent replication can help mitigate the risk of misinterpretations and improve the accuracy and reliability of biophysical modeling, toward a more accurate understanding of the BOLD signal and its relationship with neural activity.

Open challenges and future directions

Biophysical modeling of the BOLD signal stands as a vibrant and ever-evolving realm of research, replete with open challenges that, when addressed, promise to propel our comprehension of brain function to new heights while refining the precision and versatility of the models themselves. The landscape of these challenges encompasses a range of details, each representing a unique way for exploration and advancement.

Bridging scales

One pressing challenge involves the integration of microscale connectivity into biophysical models that predominantly operate at the macroscale. Existing models often distill brain regions into average properties, neglecting the intricate web of connections that individual neurons interlace. Inclusion of microscale connectivity stands to illuminate how local neural circuits contribute to the observed BOLD signal, offering insights into the fine-grained dynamics of brain function (Chen et al., 2013a,b; Kleinnijenhuis et al., 2013; Koopmans et al., 2010).

In addition to the neuronal microscale connectivity, a challenge is the integration of the microscale structure and function of the vascular tree. While significant progress has been done to this end (Boas et al. 2008, Gagnon et al. 2015, Báez-Yáñez et al., 2023, Hartung et al. 2021), an accurate representation of the human vascular tree is still limited by the lack of sufficient knowledge on the three-dimensional structure of the human vascular tree and physiological vascular responses. This insufficient knowledge is because we currently do not have noninvasive measurement techniques that can provide it directly in the human brain in vivo. Perhaps a promising approach that can enable visualization of the structure of the human vascular tree is immunohistochemistry combined with 3D microscopy (Reveles Jensen and Berg, 2017; Kirst et al., 2020), albeit postmortem. Along the same lines, the assumption of uniform vascular properties within brain regions can compromise model accuracy. Factoring in vascular diversity—embodying variations in vessel caliber or microvascular traits—promises regional precision in modeling. Crafting models that embrace such heterogeneity engenders a more authentic portrayal of the BOLD signal.

Besides the availability of knowledge that is needed to bridge realms from microscopic cellular dynamics to sweeping macroscopic networks, there is the requirement for comprehensive multiscale modeling (Stephan and Roebroeck, 2012; Marreiros et al. 2008). These complex interplays demand models that transcend traditional boundaries, capturing the intricate symbiosis between different spatial and temporal dimensions. By uniting models spanning cellular-level dynamics to extensive network interactions, a more nuanced comprehension of the BOLD signal and its interplay with neural activity can be achieved.

Incorporation of neuronal nonlinearities and astrocyte function

Another critical barrier centers on the incorporation of nonlinear dynamics into current models. Prevailing approaches frequently assume linearity, overlooking the nonlinearities inherent in neural activity or hemodynamic response (Havlicek et al., 2015). By tackling this challenge, we can enhance model fidelity substantially. Nonlinear modeling paradigms, such as dynamic causal modeling or using the power of deep learning techniques (Daunizeau et al., 2011), hold the potential to captivate this complexity, helping in a more nuanced understanding of brain function (Stephan and Roebroeck, 2012; Marreiros et al. 2008). Further, the intricate role of astrocytes and pericytes—pivotal players in neurovascular coupling and CBF regulation—beckons for illumination (Iadecola, 2017; Kaul et al. 2020). Deciphering their contributions to the BOLD signal, the interplay between neuronal and astrocytic activity emerges as a puzzle that necessitates a deeper understanding of their interactions, ideally integrated into biophysical modeling frameworks.

Accounting for individual variability

An essential challenge entails accommodating individual variability, transcending the customary reliance on average properties that assume homogeneity across individuals (Kannurpatti et al., 2010). Overcoming interindividual variability demands the creation of individual-specific models, accounting for unique distinctions in brain structure, function, and connectivity. The incorporation of personalized data, including high-resolution structural MRI and subject-specific connectivity information, is poised to increase model precision and practicality. Machine learning methodologies hold the potential for capturing individual-specific attributes, enabling tailored models aligning with each individual's neurobiological profile. Tailoring models to encapsulate individual-specific attributes—ranging from anatomy to functional connectivity—can boost both accuracy and the potential for personalized interpretation of the BOLD signal and holds the key to personalized medicine, showing individual nuances and decoding brain disorders.

Model validation

Amid these challenges, validation, replication, and comparison of diverse models remain as an important problem. Establishing standardized validation protocols and benchmark datasets and open science principles facilitate cross-group and cross-dataset comparison, reproducibility, and model generalizability. Advancements in experimental design and paradigm construction can surmount hurdles tied to data quality, experimental variability, and generalizability. Thoughtfully designed experiments tailored to specific inquiries, knowing of potential confounds, and attuned to relevant neurophysiological processes promise heightened data quality and relevance. Novel task designs, controlled environments, and scrupulous parameter consideration collectively elevate the dependability and applicability of modeling outcomes. Such endeavors, fueling comparative studies, hold promise for revealing the strengths and limitations of distinct models and can usher in a new era of clarity, enabling the systematic assessment of disparate modeling approaches.

Notably, advancements in imaging technology hold promise by boosting spatial and temporal resolution, thereby providing more precise BOLD signal measurements. Innovations such as ultra-high-field MRI, multiband imaging, and expedited acquisition techniques are poised to enhance spatial accuracy and capture swift neural dynamics (Pohmann et al., 2016). Embracing higher-resolution imaging techniques, such as ultra-high-field MRI or innovative acquisition sequences, also affords a more intricate depiction of the BOLD signal (Ehses and Scheffler, 2018). By infusing these advancements into the modeling process, we can amplify the accuracy of the models. A profound opportunity also lies in the fusion of diverse neuroimaging modalities, such as fMRI, EEG, or PET. While navigating this challenge might be intricate, the resultant synergy has the potential to give a richer comprehension of brain function. Through integrative modeling, which complements insights from different modalities, a complete grasp of neural activity, metabolism, and connectivity can be refined. Beyond the BOLD signal's conventional scope lies the tantalizing prospect of unraveling the insights ensconced within other hemodynamic signals. While the BOLD signal dominates fMRI, embracing CBV or oxygen consumption signals carries potential (Huber et al., 2015; Bhogal et al., 2014).

Computational approaches

Equally pivotal is the development of advanced computational techniques and algorithms to overcome computational complexities and elevate modeling efficiency. Drawing on high-performance computing, parallel processing, and refined optimization algorithms holds the potential to magnify the scalability and efficiency of biophysical models. The application of machine learning and deep learning methodologies can adeptly capture intricate BOLD signal relationships and nonlinearities, amplifying modeling strength. A realm of attractive exploration lies in adapting biophysical models for real-time or online applications. Facilitating real-time estimation of neural activity or online prediction of the BOLD signal stands to catalyze innovative breakthroughs in neurofeedback, brain-computer interfaces, and clinical interventions. The pursuit of models amenable to real-time settings is a dynamic frontier full with potential.

Summary

Advancements in biophysical modeling are essential to overcome existing challenges and limitations in understanding brain function. More powerful models are needed to enhance accuracy, personalize interpretations, integrate diverse data sources, ensure generalizability, support real-time applications, improve computational efficiency, ensure robustness and reproducibility, and integrate cognitive and behavioral aspects. Achieving these goals requires interdisciplinary collaboration, computational innovation, and rigorous validation, ultimately leading to better insights into the relationship between neural activity and the BOLD signal, with implications for brain function understanding, clinical applications, and personalized interventions. By pursuing these approaches, future research can address the limitations in the biophysical modeling of the BOLD signal and make significant strides toward a more accurate, comprehensive, and interpretable understanding of brain function.

Take-home points

- A biophysical model of the blood-oxygen-level-dependent (BOLD) signal aims to explain the relationship between neural activity and the observed changes in oxygenation levels in functional magnetic resonance imaging (fMRI).
- These models incorporate physiological processes such as neurovascular coupling, cerebral hemodynamics, and cerebral oxygenation dynamics to simulate the BOLD response.
- A biophysical model provides a mechanistic understanding of the BOLD signal and can offer insights into brain function, connectivity, and the effects of various experimental conditions.
- Building biophysical models often involves making assumptions about the underlying physiology, such as linear neurovascular coupling responses or homogeneous cerebrovascular hemodynamic properties.
- These models help to reveal insights into the neurovascular coupling mechanisms, regional variations in the BOLD response, and the influence of experimental or pulse sequence acquisition parameters on the observed signal.

- Biophysical models contribute to the interpretation and analysis of fMRI data, providing an understanding on the neural basis of cognitive processes, and having potential applications in clinical research and medicine.
- Challenges in biophysical modeling include the need for improved spatial and temporal resolution data for validation, accounting for interindividual variability, capturing nonlinear dynamics, integrating multiscale data, and ensuring robust validation and reproducibility.
- Overcoming these challenges requires interdisciplinary collaboration, advancements in imaging technology, computational techniques, and rigorous validation practices. By addressing these challenges and refining biophysical models, we can enhance our understanding of brain function and improve the accuracy, applicability, and interpretability of the BOLD signal in fMRI studies.

References

Aguirre, G.K., Zarahn, E., D'Esposito, M., 1998. The variability of human, BOLD hemodynamic responses. Neuroimage 8 (4), 360–369.

Akbari, A., Bollmann, S., Ali, T.S., Barth, M., 2023. Modelling the depth-dependent VASO and BOLD responses in human primary visual cortex. Hum. Brain Mapp. 44 (2), 710–726.

Angleys, H., et al., 2018. The effects of capillary transit time heterogeneity on the BOLD signal. Hum. Brain Mapp. 39, 2329–2352.

Asslaender, J., Glaser, S., Henning, J., 2017. Pseudo steady state free precession for MR-fingerprinting. Magn. Reson. Med. 77, 1151–1161.

Attwell, D., Iadecola, C., 2002. The neural basis of functional brain imaging signals. Trends Neurosci. 25 (12), 621–625.

Audoly, B., et al., 2003. Correlation functions for inhomogeneous magnetic field in random media with application to a dense random pack of spheres. J. Magn. Reson., 154–159.

Báez-Yánez, M.G., et al., 2017. The impact of vessel size, orientation and intravascular contribution on the neurovascular fingerprint of BOLD bSSFP fMRI. Neuroimage 163, 13–23.

Báez-Yáñez, M.G., Siero, J.C.W., Petridou, N., 2023. A mechanistic computational framework to investigate the hemodynamic fingerprint of the blood oxygenation level-dependent signal. NMR Biomed., e5026.

Bandettini, P.A., et al., 1994. Spin-echo and gradient-echo EPI of human brain activation using BOLD contrast: a comparative study at 1.5 T. NMR Biomed. 7, 12–20.

Bandettini, P.A., Kwong, K.K., Davis, T.L., Tootell, R.B., Wong, E.C., Fox, P.T., Belliveau, J. W., Weisskoff, R.M., Rosen, B.R., 1997. Characterization of cerebral blood oxygenation and flow changes during prolonged brain activation. Hum. Brain Mapp. 5, 93–109.

Barth, M., et al., 2007. Very high-resolution three-dimensional functional MRI of the human visual cortex with elimination of large venous vessels. NMR Biomed. 20, 477–484.

Barth, M., et al., 2010. T2-weighted 3D fMRI using S2-SSFP at 7T. Magn. Reson. Med. 63, 1015–1020.

Barzykin, A., 1998. Exact solution of the Bloch-Torrey equation for a spin echo in restricted geometries. Phys. Rev. B 58, 14171–14174.

Beaulieu, C., 2002. The basis of anisotropic water diffusion in the nervous system—a technical review. NMR Biomed. 15, 435–455.

Belliveau, J., et al., 1990. Functional cerebral imaging by susceptibility contrast NMR. Magn. Reson. Med. 14, 538–546.

Bhogal, A.A., Siero, J.C., Fisher, J.A., Froeling, M., Luijten, P., Philippens, M., Hoogduin, H., 2014. Investigating the non-linearity of the BOLD cerebrovascular reactivity response to targeted hypo/hypercapnia at 7T. Neuroimage 98, 296–305.

Bieri, O., Scheffler, K., 2007. Effect of diffusion in inhomogeneous magnetic fields on balanced steady state free precession. NMR Biomed. 20, 1–10.

Bihan, D.L., 2012. Diffusion, confusion and functional MRI. Neuroimage 62, 1131–1136.

Blinder, P., et al., 2013. The cortical angiome: an interconnected vascular network with non-columnar patterns of blood flow. Nat. Neurosci. 16, 889–897.

Bloch, F., 1946. Nuclear induction. Phys. Rev. 1, 4604–4673.

Blockley, N.P., et al., 2008. Field strength dependence of R1 and R2* relaxivities of human whole blood to ProHance, vasovist, and deoxyhemoglobin. Magn. Reson. Med. 60, 1313–1320.

Blockley, N.P., Griffeth, V.E.M., Stone, A.J., Hare, H.V., Bulte, D.P., 2015. Sources of systematic error in calibrated BOLD based mapping of baseline oxygen extraction fraction. Neuroimage 122, 105–113. ISSN 1053-8119.

Boas, D., et al., 2008. A vascular anatomical network model of the spatio-temporal response to brain activation. Neuroimage 40, 1116–1129.

Boxerman, J., et al., 1995a. MR contrast due to intravascular magnetic susceptibility perturbations. Magn. Reson. Med. 34, 555–566.

Boxerman, J., et al., 1995b. The intravascular contribution to fMRI signal change: Monte Carlo modeling and diffusion weighted studies in vivo. Magn. Reson. Med. 34, 4–10.

Buckner, R.L., 1998. Event-related fMRI and the hemodynamic response. Hum. Brain Mapp. 6 (5–6), 373–377.

Buschle, L., et al., 2015. Diffusion mediated dephasing in the dipole field around a single spherical magnetic object. Magn. Reson. Med. 33, 1126–1145.

Buxton, R., 2012. Dynamic models of BOLD contrast. Neuroimage 62, 953–961.

Buxton, R., 2013. The physics of functional magnetic resonance imaging (fMRI). Rep. Prog. Phys. 76, 1–30.

Buxton, R., et al., 1998. Dynamics of blood flow and oxygenation changes during brain activation: the balloon model. Magn. Reson. Med. 39, 855–864.

Cassot, F., et al., 2006. A novel three-dimensional computer-assisted method for a quantitative study of microvascular networks of the human cerebral cortex. Microcirculation 13, 1–18.

Chen, G., et al., 2013a. Layer-specific BOLD activation in awake monkey V1 revealed by ultra-high spatial resolution functional magnetic resonance imaging. Neuroimage 64, 147–155.

Chen, W., Foxley, S., Miller, K., 2013b. Detecting microstructural properties of white matter based on compartmentalization of magnetic susceptibility. Neuroimage 70, 1–9.

Cheng, Y., Haacke, M., Yu, Y., 2001. An exact form for the magnetic field density of states for a dipole. Magn. Reson. Imaging 19, 1017–1023.

Daunizeau, J., David, O., Stephan, K.E., 2011. Dynamic causal modelling: a critical review of the biophysical and statistical foundations. Neuroimage 58, 312–322.

Davis, T.L., Kwong, K.K., Weisskoff, R.M., Rosen, B.R., 1998. Calibrated functional MRI: mapping the dynamics of oxidative metabolism. Proc. Natl. Acad. Sci. U. S. A. 95 (4), 1834–1839. https://doi.org/10.1073/pnas.95.4.1834. PMID: 9465103; PMCID: PMC19199.

de Zwart, J.A., Silva, A.C., van Gelderen, P., Kellman, P., Fukunaga, M., Chu, R., Koretsky, A. P., Frank, J.A., Duyn, J.H., 2005. Temporal dynamics of the BOLD fMRI impulse response. Neuroimage 24 (3), 667–677.

Devor, A., Sakadžić, S., Srinivasan, V.J., et al., 2012. Frontiers in optical imaging of cerebral blood flow and metabolism. J. Cereb. Blood Flow Metab. 32 (7), 1259–1276. https://doi.org/10.1038/jcbfm.2011.195.

Dickson, J., et al., 2011. Quantitative phenomenological model of the BOLD contrast mechanism. J. Magn. Reson. 212, 17–25.

Durrant, C., Hertzberg, M., Kuchel, P., 2003. Magnetic susceptibility: further insights into macroscopic and microscopic fields and the sphere of Lorentz. Concep. Magn. Reson. 18, 72–95.

Duvernoy, H.M., et al., 1981. Cortical blood vessels of the human brain. Brain Res. Bull. 7, 519–579.

Duyn, J., 2013. MR susceptibility imaging. J. Magn. Reson. 229, 198–207.

Duyn, J., Schenck, J., 2017. Contributions to magnetic susceptibility of brain tissue. NMR Biomed. 30, 1–37.

Ehses, P., Scheffler, K., 2018. Multiline balanced SSFP for rapid functional imaging at ultrahigh field. Magn. Reson. Med. 79 (2), 994–1000. https://doi.org/10.1002/mrm.26761. Epub 2017 May 25. PMID: 28547846.

El-Bouri, W.K., et al., 2015. Multi-scale homogenization of blood flow in 3-dimensional human cerebral microvascular networks. J. Theor. Biol. 380, 40–47.

El-Bouri, W.K., MacGowan, A., Józsa, T.I., Gounis, M.J., Payne, S.J., 2021. Modelling the impact of clot fragmentation on the microcirculation after thrombectomy. PLoS Comput. Biol. 17 (3), e1008515. https://doi.org/10.1371/journal.pcbi.1008515.

Elst, L., et al., 2002. Dy-DTPA derivatives as relaxation agents for very high field MRI: the beneficial effect of slow water exchange on the transverse relaxivities. Magn. Reson. Med. 47, 1121–1130.

Fang, Q., et al., 2008. Oxygen advection and diffusion in a three dimensional vascular anatomical network. Opt. Express 16, 17530–17541.

Fisel, R., et al., 1991. MR contrast due to microscopically heterogeneous magnetic susceptibility: numerical simulations and applications to cerebral physiology. Magn. Reson. Med. 17, 336–347.

Fracasso, A., Luijten, P.R., Dumoulin, S.O., Petridou, N., 2018. Laminar imaging of positive and negative BOLD in human visual cortex at 7T. Neuroimage 164, 100–111.

Freed, D., et al., 2001. Steady state free precession experiments and exact treatment of diffusion in a uniform gradient. J. Chem. Phys. 115, 4249–4258.

Frohlich, A., et al., 2005. Theory of susceptibility induced transverse relaxation in the capillary network in the diffusion narrowing regime. Magn. Reson. Med. 53, 564–573.

Fujita, N., 2001. Extravascular contribution of blood oxygenation level dependent signal changes: a numerical analysis based on a vascular network model. Magn. Reson. Med. 46, 723–734.

Gagnon, L., et al., 2015. Quantifying the microvascular origin of BOLD fMRI from first principles with two-photon microscopy and an oxygen-sensitive nanoprobe. J. Neurosci. 35, 3663–3675.

Ganter, C., 2006a. Static susceptibility effects in balanced SSFP sequences. Magn. Reson. Med. 56, 687–691.

Ganter, C., 2006b. Steady state of gradient echo sequences with radiofrequency phase cycling: analytical solution, contrast enhancement with particular spoiling. Magn. Reson. Med. 55, 98–107.

Gauthier, C.J., Fan, A.P., 2019. BOLD signal physiology: models and applications. Neuroimage 187, 116–127 (ISSN 1053-8119.).

Goa, P., et al., 2014. BOLD fMRI signal characteristics of S1- and S2- SSFP at 7 tesla. Front. Neurosci. 49, 1–6.

Goense, J., Logothetis, N., 2006. Laminar specificity in monkey V1 using high resolution SE-fMRI. Magn. Reson. Imaging 24, 381–392.

Goense, J., et al., 2016. fMRI at high spatial resolution: implications for BOLD models. Front. Comput. Neurosci. 10, 66.

Gould, I.G., et al., 2017. The capillary bed offers the largest hemodynamic resistance to the cortical blood supply. J. Cereb. Blood Flow Metab. 37, 52–68.

Grebenkov, D.S., 2007. NMR survey of reflected Brownian motion. Rev. Mod. Phys. 69, 1077–1137.

Grgac, K., et al., 2013. Hematocrit and oxygenation dependence of blood 1H2o T1 at 7T. Magn. Reson. Med. 70, 1153–1159.

Grgac, K., et al., 2017. Transverse water relaxation in whole blood and erythrocytes at 3T, 7T, 9.4T, 11.7T and 16.4T; determination of intracellular hemoglobin and extracellular albumin relaxivities. Magn. Reson. Imaging 38, 234–249.

Haacke, M., et al., 2009. Susceptibility weighted imaging: technical aspects and clinical applications, part I. Phys. Rev. 30, 19–30.

Hahn, E., 1950. Spin echoes. Phys. Rev. 80, 580–594.

Hall, M., 2016. Continuity, the Bloch Torrey equation and diffusion MRI. Med. Phys. (arXiv:1608.02859).

Hall, M., Alexander, D., 2009. Convergence and parameter choice for Monte Carlo simulations of diffusion MRI. IEEE Trans. Med. Imaging 28, 1354–1364.

Harris, J.J., Reynell, C., Attwell, D., 2011. The physiology of developmental changes in BOLD functional imaging signals. Dev. Cogn. Neurosci. 1 (3), 199–216. https://doi.org/10.1016/j.dcn.2011.04.001. Epub 2011 Apr 27. PMID: 22436508; PMCID: PMC6987565.

Hartung, G., et al., 2021. Voxelized simulation of cerebral oxygen perfusion elucidates hypoxia in aged mouse cortex. PLOS Comp. Biol. 17 (1), e1008584.

Havlicek, M., Uludağ, K., 2020. A dynamical model of the laminar BOLD response. Neuroimage 204, 116209.

Havlicek, M., et al., 2015. Physiologically informed dynamic causal modeling of fMRI data. Neuroimage 122, 355–372.

Havlicek, M., Ivanov, D., Poser, B.A., Uludag, K., 2017. Echo-time dependence of the BOLD response transients—a window into brain functional physiology. Neuroimage 159, 355–370.

He, X., Yablonskiy, A., 2009. Biophysical mechanisms of phase contrast in gradient echo MRI. PNAS 106, 13558–13563.

Hermes, D., Nguyen, M., Winawer, J., 2017. Neuronal synchrony and the relation between the blood-oxygen-level dependent response and the local field potential. PLoS Biol. 15 (7), e2001461.

Huber, L., et al., 2015. Cortical lamina-dependent blood volume changes in human brain at 7T. Neuroimage 107, 23–33.

Huber, L., Uludağ, K., Möller, H.E., 2019. Non-BOLD contrast for laminar fMRI in humans: CBF, CBV, and CMRO2. Neuroimage 197, 742–760. https://doi.org/10.1016/j.neuroimage.2017.07.041. Epub 2017 Jul 20. PMID: 28736310.

Huppert, T., et al., 2007. A multicompartment vascular model for inferring baseline and functional changes in cerebral oxygen metabolism and arterial dilatation. J. Cereb. Blood Flow Metab. 27, 1262–1279.

Iadecola, C., 2004. Neurovascular regulation in the normal brain and in Alzheimer's disease. Nat. Rev. Neurosci. 5 (5), 347–360.

Iadecola, C., 2017. The neurovascular unit coming of age: a journey through neurovascular coupling in health and disease. Neuron 96 (1), 17–42.

Jain, V., et al., 2012. Investigating the magnetic susceptibility properties of fresh human blood for noninvasive oxygen saturation quantification. Magn. Reson. Med. 68, 863–867.

Jenkinson, M., Wilson, J., Jezzard, P., 2004. Perturbation method for magnetic field calculations of nonconductive object. Magn. Reson. Med. 52, 471–477.

Jensen, J., Chandra, R., 2000. NMR relaxation in tissue with weak magnetic inhomogeneities. Magn. Reson. Med. 44, 144–156.

Jensen, J., et al., 2006. Magnetic field correlation imaging. Magn. Reson. Med. 55, 1350–1361.

Kannurpatti, S.S., Motes, M.A., Rypma, B., Biswal, B.B., 2010. Neural and vascular variability and the fMRI-BOLD response in normal aging. Magn. Reson. Imaging 28 (4), 466–476.

Kaul, S., Methner, C., Mishra, A., 2020. The role of pericytes in hyperemia-induced capillary de-recruitment following stenosis. Curr. Tissue Microenviron. Rep. 1 (4), 163–169.

Kennan, R., Zhong, J., Gore, J., 1994. Intravascular susceptibility contrast mechanisms in tissues. Magn. Reson. Med. 31, 9–21.

Khajehim, M., et al., 2017. Investigating the spatial specificity of S2-SSFP fMRI: a Monte Carlo simulation approach. Magn. Reson. Imaging 37, 282–289.

Kim, S.G., Ogawa, S., 2012. Biophysical and physiological origins of blood oxygenation level dependent fMRI signals. J. Cereb. Blood Flow Metab. 32, 1188–1206.

Kim, T.S., et al., 2012. Analysis of the BOLD characteristics in pass-band bSSPF fMRI. Int. J. Imaging Syst. Technol. 22, 23–32.

Kirst, C., et al., 2020. Mapping the fine-scale organization and plasticity of the brain vasculature. Cell 180 (4), 780–795. e25.

Kiselev, V., 2001. On the theoretical basis of perfusion measurements by dynamic susceptibility contrast MRI. Magn. Reson. Med. 46, 1113–1122.

Kiselev, V., Novikov, D., 2002. Transverse NMR relaxation as a probe of mesoscopic structure. Phys. Rev. Lett. 89, 1–4.

Kiselev, V., Posse, S., 1998. Analytical theory of susceptibility induced NMR signal dephasing in a cerebrovascular network. Phys. Rev. 81, 5696–5699.

Kiselev, V., Posse, S., 1999. Analytical model of susceptibility induced MR signal dephasing: effect of diffusion in a microvascular network. Magn. Reson. Med. 41, 499–509.

Kjolby, B., Ostergaard, L., Kiselev, V., 2006. Theoretical model of intravascular paramagnetic tracer effect on tissue relaxation. Magn. Reson. Med. 56, 187–197.

Kleinnijenhuis, M., et al., 2013. Layer-specific diffusion weighted imaging in human primary visual cortex in vitro. Cortex 49, 2569–2582.

Koch, K., et al., 2006. Rapid calculations of susceptibility-induced magnetostatic field perturbations for in vivo magnetic resonance. Phys. Med. 51, 6381–6402.

Koopmans, P., Barth, M., Norris, D., 2010. Layer-specific BOLD activation in human V1. Hum. Brain Mapp. 31, 1297–1304.

Krieger, S., et al., 2012. Cerebral blood volume changes during brain activation. J. Cereb. Blood Flow Metab. 32, 1618–1631.

Krishnamurthy, L., et al., 2014. Dependence of blood T2 on oxygenation at 7T: in vivo calibration and in vivo application. Magn. Reson. Med. 71, 2035–2042.

Kurz, F., et al., 2014. Theoretical model of the single spin-echo relaxation time for spherical magnetic perturbers. Magn. Reson. Med. 71, 1888–1895.

Kurz, F., et al., 2016a. CPMG relaxation rate dispersion in dipole fields around capillaries. Magn. Reson. Imaging 34, 875–888.

Kurz, F., et al., 2016b. Generalized moment analysis of magnetic field correlations for accumulations of spherical and cylindrical magnetic perturbers. Front. Physiol. 4, 1–16.

Kurz, F., et al., 2016c. Spin dephasing in a magnetic dipole field around large capillaries: approximative and exact results. J. Magn. Reson. 273, 83–97.

Kurz, F.T., et al., 2017. The influence of spatial patterns of capillary networks on transverse relaxation. Magn. Reson. Imaging 40, 31–47.

Lauwers, F., et al., 2008. Morphometry of the human cerebral cortex microcirculation: general characteristics and space related profiles. Neuroimage 39, 936–948.

Lee, S.P., et al., 1999. Diffusion weighted spin echo fMRI at 9.4T: microvascular/tissue contribution to BOLD signal changes. Magn. Reson. Med. 42, 919–928.

Lee, J., et al., 2010. Sensitivity of MRI resonance frequency to the orientation of brain tissue microstructure. PNAS 107, 5130–5135.

Li, L., Leigh, J., 2004. Quantifying arbitrary magnetic susceptibility distributions with MR. Magn. Reson. Med. 51, 1077–1082.

Lipowsky, H.H., 2005. Microvascular rheology and hemodynamics. Microcirculation 12 (1), 5–15. https://doi.org/10.1080/10739680590894966. PMID: 15804970.

Liu, C., 2010. Susceptibility tensor imaging. Magn. Reson. Med. 63, 1471–1477.

Logothetis, N., et al., 2001. Neurophysiological investigation of the basis of the fMRI signal. Nature 412, 150–157.

Lorthois, S., et al., 2011. Simulation study of brain blood flow regulation by intra-cortical arterioles in an anatomically accurate large human vascular network: part I: methodology and baseline flow. Neuroimage 54, 1031–1042.

Luo, J., et al., 2010. Protein-induced water 1H MR frequency shifts: contributions from magnetic susceptibility and exchange. J. Magn. Reson. 202, 1–15.

Luo, J., He, X., Yablonskiy, A., 2014. Magnetic susceptibility induced white matter MR signal frequency shifts-experimental comparison between Lorentzian sphere and generalized Lorentzian approaches. Magn. Reson. Med. 71, 1251–1263.

Luz, Z., Meiboom, S., 1963. Nuclear magnetic resonance study of the protolysis of trimethylammonium ion in aqueous solution. J. Chem. Phys. 39, 366–370.

Lyons, D., et al., 2016. Mapping oxygen concentration in the awake mouse brain. Elife 5, 1–16.

Ma, D., et al., 2013. Magnetic resonance fingerprinting. Nature 495, 187–192.

Mandeville, J.B., Marota, J.J., Kosofsky, B.E., Keltner, J.R., Weissleder, R., Rosen, B.R., Weisskoff, R.M., 1998. Dynamic functional imaging of relative cerebral blood volume during rat forepaw stimulation. Magn. Reson. Med. 39, 615–624.

Mandeville, J., et al., 1999. Evidence of a cerebrovascular postarteriole windkessel with delayed compliance. J. Cereb. Blood Flow Metab. 19, 679–689.

Markuerkiaga, I., et al., 2016. A cortical vascular model for examining the specificity of the laminar BOLD signal. Neuroimage 132, 491–498.

Marques, J., Bowtell, R., 2004. Application of a Fourier based method for rapid calculation of field inhomogeneity due to spatial variation of magnetic susceptibility. Conc. Magn. Reson. Part B 25, 65–78.

Marques, J., Bowtell, R., 2008. Using forward calculations of the magnetic field perturbation due to a realistic vascular model to explore the BOLD effect. NMR Biomed. 21, 553–565.

Marreiros, A.C., Kiebel, S.J., Friston, K.J., 2008. Dynamic causal modelling for fMRI: a two-state model. Neuroimage 39, 269–278. https://doi.org/10.1016/j.neuroimage.2007.08.019.

Miller, K., 2010a. Asymmetries of the balanced SSFP profile. Part I: theory and observation. Magn. Reson. Med. 63, 385–395.

Miller, K., 2010b. Asymmetries of the balanced SSFP profile. Part II: white matter. Magn. Reson. Med. 63, 396–406.

Miller, K., Jezzard, P., 2008. Modeling SSFP functional MRI contrast in the brain. Magn. Reson. Med. 60, 661–673.

Mueller Bierl, B., et al., 2007. Magnetic field distribution and signal decay in functional MRI in very high fields (up to 9.4T) using Monte Carlo diffusion modeling. Int. J. Biomed. Imaging 10, 1–7.

Mueller-Bierl, B., et al., 2004. Compensation of magnetic field distortions from paramagnetic instruments by added diamagnetic material: measurements and numerical simulations. Med. Phys. 32, 76–84.

Murase, K., et al., 2011. Numerical solutions to the time dependent Bloch equations revised. Magn. Reson. Med. 29, 126–131.

Ogawa, S., Lee, T.M., Kay, A.R., Tank, D.W., 1990. Brain magnetic resonance imaging with contrast dependent on blood oxygenation. Proc. Natl. Acad. Sci. U. S. A. 87, 9868–9872.

Ogawa, S., et al., 1993. Functional brain mapping by blood oxygenation level dependent contrast magnetic resonance imaging: a comparison of signal characteristics with a biophysical model. J Biophys 64, 803–812.

Panagiotaki, E., et al., 2012. Compartment models of the diffusion MR signal in the brain white matter: a taxonomy and comparison. Neuroimage 59, 2241–2254.

Pannetier, N., et al., 2014. Numerical modeling of susceptibility related MR signal dephasing with vessel size measurement: phantom validation at 3T. Magn. Reson. Med. 72, 646–658.

Park, S.H., et al., 2011. Sensitivity and specificity of high resolution balanced steady state free precession fMRI at high field of 9.4T. Neuroimage 58, 168–176.

Pathak, A., et al., 2008. A novel technique for modeling susceptibility based contrast mechanism for arbitrary microvascular geometries: the finite perturber method. Neuroimage 40, 1130–1143.

Payne, S.J., El-Bouri, W.K., 2018. Modelling dynamic changes in blood flow and volume in the cerebral vasculature. Neuroimage 176, 124–137.

Penny, W.D., Friston, K.J., Ashburner, J.T., Kiebel, S.J., Nichols, T.E., 2006. Statistical Parametric Mapping: The Analysis of Functional Brain Images, first ed.

Petridou, N., Siero, J., 2017. Laminar fMRI: what can the time domain tell us? Neuroimage 164, 100–111.

Petridou, N., Wharton, S.J., Lotfipour, A., Gowland, P., Bowtell, R., 2010. Investigating the effect of blood susceptibility on phase contrast in the human brain. Neuroimage 50 (2), 491–498.

Pfaffenrot, V., Voelker, M.N., Kashyap, S., Koopmans, P.J., 2021. Laminar fMRI using T2-prepared multi-echo FLASH. Neuroimage 236, 118163. https://doi.org/10.1016/j.neuroimage.2021.118163. Epub 2021 May 21. PMID: 34023449.

Pflugfelder, D., et al., 2011. On the numerically predicted spatial BOLD fMRI specificity for spin echo sequences. Magn. Reson. Med. 29, 1195–1204.

Pohmann, R., Speck, O., Scheffler, K., 2016. Signal-to-noise ratio and MR tissue parameters in human brain imaging at 3, 7, and 9.4 tesla using current receive coil arrays. Magn. Reson. Med. 75 (2), 801–809.

Polimeni, J., et al., 2010. Laminar analysis of 7T BOLD using an imposed spatial activation pattern in human V1. Neuroimage 52, 1334–1346.

Pries, A.R., et al., 1990. Blood flow in microvascular networks: experiments and simulation. Circ. Res. 67, 826–834.

Puckett, A.M., et al., 2016. The spatiotemporal hemodynamic response function for depth-dependent functional imaging of human cortex. Neuroimage 139, 240–248.

Reichenbach, J., 2012. The future of susceptibility contrast for assessment of anatomy and function. Neuroimage 62, 1311–1315.

Reichold, J., et al., 2009. Vascular graph model to simulate the cerebral blood flow in realistic vascular networks. J. Cereb. Blood Flow Metab. 29, 1429–1443.

Reveles Jensen, K.H., Berg, R.W., 2017. Advances and perspectives in tissue clearing using CLARITY. J. Chem. Neuroanat. 86, 19–34.

Rozenman, Y., et al., 1990. Signal loss by superparamagnetic iron oxide particles in NMR spin echo images: the role of diffusion. Magn. Reson. Med. 14, 31–39.

Ruh, A., Scherer, H., Kiselev, V., 2018. The Larmor frequency shift in magnetically heterogeneous media depends on their mesoscopic structure. Magn. Reson. Med. 79, 1101–1110.

Schaeffer, S., Iadecola, C., 2021. Revisiting the neurovascular unit. Nat. Neurosci. 24 (9), 1198–1209. https://doi.org/10.1038/s41593-021-00904-7. Epub 2021 Aug 5. PMID: 34354283; PMCID: PMC9462551.

Scheffler, K., et al., 2001. Detection of BOLD changes by means of a frequency sensitive true-FISP technique: preliminary results. NMR Biomed. 14, 490–496.

Scheffler, K., et al., 2018. The BOLD sensitivity of rapid steady-state sequences. Magn. Reson. Med. 81, 2526–2535.

Schellekens, W., Bhogal, A.A., Roefs, E.C.A., Báez-Yáñez, M.G., Siero, J.C.W., Petridou, N., 2023. The many layers of BOLD. The effect of hypercapnic and hyperoxic stimuli on macro- and micro-vascular compartments quantified by CVR, M, and CBV across cortical depth. J. Cereb. Blood Flow Metab. 43 (3), 419–432.

Schmid, F., et al., 2017. Depth-dependent flow and pressure characteristics in cortical microvascular networks. PLoS Comput. Biol. 13, 1–22.

Schmid, F., et al., 2019. Vascular density and distribution in neocortex. Neuroimage 197, 792–805.

Secomb, T.W., et al., 2004. Green's function methods for analysis of oxygen delivery to tissue by microvascular networks. Ann. Biomed. Eng. 32, 1519–1529.

Semmineh, N., et al., 2014. An efficient computational approach to characterize DSCMRI signals arising from three dimensional heterogeneous tissue structures. PLoS One 9, 1–13.

Sen, P., 2004. Time dependent diffusion coefficient as a probe of geometry. Conc. Magn. Reson. Part A 23, 1–21.

Seppenwoolde, J., et al., 2005. Spectral characterization of local magnetic field inhomogeneities. Phys. Med. Biol. 50, 361–372.

Shmuel, A., et al., 2007. Spatio-temporal point spread function of fMRI signal in human gray matter at 7 tesla. Neuroimage 35, 539–552.

Shmueli, K., et al., 2011. The contribution of chemical exchange to MRI frequency shifts in brain tissue. Magn. Reson. Med. 65, 35–43.

Siero, J., et al., 2011. Cortical depth-dependent temporal dynamics of the BOLD response in the human brain. J. Cereb. Blood Flow Metab. 31, 1999–2008.

Stanisz, G., et al., 1998. Water dynamics in human blood via combined measurement of T2 relaxation and diffusion in the presence of gadolinium. Magn. Reson. Med. 39, 223–233.

Stephan, K.E., Roebroeck, A., 2012. A short history of causal modeling of fMRI data. Neuroimage 62, 856–863.

Stephan, K.E., Weiskopf, N., Drysdale, P.M., Robinson, P.A., Friston, K.J., 2007. Comparing hemodynamic models with DCM. Neuroimage 38, 387–401.

Stephan, K.E., Penny, W.D., Moran, R.J., den Ouden, H.E.M., Daunizeau, J., Friston, K.J., 2010. Ten simple rules for dynamic causal modeling. Neuroimage 49, 3099–3109.

Su, S., et al., 2012. The influence of network structure on the transport of blood in the human cerebral microvasculature. Microcirculation 19, 175–187.

Sukstanskii, A., Yablonskiy, D., 2001. Theory of FID NRM signal dephasing induced by mesoscopic magnetic field inhomogeneities in biological systems. J. Magn. Reson. 151, 107–117.

Sukstanskii, A., Yablonskiy, D., 2002. Effects of restricted diffusion on MR signal formation. J. Magn. Reson. 157, 92–105.

Sukstanskii, A., Yablonskiy, D., 2003. Gaussian approximation in the theory of MR signal formation in the presence of structure-specific magnetic field inhomogeneities. J. Magn. Reson. 163, 236–247.

Sun, H., et al., 2015. Steady state functional MRI using spoiled small tip fast recovery imaging. Magn. Reson. Med. 73, 536–543.

Torrey, H.C., 1956. Bloch equations with diffusion terms. Phys. Rev. 104, 563.

Tsai, P., et al., 2003. All-optical histology using ultrashort laser pulses. Neuron 39, 27–41.

Tsai, P.S., et al., 2009. Correlations of neuronal and microvascular densities in murine cortex revealed by direct counting and colocalization of nuclei and vessels. J. Neurosci. 29, 14553–14570.

Uludağ, K., 2023. Physiological modeling of the BOLD signal and implications for effective connectivity: a primer. Neuroimage 277, 120249.

Uludag, K., Blinder, P., 2018. Linking brain vascular physiology to hemodynamic response in ultra-high field MRI. Neuroimage 168 (279), 295.

Uludag, K., Mueller-Bierl, B., Ugurbil, K., 2009. An integrative model for neuronal activity-induced signal changes for gradient and spin echo functional imaging. Neuroimage 48, 150–165.

Van der Zwaag, W., et al., 2009. FMRI at 1.5, 3 and 7T: characterizing BOLD signal changes. Neuroimage 47, 1425–1434.

Vazquez, A.L., et al., 2014. Neuronal and hemodynamic response elicited by forelimb and photostimulation in channelrhodopsin-2 mice: insights into the hemodynamic point spread function. Cereb. Cortex 24, 2908–2919.

Viessmann, O., Scheffler, K., Bianciardi, M., Wald, L.L., Polimeni, J.R., 2019. Dependence of resting-state fMRI fluctuation amplitudes on cerebral cortical orientation relative to the direction of B0 and anatomical axes. Neuroimage 196, 337–350.

Vovenko, E., 1999. Distribution of oxygen tension on the surface of arterioles, capillaries and venules of brain cortex and in tissue in normoxia: an experimental study on rats. Pflugers Arch. 437 (4), 617–623. https://doi.org/10.1007/s004240050825. PMID: 10089576.

Vuong, Q., Gillis, P., Gossuin, Y., 2011. Monte Carlo simulation and theory of proton NMR transverse relaxation induced by aggregation of magnetic particles used as MRI contrast agents. J. Magn. Reson. 212, 139–148.

Weber, B., et al., 2008. The microvascular system of the striate and extricate visual cortex of the macaque. Cereb. Cortex 10, 1093.

Weisskoff, R., 2012. The characterization of dynamic susceptibility effects. Neuroimage, 1014–1016.

Weisskoff, R.M., Zuo, C.S., Boxerman, J.L., Rosen, B.R., 1994. Microscopic susceptibility variation and transverse relaxation: theory and experiment. Magn. Reson. Med. 31 (6), 601–610. https://doi.org/10.1002/mrm.1910310605. PMID: 8057812.

Wells, J., et al., 2013. Measuring biexponential tranverse relaxation of the ASL signal at 9.4T to estimate arterial oxygen saturation and the time of exchange of labeled blood water into cortical brain tissue. J. Cereb. Blood Flow Metab. 33, 215–224.

Yablonskiy, D., Haacke, M., 1994. Theory of NMR signal behavior in magnetically inhomogeneous tissues: the static dephasing regime. Magn. Reson. Med. 32, 1522–2594.

Yablonskiy, D., Sukstanskii, A., 2010. Theoretical models of the diffusion weighted MR signal. NMR Biomed. 23, 661–681.

Yablonskiy, D., Sukstanskii, A., 2015. Generalized Lorentzian tensor approach (GLTA) as a biophysical background for quantitative susceptibility mapping. Magn. Reson. Med. 73, 757–764.

Yablonskiy, D., Sukstanskii, A., He, X., 2012. Blood oxygenation level dependent based techniques for the quantification of brain hemodynamic and metabolic properties, theoretical models and experimental approaches. NMR Biomed. 26, 963–986.

Ye, F., Allen, P., 1995. Relaxation enhancement of the transverse magnetization of water protons in paramagnetic suspensions of red blood cells. Magn. Reson. Med. 34, 713–720.

Yeh, C., et al., 2013. Diffusion microscopist simulator: a general Monte Carlo simulation for diffusion magnetic resonance imaging. PLoS One 8, 1–12.

Yu, X., et al., 2012. Direct imaging of macrovascular and microvascular contributions to BOLD fMRI in layers IV-V of the rat whisker-barrel cortex. Neuroimage 59, 1451–1460.

Yu, X., et al., 2014. Deciphering laminar-specific neural inputs with line-scanning fMRI. Nat. Methods 11, 55–58.

Zaldivar, D., Rauch, A., Logothetis, N.K., Goense, J., 2018. Two distinct profiles of fMRI and neurophysiological activity elicited by acetylcholine in visual cortex. Proc. Natl. Acad. Sci. U. S. A. 115 (51), E12073–E12082. https://doi.org/10.1073/pnas.1808507115. Epub 2018 Dec 3. PMID: 30510000; PMCID: PMC6304994.

Zheng, Y., et al., 2002. A model of the hemodynamic response and oxygen delivery to brain. Neuroimage 16, 617–637.

Zheng, Y., et al., 2005. A three-compartment model of the hemodynamic response and oxygen delivery to brain. Neuroimage 28, 925–939.

Zhong, K., et al., 2007. Systematic investigation of balanced steady state free precession for functional MRI in the human visual cortex at 3T. Magn. Reson. Med. 57, 67–73.

Zhou, I., et al., 2012. Balanced steady state free precession fMRI with intravascular susceptibility contrast agent. Magn. Reson. Med. 68, 65–73.

Zielinski, L., Sen, P., 2003. Combined effects of diffusion, non-uniform gradient magnetic fields and restriction on an arbitrary coherence pathway. J. Chem. Phys. 119, 1093–1104.

Ziener, C., et al., 2006. Structure-specific magnetic field inhomogeneities and its effect on the correlation time. Magn. Reson. Med. 24, 1341–1347.

Ziener, C., et al., 2007a. Scaling laws for transverse relaxation times. J. Magn. Reson. 184, 169–175.

Ziener, C.H., et al., 2007b. Local frequency density of states around field inhomogeneities in magnetic resonance imaging: effects of diffusion. Phys. Rev. E 76, 1–16.

Zur, Y., et al., 1990. Motion insensitive, steady state free precession imaging. Magn. Reson. Med. 16, 444–459.

Biophysical modeling: Multicompartment biophysical models for brain tissue microstructure imaging

H. Farooq[a], P.K. Pisharady[a,*], and C. Lenglet[a,b,*]

[a]*Center for Magnetic Resonance Research, Department of Radiology, University of Minnesota, Minneapolis, MN, United States,* [b]*Institute for Translational Neuroscience, University of Minnesota, Minneapolis, MN, United States*

Introduction to biophysical tissue models
Theoretical context

In understanding the intricate construct of the brain's microstructure, multicompartment biophysical modeling has emerged as a transformative avenue, offering enhanced capabilities to tap into the immense potential of diffusion MRI. Many studies have explored the intricate relationship connecting white matter architecture with learning tasks, healthy aging, and neurodegenerative diseases (Fig. 1). At the heart of this modeling technique lies the phenomenon of diffusion, enabling to probe the pattern of movements of water molecules and gain invaluable insights into the complex architecture of brain tissues. Therefore, we begin with a succinct overview of how nuclear magnetic resonance (NMR) can be sensitized to detect and discern the motion of water molecules, thereby providing valuable insights into the underlying microstructure in which such diffusion occurs. Furthermore, we delve into a detailed exposition of a simple model, the diffusion tensor, which serves as a straightforward approximation for understanding the diffusion patterns in tissue, reflecting its microarchitecture, known as diffusion tensor imaging. Afterward, we highlight the inherent limitations of this model in thorough comprehension of the complex architecture of the neuronal tissue. Subsequently, we describe in detail the building of more elaborate biophysical models, which allows a detailed representation of the geometrical complexities in both healthy and diseased brain tissues.

---

*P.K. Pisharady and C. Lenglet contributed equally to this work.

Computational and Network Modeling of Neuroimaging Data. https://doi.org/10.1016/B978-0-443-13480-7.00006-5

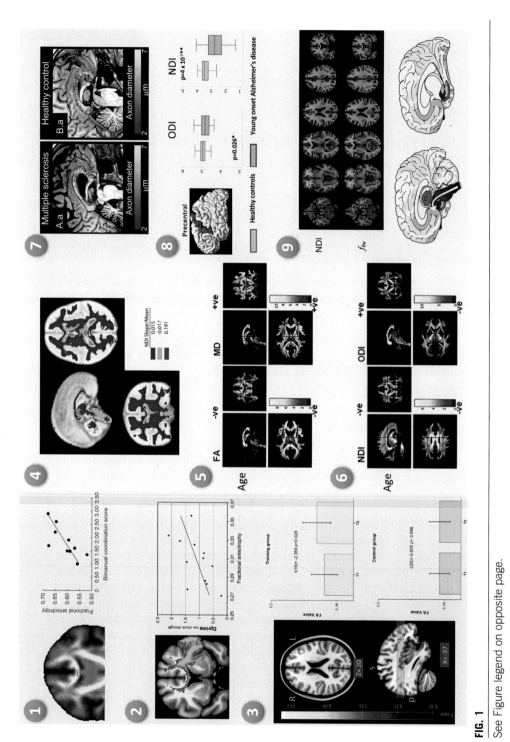

FIG. 1

See Figure legend on opposite page.

Sensitizing magnetic resonance to water diffusion

The free diffusion of water molecules through Brownian motion is significantly impeded by the geometry of neurites (Beaulieu, 2002) (e.g., neurons and dendrites). Appendix A provides a basic description of the diffusion phenomenon for the purpose of review and reference. NMR can be employed to assess the response of diffusion to the tissue microstructure by examining diffusion patterns in the millisecond range, which corresponds to a diffusion length (Eq. A3 of Appendix A) on the micrometer scale. This provides an indirect characterization of the neuronal microstructure at the scale below the resolution of diffusion MRI (Novikov et al., 2018a). The theoretical development of the process can be summarized as follows:

The influence of diffusion was initially observed on spin echo phenomena (please see Acquisition scheme for diffusion MRI data Section for details on spin echo sequence) by Hahn (Hahn, 1950) during his experiments involving NMR. Building upon Hahn's observations, Carr and Purcell (1954) devised a spin echo sequence

FIG. 1

Illustrative examples of brain tissue microstructural changes due to learning tasks, healthy aging, and diseases: *Left column—Changes in microstructure from learning tasks*: (**1**) Performance on a bimanual coordination task correlates with fractional anisotropy (FA) in the body of the corpus callosum (Johansen-Berg et al., 2007). (**2**) Acquiring grammar skills correlates with FA in pathways from Broca's area in the left hemisphere (Floel et al., 2009). (**3**) Individuals who undergo training to improve their performance in a phonetic discrimination task exhibit an increased FA in the inferior frontal gyrus area compared to the control group (Alotaibi et al., 2023). *Middle column—Changes in microstructure from healthy aging*: (**4**) Three clusters on an average brain template using k-means clustering (k = 3) shows distinct changes due to healthy aging. The red cluster shows an increase in neurite density index (NDI) across most brain regions. The green cluster indicates decreasing NDI with age in areas like the corpus callosum, corticospinal tract, thalamus, caudate, and brainstem. A smaller blue cluster in a brainstem portion exhibits substantial NDI increase (Mah et al., 2017). (**5–6**) Correlation of microstructural measures, i.e., FA, MD, NDI, and orientation dispersion index (ODI), with age (Raghavan et al., 2021). *Right column—Changes in microstructure from neurological disorders*: (**7**) Reduced axonal diameter index due to multiple sclerosis compared to healthy controls (Huang et al., 2019). (**8**) Decrease in NDI and ODI due to young-onset Alzheimer's disease (Andica et al., 2020). (**9**) Upper Image: In the cortex, reduced NDI (in blue) and increased free water fraction (ffw or isotropic volume fraction—in red) in Parkinson's disease patients compared to healthy controls (Kamiya et al., 2020). Lower Image: Depiction of Parkinson's disease-related pathology progression based on histological literature from Braak et al (Braak et al., 2003).

Right column (9) upper image adapter from Kamagata, K., Zalesky, A., Hatano, T., Ueda, R., Di Biase, M.A., Okuzumi, A., Shimoji, K., Hori, M., Caeyenberghs, K., Pantelis, C., Hattori, N., Aoki, S., 2017. Gray matter abnormalities in idiopathic Parkinson's disease: evaluation by diffusional kurtosis imaging and neurite orientation dispersion and density imaging. Hum. Brain Mapp. 38 (7), 3704–22. https://doi.org/10.1002/hbm. 23628. PubMed PMID: WOS:000404963400030.

specifically designed to measure diffusion. Torrey further expanded on this by incorporating a diffusion term into the Bloch equations, effectively treating diffusion as a relaxation process (Torrey, 1956). Stejskal and Tanner (1965) subsequently offered a solution to the Bloch-Torrey equations, revealing the relationship between the magnitude and phase of the NMR signal and the diffusivity of the system. For details on the derivation and solution of the Bloch-Torrey equations, please refer to Stejskal and Tanner's seminal work (Stejskal and Tanner, 1965).

Diffusion tensor imaging

In biological tissues, the diffusion of water molecules is hindered by surrounding membranes and structures. Specifically, in the case of brain tissues, the presence of the myelin sheath and axonal membranes contributes to the phenomenon of anisotropic water diffusion. As described in the preceding section, by exploiting the principle of diffusion, magnetic resonance (MR) signals can be sensitized to provide valuable insights into the inherent structure of those tissues. The utilization of the diffusion tensor model to capture and characterize the tissue structure is referred to as diffusion tensor imaging (DTI) (Basser et al., 1994).

With the DTI framework, the diffusion of water molecules within a given structure is assumed to follow a Gaussian distribution. More specifically, the diffusion process is characterized by the covariance matrix, referred to as the diffusion tensor, which parameterizes a Gaussian distribution at each voxel, the basic volume element. While anatomical MRI cannot reveal detailed structural information about white matter, DTI has proven to be immensely valuable in identifying such tissue geometry. By capturing anisotropic water diffusion, DTI facilitates the identification of axonal bundle orientations and enables the exploration of connectivity between (sub)cortical regions through a technique called tractography (Basser et al., 2000). The eigenvalues of the diffusion tensor and related indices such as such as fractional anisotropy (FA) and mean diffusivity (MD) (Fig. 2) provide useful markers to study the structure of white matter and provide measures to investigate fiber density or myelination (Brubaker et al., 2009). DTI finds wide-ranging applications in various clinical studies, including investigations into brain development, the effects of aging, and neurological disorders such as Alzheimer's and Parkinson's disease (Jones, 2012; Hofstetter et al., 2013).

Limitations of diffusion tensor imaging

The modeling in DTI assumes a tri-variate Gaussian dispersion pattern for the diffusion process (Ferizi et al., 2014). However, this assumption oversimplifies the true diffusive behavior, leading to limitations in sensitivity and specificity (Assaf and Basser, 2005; Assaf et al., 2019). As a result, commonly used DTI indices like FA and MD struggle to detect subtle pathological changes at the microstructural level. Various microstructural features such as axon radius index, orientation dispersion, neurite density, and existence of multiple axonal bundles in a voxel influence DTI index, making it difficult to directly associate DTI-index changes with specific

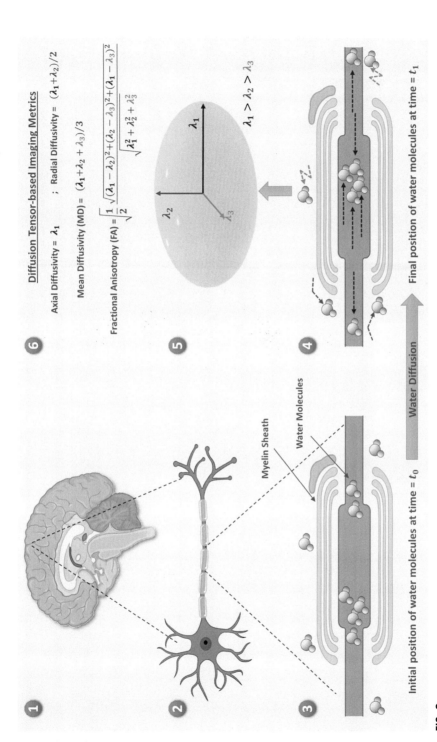

FIG. 2

Diffusion Tensor Imaging (**1–3**) Brain, axon, and water molecules in intra- and extraaxonal space and cerebrospinal fluid (CSF) areas and time t_0, (**4**) water diffusion during time $t_1 - t_0$ sensed using diffusion MRI, (**5**) water diffusion pattern approximated by a diffusion tensor, and (**6**) diffusion tensor-based imaging metrics. Here, λ_1, λ_2, and λ_3 are primary, secondary, and tertiary eigenvalues of the diffusion tensor.

Figure created with BioRender.com.

The following text appears within the figure:

Diffusion Tensor-based Imaging Metrics

Axial Diffusivity = λ_1 ; Radial Diffusivity = $(\lambda_1 + \lambda_2)/2$

Mean Diffusivity (MD) = $(\lambda_1 + \lambda_2 + \lambda_3)/3$

Fractional Anisotropy (FA) = $\sqrt{\dfrac{1}{2}}\dfrac{\sqrt{(\lambda_1 - \lambda_2)^2 + (\lambda_2 - \lambda_3)^2 + (\lambda_1 - \lambda_3)^2}}{\sqrt{\lambda_1^2 + \lambda_2^2 + \lambda_3^2}}$

$\lambda_1 > \lambda_2 > \lambda_3$

Myelin Sheath

Water Molecules

Initial position of water molecules at time = t_0

Water Diffusion

Final position of water molecules at time = t_1

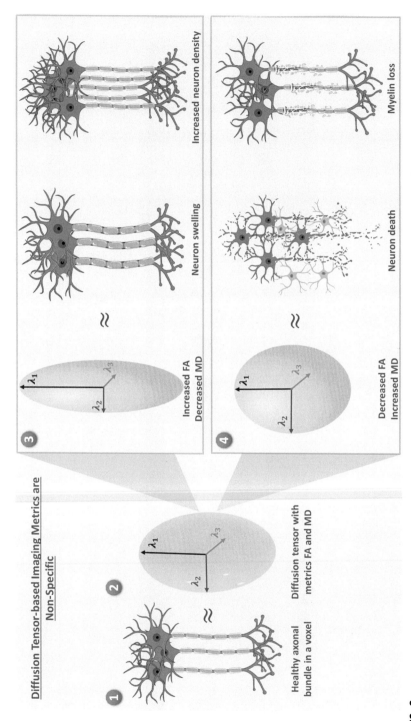

FIG. 3

The diffusion tensor model is oversimplified and provides nonspecific imaging metrics. (**1–2**) The healthy axonal bundle is represented by a diffusion tensor, characterized by metrics such as fractional anisotropy (FA) and mean diffusivity (MD). (**3**) When analyzing these metrics, an increase in FA and a decrease in MD can indicate different conditions, such as neuron swelling or increased neuron density within a voxel. (**4**) Conversely, decreased FA and increased MD are also nonspecific and can be indicative of situations like neural death or loss of myelin.

Figure created with BioRender.com.

alterations of microstructural features (Andica et al., 2020). Consequently, the interpretability of these indices is compromised, as their changes cannot be unequivocally linked to distinct microstructural modifications (Basser and Pierpaoli, 1996). Fig. 3 provides an example of the lack of sensitivity and specificity of DTI indices to alterations in tissue structure.

Multicompartment biophysical models

To overcome the limitations of DTI, an alternative approach called multicompartment biophysical modeling is employed, where the diffusion MRI signal is assumed to originate from distinct tissue "compartments" with separate microstructural properties for each compartment. In this approach, each tissue compartment represents a simplified depiction of the underlying tissue geometry. By compartmentalizing the tissue structure with the biophysical models, it becomes possible to infer micrometer-scale tissue characteristics from millimeter-scale MR data. Multicompartment biophysical models thus enable the direct measurement of microstructural properties such as average axonal radius, density, orientation, and volume fractions from diffusion MRI data across multiple compartments.

Importance and rationale for building the multicompartment biophysical models

Multicompartment biophysical models are important as these plays a crucial role in advancing our understanding of brain tissue microstructural properties and enabling improved diagnosis and treatment in neurological disorders. These models provide unique insights into tissue composition offering valuable information for studying diseases, evaluating tissue health, and tracking changes over time (Panagiotaki et al., 2015; Gardier et al., 2023). Following are important avenues where the multicompartment biophysical models contribute to the imaging process:

Improved diffusion MRI signal explanation

Multicompartment models have been extensively compared using ex vivo (Panagiotaki et al., 2012) and in vivo human brain (Ferizi et al., 2014) diffusion MRI data. These studies have shown that models with multiple compartments, specifically the three-compartment models (intraaxonal, extraaxonal, and isotropic), consistently rank higher than models with fewer compartments in explaining the diffusion MRI signals. This demonstrates that incorporating multiple compartments in the models enhances the ability to explain diffusion MRI data and provides direct microstructure information.

Clinical applications

Multicompartment models find applications in clinical settings, contributing to the diagnosis and monitoring of neurodegenerative diseases, traumatic brain injuries, and other neurological conditions (Adanyeguh et al., 2023). By more accurately characterizing the tissue microarchitecture, these models can aid in identifying abnormalities, tracking disease progression, and assessing treatment responses.

Biologically meaningful clinical metrics

Multicompartment models incorporate specific biophysical properties for different tissue compartments, providing insights into cellular properties such as axonal density, orientation, diameter, and water diffusivity indices. By incorporating specific biophysical properties of tissue compartments, these models provide quantitative measures associated with cellular changes, axonal damage, and repair processes. These clinically relevant metrics can enable the assessment of white matter health and the identification of disease conditions early. For example, a decrease in axonal density or an increase in hindered diffusivity can be indicative of axonal damage or demyelination (Adanyeguh et al., 2023).

Tissue-specific metrics

Multicompartment models facilitate the monitoring of disease progression within affected tissues, aiding in the delineation of distinct characteristics and microstructural attributes between white and gray matter. This differentiation is crucial as white matter and gray matter exhibit diverse microstructural properties and are affected by different diseases. For instance, multiple sclerosis primarily impacts the white matter, while Alzheimer's disease mainly affects gray matter. In both white and gray matter, various cellular and extracellular tissue compartments exist, each with distinct physical (and thus diffusion) properties. The application of multicompartment models permits the separation of these diverse constituents within white (Alexander et al., 2010; Zhang et al., 2012; Jelescu et al., 2015) and gray matter (Palombo et al., 2020; Jelescu et al., 2022), thereby enhancing comprehension and quantification of individual tissue properties. Such comprehension is important in studying diseases, evaluating tissue health, and tracking changes over time in both white and gray matter tissues.

Tumor-specific models

Distinct biophysical models have been created to characterize tumors using the presence of blood vessels, which can cause anisotropic diffusion, unlike what is typically observed in healthy white and gray matter (Panagiotaki et al., 2015; Gardier et al., 2023). Expanding upon the research of previous work on multicompartment biophysical models, numerous models specifically tailored to tumors have been proposed. For instance, Panagiotaki et al. introduced the Vascular, Extracellular, and Restricted Diffusion for Cytometry in Tumors (VERDICT) (Panagiotaki et al., 2015) model, while Gardier et al. recently formulated the Cellular Exchange Imaging (CEXI) (Gardier et al., 2023) model. These models hold potential for enhancing our comprehension of tumor growth and propagation.

Identification and removal of brain connectivity biases

Multicompartment models help address challenges in constructing the structural connectome better. Assessing the characteristics of white matter pathways is important, but at the same time there are challenges in accurately measuring the structural connectome using whole brain tractography (Jones et al., 2013). Simpler models like the "ball and stick" have been used in complex white matter areas with multiple orientations for probabilistic tractography (Behrens et al., 2007). However, the number of streamlines within an MRI voxel does not reflect the actual fiber density, making streamline-based density estimates biologically less relevant (Smith et al., 2013, 2015). The length of streamlines also introduces quantification bias of the connectome when using streamline density, leading to overreconstruction (Smith et al., 2013, 2015; Li et al., 2012) or underestimation (Jones, 2010). Tractography based on multicompartment models using advanced techniques called "microstructure-informed tractography" (Daducci et al., 2015a; Schiavi et al., 2020) has therefore been developed to improve the biological relevance of streamlines (Koch et al., 2022; Battocchio et al., 2022; Pestilli et al., 2014; Sreenivasan et al., 2022). Multicompartment models serve as the basis for constructing reliable macrolevel connectomes at various scales (Ocampo-Pineda et al., 2021), enabling a better understanding, identification, and monitoring of brain connectivity patterns in both healthy and pathological conditions. Microstructure information adjusts the streamlines' geometry (Battocchio et al., 2022; Yeh et al., 2021), improving the quantification of structural connectivity and mitigating some of the aforementioned biases.

Types of diffusion MRI biophysical models

While multicompartment biophysical models have been briefly introduced in the preceding subsection, it is important to note that they are just one of the types of models developed to capture various properties of tissue microstructure more comprehensively. In the following discussion, we provide a brief overview of the distinct categories of diffusion MRI biophysical models and describe their interrelationships. Broadly speaking, these models can be classified into two primary types: signal models and multicompartment models (Alexander et al., 2019).

Signal models

Signal models in diffusion MRI refer to mathematical representations that aim to elucidate the relationship between the diffusion MRI signal and the underlying tissue properties. These models typically assume that each voxel has uniform diffusion characteristics or treat the voxel as a homogeneous single compartment (Alexander et al., 2019; Afzali et al., 2021a).Conventional DTI (Basser et al., 1994) is an example of the technique, where the diffusion signal is modeled as a diffusion tensor. Other signal models include diffusion kurtosis imaging (DKI) (Jensen et al., 2005), q-space imaging (King et al., 1994), diffusion spectrum imaging (DSI) (Wedeen et al., 2005), mean apparent propagator MRI (MAP-MRI) (Ozarslan et al., 2013), and spherical mean technique (SMT) (Kaden et al., 2016a). These models primarily focus on characterizing voxel-averaged quantities and provide an overview of tissue properties.

Multicompartment models

Multicompartment models assume that the diffusion signal, within a voxel, is the result of contributions from multiple compartments, each representing a distinct cellular component with its own unique diffusion pattern. This framework enables the characterization of signal variations within the voxel and facilitates the estimation of microstructural properties specific to each tissue compartment. Multicompartment models can be further classified into two types (Alexander et al., 2019; Afzali et al., 2021a):

Multicompartment models without explicit association between diffusion signal and microstructural tissue features

These models are characterized by their compartmental nature, yet they do not establish a direct correspondence between the diffusion characteristics and specific microscopic features (Alexander et al., 2019). Rather, they describe the general diffusion behavior without offering detailed insights into the microstructure. For example, multiple-exponential decay models are frequently employed in diffusion MRI analysis, assuming that the diffusion signal in a voxel results from a combination of two or more exponential components (Niendorf et al., 1996; Zeng et al., 2018). While this allows estimation of parameters related to fast and slow diffusion processes, it does not directly reveal information about the specific microstructural features responsible for these diffusion characteristics. On the other hand, multicompartment microscopic diffusion imaging using SMT (Kaden et al., 2016b) takes into account the average behavior of the diffusion signal over a spherical region surrounding each voxel. While this approach provides an overview of the diffusion behavior within the voxel, it lacks specificity in terms of microscopic features. Similarly, DIAMOND (Scherrer et al., 2016) represents the compartments within a voxel using a finite sum of unimodal continuous distributions of diffusion tensors. Each distribution in DIAMOND (Scherrer et al., 2016) captures the microstructural diffusivity and heterogeneity of its respective compartment.

In summary, these multicompartment models describe the overall diffusion behavior without directly linking it to specific microscopic features, making them useful for capturing global diffusion characteristics but less informative in terms of detailed microstructural information.

Multicompartment models that directly estimate microscopic features of the tissue

As introduced in Multicompartment biophysical models Section, these models establish a direct association between diffusion characteristics and specific microscopic features. These models operate under the assumption that tissue is composed of multiple compartments, each possessing its own unique diffusion properties due to the underlying tissue geometry. These compartments represent distinct tissue types, such as intraaxonal (the space enclosed by neurite membranes), extraaxonal (the space surrounding neurites, occupied by various glial cells and, in gray matter, the cell bodies), and the space occupied by cerebrospinal fluid. By incorporating multiple compartments, it becomes possible to estimate the diffusion properties for multiple tissue types, enabling a more comprehensive characterization of tissue microstructure. This information facilitates the assessment of crucial microstructural features, including the diameter and orientation of axonal bundles, axon dispersion, and the volume fractions of different tissue compartments.

The subsequent sections of this chapter exclusively focus on these biophysical models that directly link tissue microstructural geometry to diffusion MRI data.

Historical background

In tracing the historical development of multicompartment biophysical tissue models, we adhere to the established naming convention of the models used in the previous literature (Ferizi et al., 2014; Panagiotaki et al., 2012; Panagiotaki, 2011), which simplifies geometrical structures of tissue, such as cylinder, stick, tensor, zeppelin, ball, dot, and more. For a detailed formal description of these simplified tissue geometries, please refer to Appendix B.

Fig. 4 offers a brief and nonexhaustive overview of the evolution of multicompartment models over the past 25 years, highlighting variations in compartment numbers and intercompartment permeability. Comprehensive guidelines for building and implementing these models are provided in Appendices B and C, respectively.

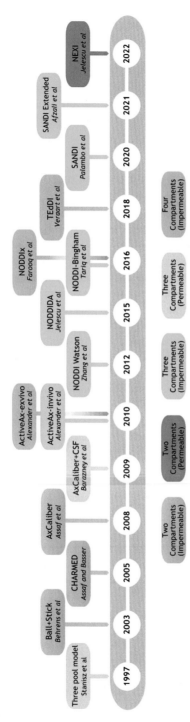

FIG. 4

Chronological development of multicompartment biophysical models over the past 25 years. The following is a nonexhaustive list of multicompartment models introduced chronologically, as depicted in the figure. Color of boxes show the number of compartments and permeability between the compartments: three pool model—Stanisz et al. (Stanisz et al., 1997); ball and stick—Behrens et al. (Behrens et al., 2003); composite hindered and restricted models of diffusion (CHARMED)—Assaf and Basser (Assaf and Basser, 2005); AxCaliber—Assaf et al. (Assaf et al., 2008; Barazany et al., 2009); AxCaliber+CSF—Barazany et al. (Barazany et al., 2009); ActiveAx—Alexander et al. (Alexander et al., 2010); neurite orientation dispersion and density imaging (NODDI) with Watson—Zhang et al. (Zhang et al., 2012); NODDI with Diffusivities Assessment (NODDIDA)—Jelescu et al. (Jelescu et al., 2015); NODDI Bingham—Tariq et al. (Tariq et al., 2016); NODDI in multiple fiber orientations (NODDIx)—Farooq et al. (Farooq et al., 2016); TE-dependent diffusion imaging (TEdDI)—Veraart et al. (Veraart et al., 2018); the soma and neurite density imaging (SANDI)—Palombo et al. (Palombo et al., 2020); extended SANDI—Afzali et al. (Afzali et al., 2021b); and neurite exchange imaging (NEXI)—Jelescu et al. (Jelescu et al., 2022).

Assumptions of biophysical tissue models

Biophysical modeling seeks to extract detailed information about neuronal tissue at the micrometer scale from diffusion MRI data, which typically has a resolution in the millimeter range. Modeling techniques address the inverse problem of inferring important tissue characteristics, such as average axonal diameter, orientation, and distribution, from the measured diffusion MRI signal. However, to make this process feasible, several assumptions are currently necessary, which are discussed as follows:

Complex tissue geometries as simplified sketches

Biophysical modeling assumes that elementary fiber segments, known as fascicles, are comprised of intra- and extraneurite compartments, while free water and cerebrospinal fluid are considered as separate compartments. These intricate structures are approximated by simplified geometrical representations of water diffusion such as sticks, cylinders, tensors, zeppelins, balls, and dots (Ferizi et al., 2014; Panagiotaki et al., 2011, 2012). Such elementary models provide a reasonable approximation of the complex tissue geometries. However, even such simplified representations involve complex mathematical functions (refer to Selection of a biophysical model Section), for example, that are challenging to take derivate (differentiate) analytically when solving optimization problems for parameter estimation or model fitting to diffusion MRI data.

Water exchange between compartments

Biophysical models have been built with the assumptions of permeability of water between the compartments, i.e., allowing water exchange between the compartments through permeable membranes (Jelescu et al., 2022; Stanisz et al., 1997; Barazany et al., 2009) and compartments with impermeable membranes (Assaf and Basser, 2005; Alexander et al., 2010; Zhang et al., 2012; Assaf et al., 2008; Barazany et al., 2009) (Fig. 4). In tissue models with impermeable membranes, the diffusion of water molecules is determined by the tissue's geometry alone. However, in tissue that allows water to exchange between compartments, both the tissue structure and the rate of exchange compete to influence the dynamics of water diffusion (Gardier et al., 2023; Olesen et al., 2022). The behavior of the diffusion signal with respect to time varies depending on the dominance of either water exchange through the membrane or tissue structure restriction. If the permeability is allowed, the Kärger model predicts a signal decrease as diffusion time increases (Fieremans et al., 2010). On the other hand, if the tissue structure imposes impermeability, the signal will be less attenuated as diffusion time increases. The suitability of the assumptions depends on the specific brain region being modeled. Assumptions related to permeability work effectively for modeling gray matter and cancer tissues, whereas traditionally, white matter models do not incorporate permeability. It is also important to acknowledge that incorporating permeability introduces significant complexity when fitting the model to diffusion MRI signals.

Fixed model constraints

Diffusivity values in diffusion MRI biophysical models quantitatively determine the diffusion of water molecules in biological tissues. These values can vary based on factors such as tissue type, pathological conditions, and model assumptions. For instance, different brain regions can exhibit distinct diffusivity values due to variations in cell density, myelination, and cellular orientation. However, several models, including neurite orientation dispersion and density imaging (NODDI) (Zhang et al., 2012), ActiveAx (Alexander et al., 2010), and AxCaliber (Assaf et al., 2008), assume fixed parallel diffusivity ($d_{\parallel}$) values (refer to Parameters of the compartment model functions Section). In these models, perpendicular diffusivity ($d_{\perp}$) is estimated from parallel diffusivity using a tortuosity model (Szafer et al., 1995). The assumptions are made to reduce the number of degrees of freedom and simplify parameter estimation problems. Nevertheless, recent studies have revealed that diffusivities tend to differ across various brain regions (Mah et al., 2017). Additionally, the validity of the tortuosity constraint (Szafer et al., 1995) is being questioned (Novikov et al., 2018a; Jelescu et al., 2015). Consequently, and also because of possible pathological changes in microarchitecture that influence parallel and perpendicular diffusivities, it is best to estimate them separately without constraints or prior assumptions to characterize the white matter microstructure more accurately (Novikov et al., 2018a; Jelescu et al., 2015).

Building, testing, interpreting biophysical tissue models

The process of building, testing, and interpreting diffusion MRI microstructural models involves several crucial steps. First, the acquired diffusion MRI data undergo preprocessing to correct for any artifacts. Next, a suitable microstructural model is chosen to describe the diffusion behavior within the tissue of interest. However, in some cases, the scanning protocols themselves can also be optimized based on the selected tissue biophysical model prior to the actual scan. In that case, the tissue biophysical model is selected prior to deciding the diffusion MRI scanning protocol. Next, the model parameters are estimated by fitting the chosen model to the diffusion MRI data. The quality of fit is typically assessed, and thorough validation and testing is conducted to evaluate the model's performance. Finally, the estimated microstructural parameters, such as axonal bundle radius, orientation, and dispersion, are interpreted within the context of tissue biology or pathology. This interpretation provides valuable insights into tissue integrity, informing both research questions and clinical decision-making. The following provides a detailed description of this process and is depicted in Fig. 5.

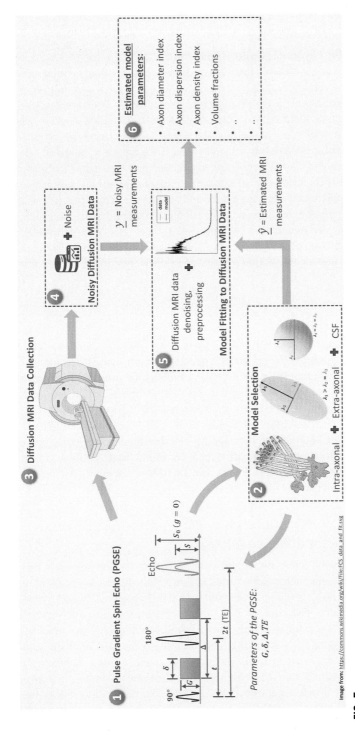

Image from: https://commons.wikimedia.org/wiki/File:FCS_data_and_fit.svg

FIG. 5

Stages of Microstructure Imaging: From dMRI data acquisition to model selection and fitting: (1, 2) Acquisition and Model Selection: Obtaining diffusion MRI data through pulse gradient spin echo (PGSE), either with single or multiple shells, precedes the selection of biophysical models aligned with the desired microscopic structures. The PGSE parameters may be fine-tuned iteratively based on the chosen model. (3, 4) Denoising and Preprocessing: Noise is inevitable during data collection, necessitating denoising and preprocessing of diffusion MRI data prior to model fitting. (5, 6) Model Fitting and Accuracy Assessment: Generally, models that closely resemble the original tissue geometry exhibit a robust correlation with the diffusion MRI data, indicating a good fit. Nonetheless, the accuracy of estimated tissue features relies on several factors, including the data quality, the type of model (including the number and type of compartments), and the specific model fitting algorithm utilized for the analysis.

Figure created with BioRencer.com.

Selection of a biophysical model

Multicompartment biophysical tissue model functions

In multicompartment models, suitable models are chosen for individual cellular volumes and integrated by considering the respective volume or signal contribution of each compartment (Ferizi et al., 2014; Panagiotaki et al., 2012). The choice of the model depends on factors such as the complexity of the tissue microstructure to be captured, the assumptions of the model, and the specific research question or clinical application. Each compartment is assumed to represent a unique normalized MR signal. These compartments are categorized as intraaxonal (restricted diffusion models, depicting diffusion inside axonal bundle), extraaxonal (including isotropic and anisotropic hindered models, depicting diffusion outside the axonal bundle), and isotropic restriction compartments depicting diffusion in extracellular structures like glial cells, trapped water on membranes, or nonparallel fibers. The overall diffusion MR signal (S) in a multicompartment model is a combination of the signals (S_i) with each compartment's volume contribution (f_i) serving as the weight in the linear combination as follows:

$$S = \sum_{i=1}^{k} f_i S_i \tag{1}$$

For k tissue compartments, $\sum_{i=1}^{k} f_i = 1$ and $f_i \geq 0$.

The mathematical equations presented in Appendix B represent tissue biophysical models shown in the Fig. 6. We offer a detailed explanation of the compartment models that are part of widely utilized biophysical models in the current research, including composite hindered and restricted models of diffusion (CHARMED) (Assaf and Basser, 2005), ActiveAx (Alexander et al., 2010), and NODDI (Zhang et al., 2012). For a more thorough understanding of other models, including their functions and implementations, information can be found in CAMINO (Cook et al., 2006), DIPY (Garyfallidis et al., 2014), and DMIPY (Fick et al., 2019) software toolboxes (Appendix C).

Acquisition scheme for diffusion MRI data

The specific parameters for acquiring the diffusion MRI data are determined in this step. This involves decisions such as the number and direction of diffusion-weighting gradients, the strength of the gradients, the number of diffusion-encoding shells, and other imaging parameters. The choices made here depend on factors such as the desired spatial resolution, signal-to-noise ratio requirements, and the specific research or clinical objectives of the study. For example, if the goal is to study in detail the white matter microstructure, a protocol with a high number of diffusion-encoding directions may be chosen. In a clinical setting, however, a protocol with a lower spatial and angular resolutions might be preferred to reduce the scan time and improve patient comfort.

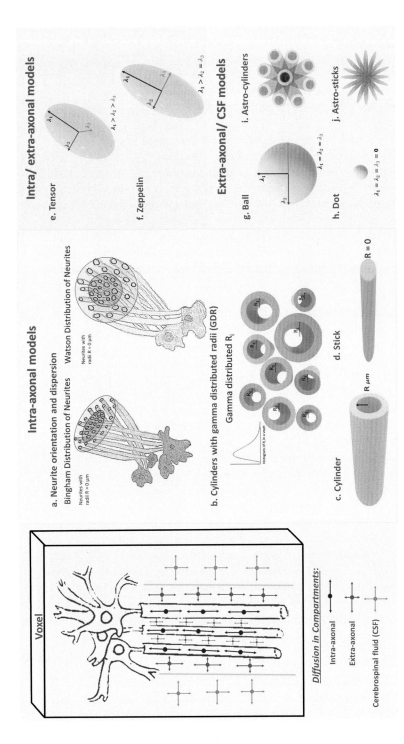

FIG. 6

Popular biophysical models for diffusion MRI. (Left) Within a voxel, a bundle of axons is illustrated. Diffusion within the axons and dendrites is restricted, enabling the determination of the axonal bundle's orientation (indicated in red). Outside the axons and dendrites, water diffusion is hindered (shown in blue), providing information about microstructures like cell bodies and glial cells. Additionally, in the cerebrospinal fluid (CSF) area, diffusion is isotropic or free (displayed in green), following the structural pattern of this space. (Center) This section presents widely used intraaxonal models. (Right-top) We show the Tensor and Zeppelin models, which are employed to characterize both restricted diffusion inside the axons and hindered diffusion outside the axons. (Right-bottom) Models such as ball, dot, astrosticks, and astrocylinders are used to describe isotropic diffusion (Ferizi et al., 2014; Panagiotaki et al., 2012; Panagiotaki, 2011).

Spin echo sequence

The spin echo sequence is widely used for diffusion-weighted imaging (DWI), to measure the diffusion properties of water molecules within biological tissues. It employs specific radiofrequency (RF) pulses and gradients. This sequence involves two primary elements: a 90-degree RF pulse followed by a 180-degree (refocusing) RF pulse.

Initially, the 90-degree RF pulse is applied to flip the magnetic spins from their equilibrium position to the transverse plane, causing them to process around the main magnetic field. However, due to magnetic field inhomogeneities, the spins experience slight variations in precession frequency, resulting in phase dispersion among the spins. This dispersion leads to the loss of coherence and dephasing of the spins over time.

To counteract this dephasing, a 180-degree RF pulse is introduced after a specific time interval. This pulse flips the spins by 180 degrees, effectively reversing their phase dispersion. Consequently, the spins begin to rephase, forming a coherent spin echo.

Selection of pulse sequence

The commonly utilized pulse sequence for diffusion modeling techniques is the pulsed gradient spin echo (PGSE), which implements single diffusion encoding (SDE). It employs a pair of pulsed magnetic field gradients surrounding the 180-degree RF pulse in a spin-echo measurement, as depicted in Supplementary Fig. B1(left) in Appendix B. There are alternative sequences like double diffusion encoding (Callaghan, 1993; Cory et al., 1990), oscillating diffusion encoding (Callaghan and Stepisnik, 1995; Topgaard et al., 2002), and the combination of double and oscillating diffusion encodings (known as double oscillating diffusion encoding (Ianus et al., 2017)) used for estimating the tissue microstructure. To the best of our knowledge, these alternative sequences have rarely been employed in combination with multicompartment tissue models; thus we exclusively examine the use of PGSE for explaining tissue microstructure estimation through multicompartment biophysical models. As an example of dMRI pulse sequence choice, we refer to the Human Connectome Project (HCP) (Sotiropoulos et al., 2013) where diffusion data were acquired on a Siemens 3 T Skyra system, utilizing a voxel size $1.25 \, \text{mm}^3$, three b-values (1000, 2000, and 3000s mm^{-2}), each with 90 directions (different across shells), and a total of 18 additional b=0 volumes. Δ/δ are fixed to 43.1/10.6 ms, while $|G|$ (Gmax $= 97.4 \, \text{mT m}^{-1}$) is varied to achieve the desired b-values. Please refer to Appendix B for description of the pulse sequence parameters.

Experiment design optimization

The parameters for the PGSE sequence (i.e., G, Δ, δ, and g) can be chosen to maximize the MR signal sensitivity to diffusion. Signal attenuation is related to the displacement of water molecules between the two pulses. Therefore, when water molecules can freely move, i.e., within an unrestricted compartment of tissue, the diffusion coefficient can be directly estimated from the signal loss using PGSE parameters. However, when diffusion is restricted and water molecules have limited movement due to the tissue geometry, the signal attenuation is less pronounced. The extent of signal attenuation in a restricted environment is intricately tied to the unique attributes of the tissue's microstructure. Factors such as the size and shape of the microstructure play a role in determining the signal attenuation. Additionally, the specific parameters of the pulse sequence, such as the duration of the pulses, time intervals between pulses, and the strength of the applied gradient, also contribute to the degree of signal attenuation. These combined factors orchestrate the nuanced interplay between the microstructure and the pulse sequence, shaping the observed signal response.

Optimal PGSE sequence parameters

Due to the intricate relationship between parameters of the PGSE sequence and the parameters of biophysical models, the optimal design of tunable parameters of the PGSE (i.e., G, Δ, and δ) ideally requires that the biophysical model be selected for scanning. The goal is to maximize sensitivity to the chosen model's parameters within a given acquisition time for MRI scanning.

Previous works have aimed at maximizing the information content that can be gathered from a diffusion MRI experiment by exploring different acquisition protocols (Alexander, 2008). A common approach in optimal protocol design is to use the Fisher information matrix. Specifically, the design process minimizes the variance of biophysical model parameter estimates, by minimizing the trace of the Fisher information matrix (besides other optimality conditions), to optimize the combination of sequence settings in the acquisition protocol. Protocols optimized for the NODDI (Zhang et al., 2012) model have been provided with the main publication (Zhang et al., 2012). These works show that designing an optimal protocol requires balancing various factors, such as increasing sensitivity with higher b-values while accommodating the lower signal-to-noise ratio associated with longer TE.

Optimal sampling of multishell schemes

The diffusion gradient directions g are crucial for sensitizing the MRI signal to the diffusion of water molecules in different orientations. By applying magnetic field gradients along specific directions, the MRI signal becomes sensitive to the physical characteristics of the tissue in those particular orientations. Caruyer et al. (2013) introduced an optimal technique for generating multishell sampling protocols. This method ensures that the samples are uniformly distributed across each shell while maintaining a globally uniform angular coverage. To achieve this, a cost function that extended the concept of electrostatic repulsion to multishell configurations was employed and can be minimized using an efficient optimization technique. The resulting sampling protocols provide an optimal uniform distribution of orientations within each shell, contributing to improved data quality, and can be readily obtained from the web application (Crauyer, 2023), provided by the authors.

Diffusion MRI data preprocessing

In the context of multicompartment modeling, the preprocessing of dMRI data serves the purpose of improving data quality, reducing artifacts, and providing accurate inputs for subsequent analysis. Key preprocessing steps for dMRI data include brain extraction, correction of susceptibility artifacts, motion-induced artifacts, and eddy current artifacts. Standardization efforts have been made to streamline dMRI data preprocessing, e.g., with the HCP developing the minimal preprocessing pipelines (MPPs) (Glasser et al., 2013). The MPPs offer a unified framework for quality control and distortion correction across various neuroimaging modalities, ensuring consistency and comparability of dMRI data in both intrastudy and interstudy analyses. For a more comprehensive understanding of dMRI data preprocessing methods and their comparison, we refer to a recent review by Tax et al. (2022). Below, we summarize the dMRI data preprocessing steps:

Diffusion data file format conversion

During data preprocessing, the initial stage involves converting the scanner-based file format, known as Digital Imaging and Communications in Medicine (DICOM), into the Neuroimaging Informatics Technology Initiative (NIfTI) format. NifTI is widely employed for storing imaging data and is essential for various dMRI data processing software packages like MRTrix3 (Tournier et al., 2019), FSL (Jenkinson et al., 2012), and Advanced Normalization Tools (ANTs) (Avants et al., 2009; Tustison et al., 2014). The preferred method for converting the file formats is the "dcm2niix (Li et al., 2016)" tool (Cai et al., 2021).

Brain extraction

Brain extraction algorithms selectively isolate the region or voxels containing solely brain data, effectively eliminating nonbrain structures such as the skull, eyes, and face muscles. This process ensures that subsequent analyses can exclusively concentrate on brain tissues, thus facilitating computationally efficient brain tissue segmentation and analysis. The brain extraction tool (bet) (Smith, 2002) from FSL is extensively utilized for its accurate results, and for not requiring a brain template during the extraction process. On the other hand, "antsBrainExtraction" from ANTs (Avants et al., 2009; Tustison et al., 2014), another brain extraction tool, relies on templates for extracting the brain area.

Artifact correction
Susceptibility-induced distortion correction

The application of echo-planar imaging (EPI) with the PGSE technique is susceptible to magnetic field inhomogeneities caused by substantial magnetic susceptibility variations within the imaged region. As a result, these inhomogeneities cause geometric distortions and signal perturbations along the phase-encode direction during the dMRI data acquisition. The most widely used toolbox to correct susceptibility-induced distortions is FSL's "topup" (Smith et al., 2004; Andersson et al., 2003).

Eddy current-induced distortion correction

During this stage, artifacts induced by eddy currents, which can result in distortions in diffusion-weighted images, are effectively rectified. Furthermore, susceptibility artifacts originating from magnetic field variations near air-tissue interfaces are also addressed. Specifically, FSL's "eddy (Jenkinson et al., 2012)" or "eddy_openmp (Andersson and Sotiropoulos, 2016)" tool is applied to the extracted brain (Smith, 2002) and the averaged output from "topup (Smith et al., 2004; Andersson et al., 2003)," to correct for eddy current-induced distortions, intervolume motion, and slice-wise signal dropout (Cai et al., 2021; Andersson et al., 2016). These corrective measures significantly enhance the accuracy of analysis performed on the corrected diffusion MRI data.

Tissue segmentation

During this step, diffusion-weighted images are divided into separate tissue types like gray matter, white matter, and CSF. This process is known as tissue segmentation and is crucial for separating distinct compartments, enabling better modeling in further analysis. FSL FAST (Zhang et al., 2001) is a reliable tool used for accurately classifying gray matter, white matter, and CSF in diffusion-weighted images.

In recent years, several comprehensive pipelines have emerged to automate the preprocessing of dMRI data by utilizing conventional toolboxes. Notably, the integrated diffusion image operator (iDIO) (Hsu et al., 2023) and PreQual (Cai et al., 2021) are examples of dedicated solutions for dMRI data preprocessing, offering an integrated pipeline that seamlessly incorporates various diffusion preprocessing tools and includes built-in quality assurance visualizations. Additionally, Connectome-Mapper-3 (Tourbier et al., 2022), Micapipe (Cruces et al., 2022), PhiPipe (Hu et al., 2023), and QuNex (Ji et al., 2023) pipelines are capable of integrating data preprocessing for multiple imaging modalities, including dMRI data, and ensure quality control throughout the process by delivering visual output analysis modules.

Model fitting

After selecting the appropriate model and acquiring the data, the subsequent step involves fitting the chosen model to the obtained data. To illustrate the challenges encountered during the model fitting process, we first introduce a standard problem formulation for fitting the model to diffusion MRI data.

Problem formulation for fitting biophysical models to diffusion MRI data

Let us define $\widehat{S}(x,f)$, the model-predicted normalized diffusion MRI signal from "k" different tissue compartments, i.e.,

$$\widehat{S}(x,f) = f_1 S_1(x_1 \cdots x_{m1}) + f_2 S_2(x_{m1+1} \cdots x_{m2}) + \cdots + f_k S_k(x_{m2+1} \cdots x_m) \tag{2}$$

where S_1, S_2, $\cdots$, S_n are selected biophysical models for intraaxonal, extraaxonal, cerebrospinal fluid and glial cell compartments, etc. (Ferizi et al., 2014; Panagiotaki et al., 2012). These models depend upon "m" different parameters given in $x = [x_1 \, x_2 \cdots x_m]$, while $f = [f_1 \, f_2 \cdots f_k]$ is the vector containing volume fractions of the "k" tissue compartments. The objective function for the model parameter estimation from diffusion MRI data, with implicit assumption of offset Gaussian noise as used in the literature (Ferizi et al., 2014; Panagiotaki et al., 2012; Daducci et al., 2015b), is as follows:

$$\min_{x,f} \left\| S - \left(\widehat{S}(x,f) \right) \right\|_2^2 \tag{3}$$

$$such \quad that \quad \sum_{i=1}^{k} f_i = 1, \quad f_i \geq 0, i = 1,2 \cdots k$$

$$x_j^{min} \leq x_j \leq x_j^{max}, \quad j = 1,2 \cdots m$$

where "S" represents the normalized diffusion MRI measurements. As defined in Eq. (2), $\widehat{S}(x,f)$ represents the estimated signal from a given multicompartment model, while f_is are the volume fractions of the "k" compartments. x_j^{min} and x_j^{max} represent lower and upper bounds for the unknown deterministic variable vector "x," respectively. Magnitude diffusion-weighted MRI data inherently have Rician noise (Pierpaoli et al., 1996). However, assuming an offset Gaussian noise model, the objective function defined in Eq. (3) becomes simple and more stable numerically than Rician log-likelihood function (Ferizi et al., 2014; Panagiotaki et al., 2012). However, at low SNR (i.e., <5) (Gudbjartsson and Patz, 1995) the assumption of the noise model can induce significant bias in the estimated model parameters (Jones and Basser, 2004).

Optimization strategies

Estimating the parameters of multicompartment models involves a nonconvex optimization. The task is further complicated by two factors: (i) the nonlinear relationships between model parameters (denoted as "x" in Eq. (2)) and diffusion MRI data, which makes the problem nonconvex and nonlinear, and (ii) the presence of noise, particularly in shorter clinical scans, which has a significant impact.

To address these challenges, various methods have been proposed. Typically, the number of nonlinear parameters ("x") is reduced by assuming a single fiber orientation within a voxel, with the orientation computed using a simpler model like the diffusion tensor. Additionally, by fixing diffusivity values (please see Glossary) for the entire brain, the optimization problem in Eq. (3) is transformed into a convex problem that searches for linearly entering parameters "f" from a dictionary of the remaining few parameters "x". This linearization of the problem greatly speeds up the process of finding a solution. Examples of such methods include (Daducci et al., 2015b) Accelerated Microstructure Imaging via Convex Optimization (AMICO) and (Novikov et al., 2015) Linearly Estimated Moments provide Orientations of Neurites And their Diffusivities Exactly (LEMONADE).

Several research works have employed gradient-based methods such as Levenberg–Marquardt and gradient descent. Unlike dictionary-based solutions, these methods do not rely on preexisting dictionaries. Instead, they search for a reliable starting point by solving many simpler models before attempting the complex multicompartment model. While this technique offers improved accuracy, it tends to be slower compared to the linear formulation. Examples are CAMINO (Ferizi et al., 2014; Panagiotaki et al., 2012; Cook et al., 2006) and NODDI (Zhang et al., 2012) model fitting (Zhang et al., 2012).

In recent studies, there has been a shift toward relaxing the constraints of single fiber orientation and fixed diffusivity values (Jelescu et al., 2016, 2020). This necessitates the use of more advanced optimization methods to better solve the model fitting problem more efficiently. One such approach is Microstructure Imaging of Crossing (MIX) White Matter Fibers from diffusion MRI (Farooq et al., 2016), which addresses the problem by dividing the optimization into two subproblems. This method enables the efficient estimation of more accurate solutions without making any assumptions about fiber orientations or solving simpler problems to obtain a reliable starting point.

For further comprehensive understanding of the parameter estimation problem and its associated solutions, we recommend referring to the relevant papers (Farooq et al., 2016; Jelescu et al., 2020; Novikov et al., 2019).

Further resources for more details

For a deeper understanding of the construction, evaluation, and interpretation of biophysical models, we recommend referring to recent review articles (Assaf et al., 2019; Alexander et al., 2019; Afzali et al., 2021a; Jelescu et al., 2020). These publications provide valuable insights and comprehensive analysis on the subject matter.

Examples of successful multicompartment biophysical tissue models

Selecting an appropriate biophysical model for microstructural imaging with diffusion MRI depends on factors such as the tissue type, complexity of microstructure, available data quality, and specific research goals. It often entails comparing different models, assessing their compatibility with observed data, and choosing the one that best captures the intricate diffusion patterns of the tissue being studied. In this section, we discuss successful multicompartment biophysical models and summarize their transformative impact in brain imaging studies:

Ball and stick

Behrens et al. (2003) proposed the ball and stick model to depict brain tissue with two distinct compartments: anisotropic diffusion within axonal fibers (sticks) and isotropic diffusion in the extracellular space (balls). Generally, the model uses multiple sticks to model different orientations of axonal bundles. Therefore, the approach has contributed significantly to our understanding of white matter connectivity and integrity in various neurological disorders, including multiple sclerosis (Snoussi et al., 2022) and Alzheimer's disease (Snoussi et al., 2023; Zhan et al., 2015).

Neurite orientation dispersion and density imaging

NODDI (Zhang et al., 2012) is also a well-established model for estimating the microstructural properties of brain tissues. It quantifies parameters such as neurite density, neurite orientation dispersion, and compartment volume fractions. NODDI (Zhang et al., 2012) has played a crucial role in studying various brain diseases, including Parkinson's disease (Kamagata et al., 2016), ischemic stroke (Wang et al., 2019), differentiation of glioma malignancy grade (Maximov et al., 2017), and the spinal cord in multiple sclerosis (By et al., 2017; Alghamdi, 2023). A recent review offers an extensive compilation of studies utilizing NODDI (Zhang et al., 2012) to investigate early brain development (DiPiero et al., 2023). Additionally, for a comprehensive understanding of the NODDI (Zhang et al., 2012) model's contributions in clinical research, we recommend referring to a recent review paper (Kamiya et al., 2020).

AxCaliber

AxCaliber (Assaf et al., 2008) focuses on estimating the distribution of axon diameters within white matter pathways. The model assumes that diffusion within axons is restricted due to their finite size, while diffusion outside the axons is hindered due to barriers formed by other cellular structures. The model estimates parameters such as the mean axon diameter and the dispersion of axon diameters within a voxel. AxCaliber (Assaf et al., 2008) has been applied in various studies to investigate changes in axons within different brain regions, such as exploring the impact of multiple sclerosis on properties (Alghamdi, 2023; De Santis et al., 2019a). Additionally, it has been utilized to study the relationship between axon diameter and the conduction velocity of axons (Horowitz et al., 2015), as well as to examine the influence of axonal properties on upper and lower limb motor performance (Gooijers et al., 2021).

Composite hindered and restricted model of diffusion

The CHARMED (Assaf and Basser, 2005) model incorporates hindered diffusion within extracellular compartments and restricted diffusion within axons. It has been employed to examine the microstructure of white matter, including factors such as axonal density and diameter. For example, it has been applied to study the rapid and dynamic changes in the microstructural architecture of the developing brain in early life (DiPiero et al., 2023), assess the dynamics of short-term plasticity (Tavor et al., 2013), and investigate multiple sclerosis-related alterations in white matter microstructure (De Santis et al., 2019b).

ActiveAx or minimal model of white matter diffusion

The ActiveAx (Alexander et al., 2010) model assumes that the brain tissue consists of four distinct compartments: an intraaxonal compartment, an extraaxonal compartment, a hindered compartment, and a free water compartment. By employing the ActiveAx (Alexander et al., 2010) model, several parameters related to the microstructure of white matter can be estimated. These parameters include the axon diameter index, which reflects the average diameter of axons within a voxel, the axon density index, which quantifies the number of axons in a voxel, and the hindered diffusion coefficient, which measures the level of water diffusion restriction within a voxel. The ActiveAx (Alexander et al., 2010) model has been widely employed in various studies to investigate the microstructural properties of white matter in both healthy and diseased brains. For instance, the model has been utilized to examine the relationship between the gut and brain (gut-brain axis) (Colon-Perez et al., 2022), explore human traumatic spinal cord injuries (Morris et al., 2022), and investigate the impact of multiple sclerosis on white matter microstructure (Huang et al., 2016). The CONNECT (Assaf et al., 2013) project leveraged the CHARMED (Assaf and Basser, 2005), AxCaliber (Assaf et al., 2008), and ActiveAx (Alexander et al., 2010) models, in combination with tractography methods, to improve structural connectivity mapping methods.

Impact of the modeling technique

The successful multicompartment biophysical models described in this section have had a transformative impact on estimating tissue microstructure. These models have been cited in about 9000 peer-reviewed articles (source Google Scholar, accessed on July 8, 2023), showing their widespread adoption in the field of brain tissue imaging in our understanding of microstructural changes associated with both disease and healthy development. This also shows a significant contribution to advancing our capabilities in microstructural imaging and enhancing our knowledge of tissue microstructure.

Limitations of multicompartment biophysical tissue modeling

Current microstructure imaging techniques hold immense promise, but they are limited by their simplified approach and well-known constraints. In addition to these limitations, studies have thoroughly examined other pitfalls, and progress in the past 15 years has been slower than anticipated (Novikov et al., 2018b). This suggests the presence of significant challenges and shortcomings that need to be identified and addressed. Following is a nonexhaustive summary of those pitfalls, discussed in detail in other reviews (Jelescu et al., 2020; Novikov et al., 2018b; Lampinen et al., 2019):

Current biophysical models ignore key physical effects

The accuracy of biophysical models in representing physical phenomena may be inadequate (Novikov et al., 2018b). For instance, recent research indicates that the time-dependent characteristics of diffusion signals might be influenced by factors outside the axonal space (Fieremans et al., 2016; Lee et al., 2018). However, earlier studies, which measured the axonal diameter index, assumed that the time-dependent behavior solely stemmed from variations in axon sizes (Alexander et al., 2010). This calls for a reassessment of the interpretation of studies measuring axonal diameter and the incorporation of this phenomenon into modeling techniques.

Biophysical model fitting techniques may provide nonoptimal solutions

The task of fitting biophysical models to diffusion MRI data is typically a nonconvex problem. This means that multiple locally optimal solutions can potentially exist, and the optimization or fitting method used may get stuck in an incorrect solution. Consequently, this can lead to an inaccurate understanding of the microstructure from a biological or pathological perspective. This challenge is exacerbated by the fact that solving the fitting problem necessitates significant computational resources. As a result, it becomes impractical to perform multiple fitting operations to achieve the globally optimal solution. Advanced optimization techniques such as Microstructure Imaging in Crossing Fibers (MIX) (Farooq et al., 2016) can overcome the problem by finding the global optima reliably while fitting the models to diffusion MRI data.

Nonoptimal scanning protocols

In order to guarantee that the obtained diffusion MRI data contain sufficient information for a desired microstructure model, it is necessary to engage in experiment design optimization. Additionally, it is important to note that a one-size-fits-all approach cannot be applied to achieve an optimal fit for different models. The complexity of the model functions further compounds the issue. Although some attempts have been made to try to find optimal scanning parameters for multicompartment models (Alexander, 2008), there is still a significant need to develop optimal scanning protocols for biophysical models that can be readily implemented in both research and clinical environments.

Open challenges and future directions
Exciting open challenges
Unveiling heterogeneity in brain tissues

Recent studies have demonstrated the necessity of employing distinct biophysical models to image white matter, gray matter, and cancer cells accurately. However, current imaging approaches utilize a single model for the entire brain, which overlooks this heterogeneity. It is therefore important to explore adaptive modeling strategies that can learn and incorporate appropriate microstructural information specific to different brain regions.

Reimagining model fitting in brain imaging: Leveraging connected brain regions

Conventional model fitting methods typically approach each voxel individually and fit the model to its data. Nonetheless, it is important to acknowledge that brain regions are interconnected. The connected structure of the brain presents an opportunity to enhance model fitting algorithms by allowing the regression process to learn from neighboring voxels. By incorporating information from adjacent regions, this approach has the potential to improve the overall regression process. Moreover, it can help identify outliers where the regression algorithm may fail or get stuck at local optima and might lead to more robust and accurate results.

Future directions
Machine learning-based methods

As discussed previously, estimation errors in advanced multicompartment models stem from assumptions and fixed parameters, suboptimal scanning protocols, nonconvex problem formulation, and noise in the data. However, machine learning-based methods offer data-driven solutions to overcome challenges in microstructure estimation. Numerous algorithms, particularly employing deep learning (DL), have been developed to enhance the performance of maximum likelihood estimation (MLE) (Alexander et al., 2019). Furthermore, the integration of natural language processing (NLP) techniques from machine learning shows great promise in improving the direct mapping of DWI signals to microstructural information (Faiyaz et al., 2023).

Machine learning-based data harmonization

Diffusion MRI data collection from multiple sites often lacks harmonization, posing challenges for consistent analysis. However, incorporating multisite data is crucial for training accurate and adaptable machine learning models (Faiyaz et al., 2023). Machine learning offers various frameworks for data harmonization, e.g., COMBAT framework (Fortin et al., 2017), presenting a potential solution (Koppers et al., 2019). These harmonized datasets serve as a fundamental basis for conducting comprehensive analyses and extracting new information using big data analytics. Through data harmonization, disparate data sources can be seamlessly integrated, leading to an increase in the volume and variety of data available for big data techniques. Ultimately, data harmonization empowers researchers and organizations to harness the full potential of big data, enabling data-driven decision-making and enabling a deeper understanding of the brain microstructure and connectivity.

Take-home points

- *Enhanced Understanding of the brain microstructure*: Biophysical multicompartment models offer valuable insights into microstructural changes caused by aging or disease.
- *Trade-offs in Time and Resources*: While providing unique information, fitting these models to data requires longer scanning time and increased computational resources.
- *Accessibility through Python Libraries and GPU Acceleration*: The implementation and utilization of these models have been made more accessible via efficient open-source Python libraries and GPU-accelerated implementations.
- *Advancing more Realistic Biophysical Models*: There is a need to develop improved biophysical models that accurately incorporate realistic axonal diameter distributions and volume fractions, enhancing sensitivity to tissue features.
- *Challenges in Experiment Design*: Designing experiments for biophysical modeling techniques remains an unaddressed and challenging aspect in the field of imaging methods.
- *Optimization for Microstructure Extraction*: Advanced optimization strategies are necessary to extract microstructure information from diffusion data by fitting multicompartment models.
- *Promising Role of Machine Learning*: Machine learning-based model-fitting and data synthesis methods hold significant promise in advancing the field and pushing its boundaries.

Appendix A: Water diffusion

Brownian motion of particles is a random process, driven by thermal energy (Brown, 1828). At a macroscopic level, the phenomenon is called diffusion. The rate of change of concentration $\frac{\partial \psi}{\partial t}$ is equal to the net flux (J_f) across boundaries of an area of interest. That is:

$$\frac{\partial \psi(t, s)}{\partial t} = -\nabla J_f \tag{A1}$$

From Fick's first law of diffusion in a unidimensional context:

$$J_f = -D\frac{\partial \psi}{\partial s} \tag{A2}$$

where s is the distance and D is the diffusion co-efficient defined by Einstein (Einstein, 1905, 1956) in one-dimensional case as follows:

$$D = \frac{\Delta s^2}{2\Delta t} \tag{A3}$$

Δs is the mean squared displacement of the molecules in time Δt. For three-dimensional case, let $\Delta s = s - s_0$, where s_0 is the original position and s is the final position of a particle after diffusion for time Δt, Eq. (A3) can be written as follows:

$$\boldsymbol{D} = \frac{\langle \Delta s^T \Delta s \rangle}{6\Delta t} \tag{A4}$$

where $\langle\rangle$ is the ensemble average. Finally, from Eqs. (A1) and (A4), we have:

$$\frac{\partial \psi}{\partial t} = -\nabla(-\boldsymbol{D}\nabla\psi) = \boldsymbol{D}\nabla^2\psi \tag{A5}$$

Note that the diffusion coefficient D ($length^2 . time^{-1}$) is a scalar constant, which describes the mobility of molecules in the isotropic case and depends on the molecule as well as the medium in which the diffusion occurs, However, it cannot give the direction of the diffusion (i.e., antipodal symmetric).

Appendix B: Multicompartment biophysical model functions
parameters of the compartment model functions

Following is a detailed description of the common parameters used in model functions of tissue compartments:

a. **Fiber orientation vector:** Fiber (axon) orientation vector "n" is given by Eq. (B1), where θ, φ, and α are angles (in radians) as shown in Supplementary Fig. B1 in the online version at https://doi.org/10.1016/B978-0-443-13480-7.00006-5(Right)

$$n = [\cos\varphi\sin\theta; \; \sin\varphi\cos\theta; \; \cos\theta] \tag{B1}$$

b. $d_{\parallel}$ ($\mu m^2 . sec^{-1}$) = Diffusivity in the direction parallel to "n" (intrinsic free diffusivity), fixed to $0.6e^3$ ($\mu m^2 . sec^{-1}$) for ex vivo and $1.7e^3$($\mu m^2 . sec^{-1}$) for in vivo

c. $d_{\perp 1}$ ($\mu m^2 . sec^{-1}$) = Diffusivity in the direction perpendicular to "n"

d. $d_{\perp 2}$ ($\mu m^2 . sec^{-1}$) = Diffusivity in the direction perpendicular to "n" and $d_{\perp 1}$

e. d_{iso} ($\mu m^2 . sec^{-1}$) = Isotropic diffusivity, fixed to $2.0e^3$($\mu m^2 . sec^{-1}$) for ex vivo and $3.0e^3$ ($\mu m^2 . sec^{-1}$) for in vivo

f. R (μm) = Axon radius index

g. $R_s(\mu m)$ = Soma, extraaxonal spherical diffusion radius index

h. v_{ic} = Intracellular volume fraction

i. v_{iso} = Isotropic volume fraction or free water fraction (Ffw)

j. f_i = Volume fraction for the ith tissue compartment

k. ODI = Orientation dispersion index

l. NDI = Neurite density index

m. Ffw = Free water fraction or v_{iso} (isotropic volume fraction)

n. μ (rad) = Mean orientation of axonal/ neurite bundle

o. S = Normalized diffusion signal

Parameters of the pulse gradient spin echo sequence

The commonly used pulse sequence for biophysical modeling technique is the PGSE, which implements SDE. As illustrated in Supplementary Fig. B1 in the online version at https://doi.org/10.1016/B978-0-443-13480-7.00006-5(center and left), the parameters of the PGSE for SDE are as follows:

a. G (Tesla. μm^{-1}) = Gradient strength

b. $g = [g_x; g_y; g_z]$ Gradient direction (unit) vector

c. δ (sec) = Diffusion pulse duration (width)

d. $\Delta(sec)$ = Diffusion time (t)

e. γ (rad(sec. Tesla)$^{-1}$) = Gyro-magnetic ratio

f. b (sec. μm^{-2}) = $\Delta - (\delta/3)(\gamma\delta G)^2$, Diffusion weighting factor, also called b-value

g. TE (sec) = Echo time

Intracellular compartment model functions

Cylinder

The cylinder model for the intraaxonal compartment assumes a nonzero pore size of the axonal bundle. The signal model uses the Gaussian phase distribution approximation (Balinov et al., 1993; Douglass and McCall, 1958) for the signal in cylinders of radius (Vangelderen et al., 1994; Stepisnik, 1993) R expressed as follows:

$$S_{cylinder} = e^{S_{cyl\|}}.e^{S_{cyl\perp}} \tag{B2}$$

where

$$e^{S_{cyl\|}} = e^{-(\Delta-(\delta/3))(\gamma\delta G)^2(g^T.n)^2 d_\|} \tag{B2a}$$

$$e^{S_{cyl\perp}} = e^{-2\gamma^2\left((G)-(G.n)^2\right)\sum_{i=1}^{\infty}\frac{2d_\|\beta_i^2\delta-2+2e^{-d_\|\beta_i^2\Delta}-e^{-d_\|\beta_i^2(\Delta-\delta)}-e^{-d_\|\beta_i^2(\Delta+\delta)}+2e^{-d_\|\beta_i^2\delta}}{d_\|^2\beta_i^6\left(\beta_i^2R^2-1\right)}} \tag{B2b}$$

with β_i as the ith root of the equation, $J'(\beta_i R) = 0$, where J' is the derivative of the Bessel function of the first kind and first order.

Stick

Behrens et al. (2003) proposed the stick model, depicting a cylinder where the diameter is considered negligible and can effectively be regarded as zero. Therefore, with $R = 0$ in Eq. (B2), the signal from a stick compartment is given as follows:

$$S_{stick} = e^{S_{cyl\parallel}} = e^{-(\Delta - (\delta/3))(\gamma\delta G)^2 (g^T.n)^2 d_\parallel} \tag{B4}$$

Neurite orientation and dispersion

By utilizing a set of sticks (Eq. B4) and the zeppelin (Eq. B10) model functions, diffusion in the intracellular and extracellular compartments can been effectively modeled with Watson (Zhang et al., 2012) and Bingham (Tariq et al., 2016) distributions. These models successfully capture the two perpendicular directions of diffusion, i.e., the restricted diffusion along neurites and the hindered diffusion perpendicular to them. This configuration encompasses a broad range of neurite orientation patterns observed in brain tissues, spanning from highly parallel alignments to extensively dispersed orientations. These patterns encompass diverse white matter structures, such as the corpus callosum with its remarkably coherent orientations, the centrum semiovale characterized by axons that bend and fan out, as well as cortical and subcortical gray matter structures where dendrites exhibit a dispersed pattern (Zhang et al., 2012).

In a general setting, the signal originating from a dispersed compartment model (S_C), i.e., core diffusion signal integrated over a sphere, takes the following form:

$$S_{OD-C} = \int_{\mathbb{S}^2} f(n) \, S_C \, dn \tag{B5}$$

Here, S_C is the normalized diffusion signal from the intracellular or extracellular compartment and $f(n)dn$ is the probability of finding neurite/ axonal bundle parallel to n. In the literature, the following probability functions for $f(n)$ have been used:

Watson distribution

The Watson distribution is a parametric distribution, defined as an isotropic Gaussian distribution on the sphere and characterized by orientation μ and concentration parameter κ. It serves as the simplest distribution capable of capturing the dispersion of bundle of neurites, such as axons and dendrites (Zhang et al., 2012; Mardia et al., 2000). In Eq. (B6a), $f(n)dn$ for the Watson distribution is formally expressed as follows:

$$f(n)dn = M\left(\frac{1}{2}, \frac{3}{2}, \kappa\right)^{-1} e^{-\kappa(\mu.n)^2} \tag{B6a}$$

where M is a confluent hypergeometric function, μ is the mean orientation, and κ is the concentration parameter that measures the extent of orientation dispersion around μ. Zhang et al. (2012) introduced the parameter called "orientation dispersion index" (ODI), calculated from κ as follows:

$$ODI = \frac{2}{\pi} \; arctan \left(\frac{1}{\kappa} \right)$$ (B6b)

Here, ODI value of 0 corresponds to a spike function along μ (the mean orientation of axonal bundle), while an ODI value of 1 represents an isotropically dispersed profile of S_C on the sphere.

Bingham distribution

Similar to the Watson distribution, the Bingham distribution is a parametric orientation distribution, defined as an anisotropic Gaussian distribution on the sphere with orientation μ as defined as follows:

$$f(n)dn = \frac{1}{{}_1^0F_1 \left(\frac{1}{2}; \frac{3}{2}; \boldsymbol{B} \right)} \; e^{(n^T Bn)}$$ (B6c)

${}_1^0F_1$ is a confluent hypergeometric function of the first kind, $B = R^T BR$, where R is a rotation matrix aligning the orientation distribution with the mean orientation of neurites μ and B is given by:

$$B = \begin{bmatrix} \kappa_1 & 0 & 0 \\ 0 & \kappa_2 & 0 \\ 0 & 0 & \kappa_3 \end{bmatrix}$$ (B6d)

κ_1, κ_2, and κ_3 are the primary, secondary, and tertiary concentration parameters, respectively. Tariq et al. (2016) suggested an inverse relationship between the concentration parameters κ_1 and κ_2, and the dispersion of neurites. Measure of the dispersion anisotropy of neurites with Bingham–NODDI (Tariq et al., 2016) (DA_B) is given as follows:

$$DA_B = \frac{2}{\pi} \; arctan \left(\frac{\kappa_2 - \kappa_3}{\kappa_1 - \kappa_3} \right)$$ (B6e)

Extracellular and cerebrospinal fluid models
Sphere

To model the diffusion of water particles confined within a "spherically" shaped tissue, the following expression offers an approximate measure of the signal attenuation. It utilizes a Gaussian phase distribution (Balinov et al., 1993; Douglass and McCall, 1958) approximation:

$$S_{sphere} = e^{-2\Upsilon^2\hat{G}^2 \sum\limits_{i=1}^{\infty} \frac{2d_\| \alpha_i^2 \delta - 2 + 2e^{-d_\| \beta_i^2 \Delta} - e^{-d_\| \alpha_i^2 (\Delta - \delta)} - e^{-d_\| \alpha_i^2 (\Delta + \delta)} + 2e^{-d_\| \alpha_i^2 \delta}}{d_\|^2 \alpha_i^6 \left(\alpha_i^2 R_s^2 - 1 \right)}}$$ (B7)

where α_i is the ith positive root of the equation, $(\alpha_i R_s) J'_{3/2}(\alpha_i R_s) - \frac{1}{2} J_{3/2}(\alpha_i R_s) = 0$ (or $(\alpha_i R_s)^{-1} J_{3/2}(\alpha_i R_s) = \frac{1}{2} J_{5/2}(\alpha_i R_s)$) and J' is the derivative of Bessel function of the first kind.

Dot

Similar to zero radius cylinders modeled as sticks, zero radius spheres are modeled as nondiffusion dots.

$$S_{dot} = e^{-b_0 = 0} = 1 \tag{B8}$$

Tensor

Within the tensor model (Basser et al., 1994), three diffusivity parameters are present: the parallel diffusivity ($d_{\parallel}$) and two perpendicular diffusivities ($d_{\perp 1}$ and $d_{\perp 2}$) with $d_{\perp 1}$ and $d_{\perp 2}$ being distinct from each other. Additionally, the model incorporates three degrees of freedom for the orthogonal eigenvectors. For details, please refer to Fig. 2 and Diffusion tensor imaging Section.

$$S_{tensor} = e^{-bg' Dg} \tag{B9}$$

where D is a symmetric positive definite 3×3 matrix.

Zeppelin

The zeppelin is a tensor with axial symmetry and is aligned along the orientation μ. It possesses parallel and perpendicular diffusivity, where the parallel diffusivity ($d_{\parallel}$) is greater than or equal to the perpendicular diffusivity ($d_{\perp 1} = d_{\perp 2}$). This distribution is commonly employed to depict the diffusion signal arising from the extraaxonal space.

$$S_{zeppelin} = e^{-b\left(d_{\parallel} - d_{\perp}\right)\left(g^T . n\right)^2 + d_{\perp}} \tag{B10}$$

Ball

The "ball" model, as described in Behrens et al. (2003), is isotropic and characterized by a single parameter, which is the isotropic diffusivity d_{iso}.

$$S_{ball} = e^{-bd_{iso}} \tag{B11}$$

For a comprehensive list of model functions including cylinders with Gamma distributed radii, astrosticks, and astrocylinders, please refer to Panagiotaki et al. (2012).

Examples to build commonly used multicompartment models

Within this section, we demonstrate how commonly employed multicompartment models can be constructed by utilizing the "building block" model functions discussed earlier. To avoid redundancy, we refer readers to Multicompartment biophysical tissue model functions to Extracellular and cerebrospinal fluid models Sections for a detailed explanation of the variables and parameters found in the subsequent functions.

Ball and stick

Behrens et al. (2003) introduced the ball and stick model that comprises of two compartments. The model, however, considers only a discrete set of axon orientations, neglecting orientation dispersion:

$$S_{ball\ and\ stick} = f_1 S_{stick} + f_2 S_{ball} \tag{B12}$$

NODDI/Bingham-NODDI

The NODDI (Zhang et al., 2012) model uses normalized signals from the following three compartments:

$$S_{Noddi} = (1 - v_{iso})\left(S_{OD-C}(v_{ic}S_{stick} + (1 - v_{ic})S_{zeppelin})\right) + v_{iso}S_{dot} \tag{B13}$$

where S_{OD-C} is Watson distribution given by Eqs. (B5) and (B6a). For Bingham-NODDI (Tariq et al., 2016), S_{OD-C} is the Bingham distribution as described in Eqs. (B5) and (B6c). For NODDI (Zhang et al., 2012), there are six parameters $(v_{ic}, v_{iso}, ODI, d_\perp, \theta,$ and $\varphi)$ that explain the neurite orientation, dispersion, and volume fractions of the intracellular and isotropic compartments. Neurite dispersion index (NDI) is calculated using estimated $d_\perp$ as $d_\perp = (1 - NDI)d_\parallel$, where $d_\parallel = 1.7e^3(\mu m^2. \ sec^{-1})$ fixed in vivo.

ActiveAx

ActiveAx (Alexander et al., 2010) for ex vivo data is a four-compartment model, i.e., Zeppelin-Cylinder-Ball-Dot (Alexander et al., 2010). It uses eight parameters to describe the tissue properties $(f_1, f_2, f_3, f_4, R, d_\perp, \theta,$ and $\varphi)$. For the in vivo case, the dot component is not considered (Ferizi et al., 2014; Alexander et al., 2010).

$$S_{ActiveAx_e xvivo} = f_1 S_{cylinder} + f_2 S_{zeppelin} + f_3 S_{ball} + f_4 S_{dot} \tag{B14}$$

$$S_{ActiveAx_i nvivo} = f_1 S_{cylinder} + f_2 S_{zeppelin} + f_3 S_{dot} \tag{B15}$$

Similarly, recently proposed models like SANDI (Palombo et al., 2020) can be built with stick, sphere, ball and extended SANDI (Palombo et al., 2020) with cylinder, sphere, and ball.

Multiple-orientation extensions of the multicompartment models

Extending the multicompartment models discussed earlier to multiple orientations is possible. However, this expansion introduces greater complexity in the modeling process and necessitates a larger number of MR measurements at multiple b-values to ensure the accurate estimation of tissue microstructural properties. Following are a few examples from previous work:

Ball and sticks

The ball and sticks model utilizes multiple sticks to capture the presence and the orientation of multiple fiber bundles within a voxel. This information is highly valuable and has been widely employed in probabilistic tractography techniques (Behrens et al., 2007). By leveraging the orientation data from the ball and stick model, probabilistic tractography can be used to infer and visualize the connectivity patterns of neural pathways in the brain.

NODDIx

To incorporate multiple directions of fiber orientation, the NODDI (Zhang et al., 2012) model can also be extended. In the case of two fiber orientations, this expansion involves considering normalized signals from five distinct compartments (Farooq et al., 2016). In particular, S_{stick^k} represents the normalized signal from the intracellular compartment, while $S_{zeppelin^k}$ represents the normalized signal from the extracellular compartment in the kth fiber orientation. The model incorporates 13 parameters that define the geometric properties of the tissue (i.e., $v_{ic1}, v_{ec1}, v_{ic2}, v_{ec2}, v_{iso}, OD_1, OD_2, \theta_1, \varphi_1, \theta_2, \varphi_2, d_{\perp 1}$, and $d_{\perp 2}$). The full model function is given as follows:

$$S_{Noddix} = \sum_{k=1}^{2}(1 - v_{iso})\left(S_{OD-C}\left(v_{ick}S_{stick^k} + (1 - v_{ick})S_{zeppelin^k}\right)\right) + v_{iso}S_{dot} \qquad (B16)$$

ActiveAx in multiple orientations

An illustrative instance of expanding the ActiveAx (Alexander et al., 2010) model involves the inclusion of three distinct fiber orientations within a voxel. Specifically, we assume that a zeppelin and a cylinder represent the extraaxonal and intraaxonal signal, respectively, in each of the three orientations (Farooq et al., 2016). The cumulative diffusion MRI signal is therefore assumed to be composed of seven compartments:

$$S_{ZCD} = \underbrace{\overbrace{f_{11}S_{cylinder1}}^{Intra-axonal1} + \overbrace{f_{12}S_{zeppelin1}}^{Extra-axonal1}}_{Primary\ Orientation} + \underbrace{\overbrace{f_{21}S_{cylinder2}}^{Intra-axonal2} + \overbrace{f_{22}S_{zeppelin2}}^{Extra-axonal2}}_{Secondary\ Orientation}$$

$$+ \underbrace{\overbrace{f_{31}S_{cylinder3}}^{Intra-axonal3} + \overbrace{f_{32}S_{zeppelin3}}^{Extra-axonal3}}_{Tertiary\ Orientation} + \underbrace{f_3 S_{dot}}_{CSF\ Compartmet\ (Dot)} \qquad (B17)$$

such that $\sum_{i=1}^{3}\sum_{k=1}^{4}f_{ik} + f_3 = 1$, $d_{\|k} = 1.7e^3$ ($\mu m^2.sec^{-1}$), and $d_{\perp k} = (1 - f_{1k})\,d_{\|k}$ ($\mu m^2.sec^{-1}$) are fixed in vivo, with $k = 1, 2, 3$ representing the fiber orientations inside the voxel. There are 19 model parameters that describe the tissue geometry in each fiber orientation (i.e., $f_{11}, f_{12}, f_{21}, f_{22}, f_{31}, f_{32}, f_3, R_1, d_{\perp 1}, \theta_1, \emptyset_1, R_2, d_{\perp 2}, \theta_2, \emptyset_2, R_3, d_{\perp 3}, \theta_3$, and $\emptyset_3$).

AxCaliber3D

Barazany et al. (2011) expanded the AxCaliber (Assaf et al., 2008) framework to AxCaliber3D (Barazany et al., 2011), allowing for the estimation axon diameter distribution (ADD) in any arbitrary fiber orientations, within a voxel. The technique enabled analysis of complex fiber systems, such as those found in the porcine spinal cord, distinguishing between different fiber fascicles and calculating their ADDs. This extension not only provided valuable insights into complex fiber structures but also enhanced tractography and brain connectivity assessment by offering a comprehensive three-dimensional perspective on the ADD along multiple pathways.

Appendix C: Biophysical model implementation libraries

There are several libraries that can be used to implement multicompartment biophysical models for diffusion MRI data analysis. These libraries provide prebuilt tools and functions that make the analysis process simpler, efficient, and more accurate. They also ensure that the results are consistent and reliable, which frees up researchers to focus on interpreting the data and gaining insights into tissue microstructure. Following is a nonexhaustive list:

Diffusion microstructure imaging in python (Dmipy)

Diffusion Microstructure Imaging in Python (Dmipy) (Fick et al., 2019) https://github.com/AthenaEPI/dmipy, an open-source Python toolbox, is an implementation of PGSE-based multicompartment modeling in a comprehensive and flexible framework. It employs an elementary component approach, enabling effortless integration of diverse biophysical model blocks into a unified multicompartment model. This feature enhances its value as an advanced tool for diffusion MRI microstructure imaging, providing researchers with cutting-edge capabilities. Moreover, Dmipy builds upon the functionality of Diffusion Imaging in Python (DIPY) (Garyfallidis et al., 2014) https://dipy.org/, another open-source software project dedicated to processing, modeling, and visualization of diffusion MRI data, thereby facilitating the analysis of diffusion MRI data. Code snippets for easy and efficient implementation of biophysical modeling can be accessed at https://www.ncbi.nlm.nih.gov/pmc/articles/PMC6803556/ and https://dmipy.readthedocs.io/en/latest/index.html.

Accelerated microstructure imaging via convex optimization (AMICO)

Accelerated Microstructure Imaging via Convex Optimization (AMICO) (Daducci et al., 2015b) https://github.com/daducci/AMICO/wiki is an open-source python library specifically designed for the analysis and fitting of multicompartment models to diffusion MRI data. It focuses on estimating microstructural properties of brain tissue, such as intra- and extracellular volume fractions, diffusivities, and fiber orientations by utilizing a convex optimization framework. It has implementations for

NODDI (Zhang et al., 2012), SANDI (Palombo et al., 2020), and ActiveAx (Alexander et al., 2010).

CAMINO

CAMINO (Cook et al., 2006), a widely used software package for diffusion MRI analysis, is developed by the University College London. It provides an extensive collection of tools for processing, modeling, and analyzing diffusion MRI data, incorporating various biophysical models. The toolkit is utilized in research related to multicompartment model taxonomy to fit models accurately to the data.

GPU accelerated implementations

Disimpy

Disimpy (Kerkelä et al., 2020) https://github.com/kerkelae/disimpy,a Python package, allows the generation of synthetic diffusion-weighted MR signals from various geometrical compartments such as spheres and zeppelins, commonly employed in multicompartment modeling technique. The simulated data prove valuable for conducting synthetic data experiments, aiding in the development and validation of data acquisition and analysis methods. Monte Carlo random walk simulations power the generation of this data, leveraging the massive parallel computing capabilities of Nvidia CUDA-capable GPUs.

GPU-accelerated computational diffusion MRI

Hernandez-Fernandez et al. (2019) introduced GPU-based frameworks to accelerate various diffusion MRI computations. These frameworks enable fast processing of tasks such as biophysical modeling, microstructure estimation, and probabilistic tractography. The authors conducted validation experiments comparing the GPU-based implementations with traditional CPU-based approaches, demonstrating that a single GPU can outperform the performance of 200 CPU cores. This superior performance is attributed to the efficient parallelization capabilities of GPUs, which enable more effective utilization of computational resources.

Estimation of diffusion MRI microstructure model parameters

Harms and Roebroeck (2018) introduced an efficient GPU-based Markov chain Monte Carlo (MCMC) sampling method for multicompartment biophysical models. Parameter estimation in these models can be accomplished using maximum likelihood estimation (MLE) or MCMC sampling. While MLE provides a single-point estimate for model parameters, MCMC sampling allows for the recovery of the complete posterior distribution, providing additional insights into parameter uncertainty and correlations. However, MCMC sampling is not widely utilized in diffusion MRI microstructure modeling due to the time-consuming nature of model-specific adjustments and tuning. The study demonstrated that by leveraging an efficient GPU-based implementation, the time constraints associated with MCMC sampling for diffusion multicompartment models can be significantly reduced.

References

Adanyeguh, I.M., Joers, J.M., Deelchand, D.K., Hutter, D.H., Eberly, L.E., Iltis, I., Bushara, K. O., Lenglet, C., Henry, P.-G., 2023. Brain MRI detects early-stage alterations and disease progression in Friedreich ataxia. Brain Commun.

Afzali, M., Pieciak, T., Newman, S., Garyfallidis, E., Ozarslan, E., Cheng, H., Jones, D.K., 2021a. The sensitivity of diffusion MRI to microstructural properties and experimental factors. J. Neurosci. Methods, 347. PubMed PMID: WOS:000600848600008.

Afzali, M., Nilsson, M., Palombo, M., Jones, D.K., 2021b. SPHERIOUSLY? The challenges of estimating sphere radius non-invasively in the human brain from diffusion MRI. Neuroimage, 237. https://doi.org/10.1016/j.neuroimage.2021.118183. PubMed PMID: WOS:000671132300002.

Alexander, D.C., 2008. A general framework for experiment design in diffusion MRI and its application in measuring direct tissue-microstructure features. Magn. Reson. Med. 60 (2), 439–448. https://doi.org/10.1002/mrm.21646. PubMed PMID: WOS:000258105800025.

Alexander, D.C., Hubbard, P.L., Hall, M.G., Moore, E.A., Ptito, M., Parker, G.J.M., Dyrby, T. B., 2010. Orientationally invariant indices of axon diameter and density from diffusion MRI. Neuroimage 52 (4), 1374–1389. https://doi.org/10.1016/j.neuroimage.2010.05.043. PubMed PMID: WOS:000280695200026.

Alexander, D.C., Dyrby, T.B., Nilsson, M., Zhang, H., 2019. Imaging brain microstructure with diffusion MRI: practicality and applications. NMR Biomed. 32 (4). https://doi.org/10.1002/nbm.3841. PubMed PMID: WOS:000461892300008.

Alghamdi, A.J., 2023. The value of various post-processing modalities of diffusion weighted imaging in the detection of multiple sclerosis. Brain Sci. 13 (4). https://doi.org/10.3390/brainsci13040622. PubMed PMID: WOS:000977417800001.

Alotaibi, S., Alsaleh, A., Wuerger, S., Meyer, G., 2023. Rapid neural changes during novel speech-sound learning: an fMRI and DTI study. Brain Lang., 245. https://doi.org/10.1016/j.bandl.2023.105324. PubMed PMID: WOS:001080225600001.

Andersson, J.L.R., Sotiropoulos, S.N., 2016. An integrated approach to correction for off-resonance effects and subject movement in diffusion MR imaging. Neuroimage 125, 1063–1078. https://doi.org/10.1016/j.neuroimage.2015.10.019. PubMed PMID: WOS:000366647500095.

Andersson, J.L.R., Skare, S., Ashburner, J., 2003. How to correct susceptibility distortions in spin-echo echo-planar images: application to diffusion tensor imaging. Neuroimage 20 (2), 870–888. https://doi.org/10.1016/s1053-8119(03)00336-7. PubMed PMID: WOS:000186260700020.

Andersson, J.L.R., Graham, M.S., Zsoldos, E., Sotiropoulos, S.N., 2016. Incorporating outlier detection and replacement into a non-parametric framework for movement and distortion correction of diffusion MR images. Neuroimage 141, 556–572. https://doi.org/10.1016/j.neuroimage.2016.06.058. PubMed PMID: WOS:000384074500047.

Andica, C., Kamagata, K., Hatano, T., Saito, Y., Ogaki, K., Hattori, N., Aoki, S., 2020. MR biomarkers of degenerative brain disorders derived from diffusion imaging. J. Magn. Reson. Imaging 52 (6), 1620–1636. https://doi.org/10.1002/jmri.27019. PubMed PMID: WOS:000502465300001.

Assaf, Y., Basser, P.J., 2005. Composite hindered and restricted model of diffusion (CHARMED) MR imaging of the human brain. Neuroimage 27 (1), 48–58. https://doi.org/10.1016/j.neuroimage.2005.03.042. PubMed PMID: WOS:000230701200005.

Assaf, Y., Blumenfeld-Katzir, T., Yovel, Y., Basser, P.J., 2008. AxCaliber: a method for measuring axon diameter distribution from diffusion MRI. Magn. Reson. Med. 59 (6), 1347–1354. https://doi.org/10.1002/mrm.21577. PubMed PMID: WOS: 000256266400017.

Assaf, Y., Alexander, D.C., Jones, D.K., Bizzi, A., TEJ, B., Clark, C.A., Cohen, Y., Dyrby, T.B., Huppi, P.S., Knoesche, T.R., LeBihan, D., GJM, P., Poupon, C., CONNECT Consortium, 2013. The CONNECT project: combining macro- and micro-structure. Neuroimage 80, 273–282. https://doi.org/10.1016/j.neuroimage.2013.05.055. PubMed PMID: WOS:000322416000021.

Assaf, Y., Johansen-Berg, H., de Schotten, M.T., 2019. The role of diffusion MRI in neuroscience. NMR Biomed. 32 (4). https://doi.org/10.1002/nbm.3762. PubMed PMID: WOS:000461892300006.

Avants, B.B., Tustison, N., Song, G., 2009. Advanced normalization tools (ANTS). Insight J. 2 (365), 1–35.

Balinov, B., Jonsson, B., Linse, P., Soderman, O., 1993. The NMR self-diffusion method applied to restricted diffusion—simulation of echo attenuation from molecules in spheres and between planes. J. Magn. Reson. A 104 (1), 17–25. https://doi.org/10.1006/jmra.1993.1184. PubMed PMID: WOS:A1993LQ74500003.

Barazany, D., Basser, P.J., Assaf, Y., 2009. In vivo measurement of axon diameter distribution in the corpus callosum of rat brain. Brain 132, 1210–1220. https://doi.org/10.1093/brain/awp042. PubMed PMID: WOS:000265950900010.

Barazany, D., Jones, D., Assaf, Y. (Eds.), 2011. AxCaliber 3D. Proc. Int. Soc. Magn. Reson. Med.

Basser, P.J., Pierpaoli, C., 1996. Microstructural and physiological features of tissues elucidated by quantitative-diffusion-tensor MRI. J. Magn. Reson. B 111 (3), 209–219. https://doi.org/10.1006/jmrb.1996.0086. PubMed PMID: WOS:A1996UQ67800001.

Basser, P.J., Mattiello, J., Lebihan, D., 1994. MR diffusion tensor spectroscopy and imaging. Biophys. J. 66 (1), 259–267. PubMed PMID: WOS:A1994MP04100030.

Basser, P.J., Pajevic, S., Pierpaoli, C., Duda, J., Aldroubi, A., 2000. In vivo fiber tractography using DT-MRI data. Magn. Reson. Med. 44 (4), 625–632. https://doi.org/10.1002/1522-2594(200010)44:4<625::aid-mrm17>3.0.co;2-o. PubMed PMID: WOS:000089671300017.

Battocchio, M., Schiavi, S., Descoteaux, M., Daducci, A., 2022. Bundle-o-graphy: improving structural connectivity estimation with adaptive microstructure-informed tractography. Neuroimage, 263. https://doi.org/10.1016/j.neuroimage.2022.119600. PubMed PMID: WOS:000856034500010.

Beaulieu, C., 2002. The basis of anisotropic water diffusion in the nervous system—a technical review. NMR Biomed. 15 (7–8), 435–455. https://doi.org/10.1002/nbm.782. PubMed PMID: WOS:000180166100002.

Behrens, T.E.J., Woolrich, M.W., Jenkinson, M., Johansen-Berg, H., Nunes, R.G., Clare, S., Matthews, P.M., Brady, J.M., Smith, S.M., 2003. Characterization and propagation of uncertainty in diffusion-weighted MR imaging. Magn. Reson. Med. 50 (5), 1077–1088. https://doi.org/10.1002/mrm.10609. PubMed PMID: WOS:000186326400022.

Behrens, T.E.J., Berg, H.J., Jbabdi, S., Rushworth, M.F.S., Woolrich, M.W., 2007. Probabilistic diffusion tractography with multiple fibre orientations: what can we gain? Neuroimage 34 (1), 144–155. https://doi.org/10.1016/j.neuroimage.2006.09.018. PubMed PMID: WOS:000242735300015.

Braak, H., Del Tredici, K., Rub, U., de Vos, R.A.I., Steur, E., Braak, E., 2003. Staging of brain pathology related to sporadic Parkinson's disease. Neurobiol. Aging 24 (2), 197–211. https://doi.org/10.1016/s0197-4580(02)00065-9. PubMed PMID: WOS:000180616100001.

Brown, R., 1828. XXVII. A brief account of microscopical observations made in the months of June, July and August 1827, on the particles contained in the pollen of plants; and on the general existence of active molecules in organic and inorganic bodies. Philos. Mag. 4 (21), 161–173. https://doi.org/10.1080/14786442808674769.

Brubaker, C.J., Schmithorst, V.J., Haynes, E.N., Dietrich, K.N., Egelhoff, J.C., Lindquist, D. M., Lanphear, B.P., Cecil, K.M., 2009. Altered myelination and axonal integrity in adults with childhood lead exposure: a diffusion tensor imaging study. Neurotoxicology 30 (6), 867–875. https://doi.org/10.1016/j.neuro.2009.07.007. PubMed PMID: WOS:000272873700001.

By, S., Xu, J.Z., Box, B.A., Bagnato, F.R., Smith, S.A., 2017. Application and evaluation of NODDI in the cervical spinal cord of multiple sclerosis patients. Neuroimage-Clinical. 15, 333–342. https://doi.org/10.1016/j.nicl.2017.05.010. PubMed PMID: WOS:000410067200036.

Cai, L.Y., Yang, Q., Hansen, C.B., Nath, V., Ramadass, K., Johnson, G.W., Conrad, B.N., Boyd, B.D., Begnoche, J.P., Beason-Held, L.L., Shafer, A.T., Resnick, S.M., Taylor, W.D., Price, G.R., Morgan, V.L., Rogers, B.P., Schilling, K.G., Landman, B.A., 2021. PreQual: an automated pipeline for integrated preprocessing and quality assurance of diffusion weighted MRI images. Magn. Reson. Med. 86 (1), 456–470. https://doi.org/10.1002/mrm.28678. PubMed PMID: WOS:000614037300001.

Callaghan, P.T., 1993. Principles of Nuclear Magnetic Resonance Microscopy. Clarendon Press.

Callaghan, P.T., Stepisnik, J., 1995. Frequency-domain analysis of spin motion using modulated-gradient NMR. J. Magn. Reson. A 117 (1), 118–122. https://doi.org/10.1006/jmra.1995.9959. PubMed PMID: WOS:A1995TE63700020.

Carr, H.Y., Purcell, E.M., 1954. Effects of diffusion on free precession in nuclear magnetic resonance experiments. Phys. Rev. 94 (3), 630–638. https://doi.org/10.1103/PhysRev.94.630. PubMed PMID: WOS:A1954UB48000018.

Caruyer, E., Lenglet, C., Sapiro, G., Deriche, R., 2013. Design of multishell sampling schemes with uniform coverage in diffusion MRI. Magn. Reson. Med. 69 (6), 1534–1540. https://doi.org/10.1002/mrm.24736.

Colon-Perez, L., Montesinos, J., Monsivais, M., 2022. The future of neuroimaging and gut-brain axis research for substance use disorders. Brain Res. 1781. https://doi.org/10.1016/j.brainres.2022.147835. PubMed PMID: WOS:000793392300002.

Cook, P.A., Bai, Y., Nedjati-Gilani, S., Seunarine, K.K., Hall, M.G., Parker, G.J., Alexander, D.C., 2006. Camino: open-source diffusion-MRI reconstruction and processing. In: 14th Scientific Meeting of the International Society for Magnetic Resonance in Medicine; Seattle, WA, USA, p. 2759.

Cory, D.G., Garroway, A.N., Miller, J.B. (Eds.), 1990. Applications of Spin Transport as a Probe of Local Geometry. Abstracts of Papers of the American Chemical Society. AMER Chemical SOC 1155 16TH St, NW, WASHINGTON, DC 20036.

Crauyer, E., 2023. Web Application for Multishell Protocol Design. Available from: http://www.emmanuelcaruyer.com/q-space-sampling.php.

Cruces, R.R., Royer, J., Herholz, P., Lariviere, S., De Wael, R.V., Paquola, C., Benkarim, O., Park, B.Y., Degre-Pelletier, J., Nelson, M.C., DeKraker, J., Leppert, I.R., Tardif, C., Poline, J.B., Concha, L., Bernhardt, B.C., 2022. Micapipe: a pipeline for multimodal neuroimaging and connectome analysis. Neuroimage, 263. https://doi.org/10.1016/j.neuroimage.2022.119612. PubMed PMID: WOS:000856034500006.

Daducci, A., Dal Palu, A., Lemkaddem, A., Thiran, J.P., 2015a. COMMIT: convex optimization modeling for microstructure informed tractography. IEEE Trans. Med. Imaging 34 (1), 246–257. https://doi.org/10.1109/tmi.2014.2352414. PubMed PMID: WOS:000346975900023.

Daducci, A., Canales-Rodriguez, E.J., Zhang, H., Dyrby, T.B., Alexander, D.C., Thiran, J.-P., 2015b. Accelerated microstructure imaging via convex optimization (AMICO) from diffusion MRI data. Neuroimage 105, 32–44. https://doi.org/10.1016/j.neuroimage.2014.10.026. PubMed PMID: WOS:000346050300004.

De Santis, S., Herranz, E., Treaba, C.A., Barletta, V., Mehndiratta, A., Mainero, C., Toschi, N., 2019a. Whole brain in vivo axonal diameter mapping in multiple sclerosis. In: 41st Annual International Conference of the IEEE Engineering in Medicine and Biology Society (EMBC); 2019 Jul 23-27; Berlin, GERMANY.

De Santis, S., Bastiani, M., Droby, A., Kolber, P., Zipp, F., Pracht, E., Stoecker, T., Groppa, S., Roebroeck, A., 2019b. Characterizing microstructural tissue properties in multiple sclerosis with diffusion MRI at 7 T and 3 T: the impact of the experimental design. Neuroscience 403, 17–26. https://doi.org/10.1016/j.neuroscience.2018.03.048. PubMed PMID: WOS:000462916000003.

DiPiero, M., Rodrigues, P.G., Gromala, A., Dean, D.C., 2023. Applications of advanced diffusion MRI in early brain development: a comprehensive review. Brain Struct. Funct. 228 (2), 367–392. https://doi.org/10.1007/s00429-022-02605-8. PubMed PMID: WOS:000906295300001.

Douglass, D.C., McCall, D.W., 1958. Diffusion in paraffin hydrocarbons. J. Phys. Chem. 62 (9), 1102–1107. https://doi.org/10.1021/j150567a020. PubMed PMID: WOS:A1958WK47500020.

Einstein, A., 1905. Über die von der molekularkinetischen Theorie der Wärme geforderte Bewegung von in ruhenden Flüssigkeiten suspendierten Teilchen. Ann. Phys. 322 (8), 549–560. https://doi.org/10.1002/andp.19053220806.

Einstein, A., 1956. Investigations on the Theory of the Brownian Movement. Courier Corporation.

Faiyaz, A., Doyley, M.M., Schifitto, G., Uddin, M.N., 2023. Artificial intelligence for diffusion MRI-based tissue microstructure estimation in the human brain: an overview. Front. Neurol., 14. https://doi.org/10.3389/fneur.2023.1168833. PubMed PMID: WOS:000980385000001.

Farooq, H., Xu, J.Q., Nam, J.W., Keefe, D.F., Yacoub, E., Georgiou, T., Lenglet, C., 2016. Microstructure imaging of crossing (MIX) white matter fibers from diffusion MRI. Sci. Rep., 6. https://doi.org/10.1038/srep38927. PubMed PMID: WOS:000389881600001.

Ferizi, U., Schneider, T., Panagiotaki, E., Nedjati-Gilani, G., Zhang, H., Wheeler-Kingshott, C.A.M., Alexander, D.C., 2014. A ranking of diffusion MRI compartment models with in vivo human brain data. Magn. Reson. Med. 72 (6), 1785–1792. https://doi.org/10.1002/mrm.25080. PubMed PMID: WOS:000344798300030.

Fick, R.H.J., Wassermann, D., Deriche, R., 2019. The Dmipy toolbox: diffusion MRI multicompartment modeling and microstructure recovery made easy. Front. Neuroinform., 13. https://doi.org/10.3389/fninf.2019.00064. PubMed PMID: WOS:000535812800001.

Fieremans, E., Novikov, D.S., Jensen, J.H., Helpern, J.A., 2010. Monte Carlo study of a two-compartment exchange model of diffusion. NMR Biomed. 23 (7), 711–724. https://doi.org/10.1002/nbm.1577. PubMed PMID: WOS:000283014300005.

Fieremans, E., Burcaw, L., Lee, H.H., Lemberskiy, G., Veraart, J., Novikov, D.S., 2016. In vivo observation and biophysical interpretation of time-dependent diffusion in human white matter. Neuroimage 129, 414–427. https://doi.org/10.1016/j.neuroimage. 2016.01.018. PubMed PMID: WOS:000372745300035.

Floel, A., de Vries, M.H., Scholz, J., Breitenstein, C., Johansen-Berg, H., 2009. White matter integrity in the vicinity of Broca's area predicts grammar learning success. Neuroimage 47 (4), 1974–1981. https://doi.org/10.1016/j.neuroimage.2009.05.046. PubMed PMID: WOS:000269035100085.

Fortin, J.P., Parker, D., Tunc, B., Watanabe, T., Elliott, M.A., Ruparel, K., Roalf, D.R., Satterthwaite, T.D., Gur, R.C., Gur, R.E., Schultz, R.T., Verma, R., Shinohara, R.T., 2017. Harmonization of multi-site diffusion tensor imaging data. Neuroimage 161, 149–170. https://doi.org/10.1016/j.neuroimage.2017.08.047. PubMed PMID: WOS:000415673100013.

Gardier, R., Haro, J.L.V., Canales-Rodriguez, E.J., Jelescu, I.O., Girard, G., Rafael-Patino, J., Thiran, J.P., 2023. Cellular exchange imaging (CEXI): evaluation of a diffusion model including water exchange in cells using numerical phantoms of permeable spheres. Magn. Reson. Med. https://doi.org/10.1002/mrm.29720. PubMed PMID: WOS:001001951600001.

Garyfallidis, E., Brett, M., Amirbekian, B., Rokem, A., van der Walt, S., Descoteaux, M., Nimmo-Smith, I., Dipy Contributors, 2014. Dipy, a library for the analysis of diffusion MRI data. Front. Neuroinform., 8. https://doi.org/10.3389/fninf.2014.00008. PubMed PMID: WOS:000348105700001.

Glasser, M.F., Sotiropoulos, S.N., Wilson, J.A., Coalson, T.S., Fischl, B., Andersson, J.L., Xu, J.Q., Jbabdi, S., Webster, M., Polimeni, J.R., Van Essen, D.C., Jenkinson, M., Consortium WU-MH, 2013. The minimal preprocessing pipelines for the human connectome project. Neuroimage 80, 105–124. https://doi.org/10.1016/j.neuroimage.2013.04.127. PubMed PMID: WOS:000322416000011.

Gooijers, J., De Luca, A., Adab, H.Z., Leemans, A., Roebroeck, A., Swinnen, S.P., 2021. Indices of callosal axonal density and radius from diffusion MRI relate to upper and lower limb motor performance. Neuroimage, 241. https://doi.org/10.1016/j.neuroimage.2021. 118433. PubMed PMID: WOS:000693400400001.

Gudbjartsson, H., Patz, S., 1995. The rician distribution of noisy MRI data. Magn. Reson. Med. 34 (6), 910–914. https://doi.org/10.1002/mrm.1910340618. PubMed PMID: WOS: A1995TH39600017.

Hahn, E.L., 1950. Spin echoes. Phys. Rev. 80 (4), 580–594. https://doi.org/10.1103/PhysRev.80.580. PubMed PMID: WOS:A1950UB34300017.

Harms, R.L., Roebroeck, A., 2018. Robust and fast Markov chain Monte Carlo sampling of diffusion MRI microstructure models. Front. Neuroinform., 12. https://doi.org/10.3389/fninf.2018.00097. PubMed PMID: WOS:000453748800001.

Hernandez-Fernandez, M., Reguly, I., Jbabdi, S., Giles, M., Smith, S., Sotiropoulos, S.N., 2019. Using GPUs to accelerate computational diffusion MRI: from microstructure estimation to tractography and connectomes. Neuroimage 188, 598–615. https://doi.org/ 10.1016/j.neuroimage.2018.12.015. PubMed PMID: WOS:000460064700053.

Hofstetter, S., Tavor, I., Moryosef, S.T., Assaf, Y., 2013. Short-term learning induces white matter plasticity in the fornix. J. Neurosci. 33 (31), 12844–12850. https://doi.org/10.1523/ jneurosci.4520-12.2013. PubMed PMID: WOS:000322514500028.

Horowitz, A., Barazany, D., Tavor, I., Bernstein, M., Yovel, G., Assaf, Y., 2015. In vivo correlation between axon diameter and conduction velocity in the human brain. Brain Struct. Funct. 220 (3), 1777–1788. https://doi.org/10.1007/s00429-014-0871-0. PubMed PMID: WOS:000353515200036.

Hsu, C.C.H., Chong, S.T., Kung, Y.C., Kuo, K.T., Huang, C.C., Lin, C.P., 2023. Integrated diffusion image operator (iDIO): a pipeline for automated configuration and processing of diffusion MRI data. Hum. Brain Mapp. https://doi.org/10.1002/hbm.26239. PubMed PMID: WOS:000935793600001.

Hu, Y., Li, Q.F., Qiao, K.N., Zhang, X.C., Chen, B., Yang, Z., 2023. PhiPipe: a multi-modal MRI data processing pipeline with test-retest reliability and predicative validity assessments. Hum. Brain Mapp. 44 (5), 2062–2084. https://doi.org/10.1002/hbm.26194. PubMed PMID: WOS:000905768400001.

Huang, S.Y., Tobyne, S.M., Nummenmaa, A., Witzel, T., Wald, L.L., McNab, J.A., Klawiter, E.C., 2016. Characterization of axonal disease in patients with multiple sclerosis using high-gradient-diffusion MR imaging. Radiology 280 (1), 244–251. https://doi.org/10.1148/radiol.2016151582. PubMed PMID: WOS:000378721900027.

Huang, S.Y., Fan, Q.Y., Machado, N., Eloyan, A., Bireley, J.D., Russo, A.W., Tobyne, S.M., Patel, K.R., Brewer, K., Rapaport, S.F., Nummenmaa, A., Witzel, T., Sherman, J.C., Wald, L.L., Klawiter, E.C., 2019. Corpus callosum axon diameter relates to cognitive impairment in multiple sclerosis. Ann. Clin. Transl. Neurol. 6 (5), 882–892. https://doi.org/10.1002/acn3.760. PubMed PMID: WOS:000468630000007.

Ianus, A., Shemesh, N., Alexander, D.C., Drobnjak, I., 2017. Double oscillating diffusion encoding and sensitivity to microscopic anisotropy. Magn. Reson. Med. 78 (2), 550–564. https://doi.org/10.1002/mrm.26393. PubMed PMID: WOS:000405637000014.

Jelescu, I.O., Veraart, J., Adisetiyo, V., Milla, S.S., Novikov, D.S., Fieremans, E., 2015. One diffusion acquisition and different white matter models: how does microstructure change in human early development based on WMTI and NODDI? Neuroimage 107, 242–256. https://doi.org/10.1016/j.neuroimage.2014.12.009. PubMed PMID: WOS:000348043100026.

Jelescu, I.O., Veraart, J., Fieremans, E., Novikov, D.S., 2016. Degeneracy in model parameter estimation for multi-compartmental diffusion in neuronal tissue. NMR Biomed. 29 (1), 33–47. https://doi.org/10.1002/nbm.3450. PubMed PMID: WOS:000367316200005.

Jelescu, I.O., Palombo, M., Bagnato, F., Schilling, K.G., 2020. Challenges for biophysical modeling of microstructure. J. Neurosci. Methods, 344. https://doi.org/10.1016/j.jneumeth.2020.108861. PubMed PMID: WOS:000557763400005.

Jelescu, I.O., de Skowronski, A., Geffroy, F., Palombo, M., Novikov, D.S., 2022. Neurite exchange imaging (NEXI): a minimal model of diffusion in gray matter with inter-compartment water exchange. Neuroimage, 256. https://doi.org/10.1016/j.neuroimage.2022.119277. PubMed PMID: WOS:000830858700008.

Jenkinson, M., Beckmann, C.F., Behrens, T.E., Woolrich, M.W., Smith, S.M., 2012. FSL. Neuroimage. 62 (2), 782–790. https://doi.org/10.1016/j.neuroimage.2011.09.015. PubMed PMID: WOS:000306390600032.

Jensen, J.H., Helpern, J.A., Ramani, A., Lu, H.Z., Kaczynski, K., 2005. Diffusional kurtosis imaging: the quantification of non-Gaussian water diffusion by means of magnetic resonance imaging. Magn. Reson. Med. 53 (6), 1432–1440. https://doi.org/10.1002/mrm.20508. PubMed PMID: WOS:000229468200025.

Ji, J.L., Demsar, J., Fonteneau, C., Tamayo, Z., Pan, L.N., Kraljic, A., Matkovic, A., Purg, N., Helmer, M., Warrington, S., Winkler, A., Zerbi, V., Coalson, T.S., Glasser, M.F., Harms, M.P., Sotiropoulos, S.N., Murray, J.D., Anticevic, A., Repovs, G., 2023. QuNex-an integrative platform for reproducible neuroimaging analytics. Front. Neuroinform., 17. https://doi.org/10.3389/fninf.2023.1104508. PubMed PMID: WOS:000970089800001.

Johansen-Berg, H., Della-Maggiore, V., Behrens, T.E.J., Smith, S.M., Paus, T., 2007. Integrity of white matter in the corpus callosum correlates with bimanual co-ordination skills. Neuroimage 36, T16–T21. https://doi.org/10.1016/j.neuroimage.2007.03.041. PubMed PMID: WOS:000246911400003.

Jones, D.K., 2010. Challenges and limitations of quantifying brain connectivity in vivo with diffusion MRI. Imaging Med. 2 (3), 341.

Jones, D.K., 2012. In: PDK, J. (Ed.), Diffusion MRI: Theory, Methods, and Applications. Oxford University Press.

Jones, D.K., Basser, P.J., 2004. "Squashing peanuts and smashing pumpkins": how noise distorts diffusion-weighted MR data. Magn. Reson. Med. 52 (5), 979–993. https://doi.org/10.1002/mrm.20283. PubMed PMID: WOS:000224948700003.

Jones, D.K., Knosche, T.R., Turner, R., 2013. White matter integrity, fiber count, and other fallacies: the do's and don'ts of diffusion MRI. Neuroimage 73, 239–254. https://doi.org/10.1016/j.neuroimage.2012.06.081. PubMed PMID: WOS:000317084500024.

Kaden, E., Kruggel, F., Alexander, D.C., 2016a. Quantitative mapping of the per-axon diffusion coefficients in brain white matter. Magn. Reson. Med. 75 (4), 1752–1763. https://doi.org/10.1002/mrm.25734. PubMed PMID: WOS:000372910900037.

Kaden, E., Kelm, N.D., Carson, R.P., Does, M.D., Alexander, D.C., 2016b. Multi-compartment microscopic diffusion imaging. Neuroimage 139, 346–359. https://doi.org/10.1016/j.neuroimage.2016.06.002. PubMed PMID: WOS:000381583500034.

Kamagata, K., Hatano, T., Aoki, S., 2016. What is NODDI and what is its role in Parkinson's assessment? Expert Rev. Neurother. 16 (3), 241–243. https://doi.org/10.1586/14737175.2016.1142876. PubMed PMID: WOS:000371335900001.

Kamiya, K., Hori, M., Aoki, S., 2020. NODDI in clinical research. J. Neurosci. Methods 346. https://doi.org/10.1016/j.jneumeth.2020.108908. PubMed PMID: WOS:000580629300009.

Kerkelä, L., Nery, F., Hall, M.G., Clark, C.A., 2020. Disimpy: a massively parallel Monte Carlo simulator for generating diffusion-weighted MRI data in Python. J. Open Source Softw. 5 (52), 2527.

King, M.D., Houseman, J., Roussel, S.A., Vanbruggen, N., Williams, S.R., Gadian, D.G., 1994. Q-space imaging of the brain. Magn. Reson. Med. 32 (6), 707–713. https://doi.org/10.1002/mrm.1910320605. PubMed PMID: WOS:A1994PV32900004.

Koch, P.J., Girard, G., Brugger, J., Cadic-Melchior, A.G., Beanato, E., Park, C.H., Morishita, T., Wessel, M.J., Pizzolato, M., Canales-Rodriguez, E.J., Fischi-Gomez, E., Schiavi, S., Daducci, A., Piredda, G.F., Hilbert, T., Kober, T., Thiran, J.P., Hummel, F.C., 2022. Evaluating reproducibility and subject-specificity of microstructure-informed connectivity. Neuroimage, 258. https://doi.org/10.1016/j.neuroimage.2022.119356. PubMed PMID: WOS:000814757600012.

Koppers, S., Bloy, L., Berman, J.I., Tax, C.M., Edgar, J.C., Merhof, D., 2019. Spherical harmonic residual network for diffusion signal harmonization. In: Computational Diffusion MRI: International MICCAI Workshop, Granada, Spain, September 2018 22. Springer.

Kraguljac, N.V., Guerreri, M., Strickland, M.J., Zhang, H., 2023. Neurite orientation dispersion and density imaging in psychiatric disorders: a systematic literature review and a technical note. Biol. Psychiatry Glob. Open Sci. 3 (1), 10–21. https://doi.org/10.1016/j.bpsgos.2021.12.012.

Lampinen, B., Szczepankiewicz, F., Noven, M., van Westen, D., Hansson, O., Englund, E., Martensson, J., Westin, C.F., Nilsson, M., 2019. Searching for the neurite density with diffusion MRI: challenges for biophysical modeling. Hum. Brain Mapp. 40 (8), 2529–2545. https://doi.org/10.1002/hbm.24542. PubMed PMID: WOS:000466605100019.

Lee, H.H., Fieremans, E., Novikov, D.S., 2018. What dominates the time dependence of diffusion transverse to axons: intra- or extra-axonal water? Neuroimage 182, 500–510. https://doi.org/10.1016/j.neuroimage.2017.12.038. PubMed PMID: WOS:000446316400039.

Li, L.C., Rilling, J.K., Preuss, T.M., Glasser, M.F., Hu, X.P., 2012. The effects of connection reconstruction method on the interregional connectivity of brain networks via diffusion tractography. Hum. Brain Mapp. 33 (8), 1894–1913. https://doi.org/10.1002/hbm.21332. PubMed PMID: WOS:000306409400012.

Li, X.R., Morgan, P.S., Ashburner, J., Smith, J., Rorden, C., 2016. The first step for neuroimaging data analysis: DICOM to NIfTI conversion. J. Neurosci. Methods 264, 47–56. https://doi.org/10.1016/j.jneumeth.2016.03.001. PubMed PMID: WOS:000375164400007.

Mah, A., Geeraert, B., Lebel, C., 2017. Detailing neuroanatomical development in late childhood and early adolescence using NODDI. PLoS One 12 (8). https://doi.org/10.1371/journal.pone.0182340. PubMed PMID: WOS:000407856600023.

Mardia, K.V., Jupp, P.E., Mardia, K., 2000. Directional Statistics. Wiley Online Library.

Maximov, I.I., Tonoyan, A.S., Pronin, I.N., 2017. Differentiation of glioma malignancy grade using diffusion MRI. Phys. Med. 40, 24–32. https://doi.org/10.1016/j.ejmp.2017.07.002. PubMed PMID: WOS:000410701400004.

Morris S.R., Swift-LaPointe T., Yung A., Prevost V., George S., Bauman A., Kozlowski P., Samadi F., Fournier C., Parker L. (2022). Identifying diffusion model biomarkers for inflammation in human traumatic spinal cord injury.

Niendorf, T., Dijkhuizen, R.M., Norris, D.G., Campagne, M.V., Nicolay, K., 1996. Biexponential diffusion attenuation in various states of brain tissue: implications for diffusion-weighted imaging. Magn. Reson. Med. 36 (6), 847–857. https://doi.org/10.1002/mrm.1910360607. PubMed PMID: WOS:A1996VU89000006.

Novikov, D.S., Jelescu, I.O., Fieremans, E. (Eds.), 2015. From diffusion signal moments to neurite diffusivities, volume fraction and orientation distribution: an exact solution. Proceedings of the International Society of Magnetic Resonance in Medicine.

Novikov, D.S., Veraart, J., Jelescu, I.O., Fieremans, E., 2018a. Rotationally-invariant mapping of scalar and orientational metrics of neuronal microstructure with diffusion MRI. Neuroimage 174, 518–538. https://doi.org/10.1016/j.neuroimage.2018.03.006. PubMed PMID: WOS:000438609100045.

Novikov, D.S., Kiselev, V.G., Jespersen, S.N., 2018b. On modeling. Magn. Reson. Med. 79 (6), 3172–3193. https://doi.org/10.1002/mrm.27101. PubMed PMID: WOS:000426866400031.

Novikov, D.S., Fieremans, E., Jespersen, S.N., Kiselev, V.G., 2019. Quantifying brain microstructure with diffusion MRI: theory and parameter estimation. NMR Biomed. 32 (4). https://doi.org/10.1002/nbm.3998. PubMed PMID: WOS:000461892300010.

Ocampo-Pineda, M., Schiavi, S., Rheault, F., Girard, G., Petit, L., Descoteaux, M., Daducci, A., 2021. Hierarchical microstructure informed tractography. Brain Connect. 11 (2), 75–88. https://doi.org/10.1089/brain.2020.0907. PubMed PMID: WOS:000630161200002.

Olesen, J.L., Ostergaard, L., Shemesh, N., Jespersen, S.N., 2022. Diffusion time dependence, power-law scaling, and exchange in gray matter. Neuroimage, 251. https://doi.org/10.1016/j.neuroimage.2022.118976. PubMed PMID: WOS:000766358200001.

Ozarslan, E., Koay, C.G., Shepherd, T.M., Komlosh, M.E., Irfanoglu, M.O., Pierpaoli, C., Basser, P.J., 2013. Mean apparent propagator (MAP) MRI: a novel diffusion imaging method for mapping tissue microstructure. Neuroimage 78, 16–32. https://doi.org/10.1016/j.neuroimage.2013.04.016. PubMed PMID: WOS:000320488900003.

Palombo, M., Ianus, A., Guerreri, M., Nunes, D., Alexander, D.C., Shemesh, N., Zhang, H., 2020. SANDI: a compartment-based model for non-invasive apparent soma and neurite imaging by diffusion MRI. Neuroimage, 215. https://doi.org/10.1016/j.neuroimage.2020.116835. PubMed PMID: WOS:000539990200004.

Panagiotaki, E., 2011. Geometric Models of Brain White Matter for Microstructure Imaging With Diffusion MRI. University College London.

Panagiotaki, E., Schneider, T., Siow, B., Hall, M.G., Lythgoe, M.F., Alexander, D.C., 2012. Compartment models of the diffusion MR signal in brain white matter: a taxonomy and comparison. Neuroimage 59 (3), 2241–2254. https://doi.org/10.1016/j.neuroimage.2011.09.081. PubMed PMID: WOS:000299494000026.

Panagiotaki, E., Chan, R.W., Dikaios, N., Ahmed, H.U., O'Callaghan, J., Freeman, A., Atkinson, D., Punwani, S., Hawkes, D.J., Alexander, D.C., 2015. Microstructural characterization of normal and malignant human prostate tissue with vascular, extracellular, and restricted diffusion for cytometry in Tumours magnetic resonance imaging. Invest. Radiol. 50 (4), 218–227. https://doi.org/10.1097/rli.0000000000000115. PubMed PMID: WOS:000350915500006.

Pestilli, F., Yeatman, J.D., Rokem, A., Kay, K.N., Wandell, B.A., 2014. Evaluation and statistical inference for human connectomes. Nat. Methods 11 (10), 1058–1063. https://doi.org/10.1038/nmeth.3098. PubMed PMID: WOS:000342719100026.

Pierpaoli, C., Jezzard, P., Basser, P.J., Barnett, A., DiChiro, G., 1996. Diffusion tensor MR imaging of the human brain. Radiology 201 (3), 637–648. PubMed PMID: WOS: A1996VU50000010.

Raghavan, S., Reid, R.I., Przybelski, S.A., Lesnick, T.G., Graff-Radford, J., Schwarz, C.G., Knopman, D.S., Mielke, M.M., Machulda, M.M., Petersen, R.C., Jack, C.R., Vemuri, P., 2021. Diffusion models reveal white matter microstructural changes with ageing, pathology and cognition. Brain. I.D.A.A. Commun. 3 (2). https://doi.org/10.1093/braincomms/fcab106. PubMed PMID: WOS:000667313900024.

Scherrer, B., Schwartzman, A., Taquet, M., Sahin, M., Prabhu, S.P., Warfield, S.K., 2016. Characterizing brain tissue by assessment of the distribution of anisotropic microstructural environments in diffusion-compartment imaging (DIAMOND). Magn. Reson. Med. 76, 963–977. https://doi.org/10.1002/mrm.25912.

Schiavi, S., Ocampo-Pineda, M., Barakovic, M., Petit, L., Descoteaux, M., Thiran, J.-P., Daducci, A., 2020. A new method for accurate in vivo mapping of human brain connections using microstructural and anatomical information. Sci. Adv. 6 (31), eaba8245. https://doi.org/10.1126/sciadv.aba8245.

Smith, S.M., 2002. Fast robust automated brain extraction. Hum. Brain Mapp. 17 (3), 143–155. https://doi.org/10.1002/hbm.10062. PubMed PMID: WOS:000178994100001.

Smith, S.M., Jenkinson, M., Woolrich, M.W., Beckmann, C.F., Behrens, T.E.J., Johansen-Berg, H., Bannister, P.R., De Luca, M., Drobnjak, I., Flitney, D.E., Niazy, R.K., Saunders, J., Vickers, J., Zhang, Y.Y., De Stefano, N., Brady, J.M., Matthews, P.M., 2004. Advances in functional and structural MR image analysis and implementation as FSL. Neuroimage 23, S208–S219. https://doi.org/10.1016/j.neuroimage.2004.07.051. PubMed PMID: WOS:000225374100020.

Smith, R.E., Tournier, J.D., Calamante, F., Connelly, A., 2013. SIFT: spherical-deconvolution informed filtering of tractograms. Neuroimage 67, 298–312. https://doi.org/10.1016/j.neuroimage.2012.11.049. PubMed PMID: WOS:000314144600028.

Smith, R.E., Tournier, J.D., Calamante, F., Connelly, A., 2015. The effects of SIFT on the reproducibility and biological accuracy of the structural connectome. Neuroimage 104, 253–265. https://doi.org/10.1016/j.neuroimage.2014.10.004. PubMed PMID: WOS:00345393800025.

Snoussi, H., Combes, B., Commowick, O., Bannier, E., Kerbrat, A., Cohen-Adad, J., Barillot, C., Caruyer, E., 2022. Reproducibility and evolution of diffusion MRI measurements within the cervical spinal cord in multiple sclerosis. In: 2022 IEEE 19th International Symposium on Biomedical Imaging (ISBI), India., https://doi.org/10.1109/isbi52829.2022.9761680. PubMed PMID: WOS:000836243800276.

Snoussi, H., Rashid, T., Seshadri, S., Habes, M., Satizabal, C.L., 2023. Diffusion MRI indices of brain microstructure: evaluating the ball-and-stick model in distinguishing Alzheimer's disease. Alzheimers Dement. 19 (S3), e066019. https://doi.org/10.1002/alz.066019.

Sotiropoulos, S.N., Jbabdi, S., Xu, J.Q., Andersson, J.L., Moeller, S., Auerbach, E.J., Glasser, M.F., Hernandez, M., Sapiro, G., Jenkinson, M., Feinberg, D.A., Yacoub, E., Lenglet, C., Van Essen, D.C., Ugurbil, K., TEJ, B., Consortium WU-MH, 2013. Advances in diffusion MRI acquisition and processing in the human connectome project. Neuroimage 80, 125–143. https://doi.org/10.1016/j.neuroimage.2013.05.057. PubMed PMID: WOS:000322416000012.

Sreenivasan, V., Kumar, S., Pestilli, F., Talukdar, P., Sridharan, D., 2022. GPU-accelerated connectome discovery at scale. Nat. Comput. Sci. 2 (5), 298. https://doi.org/10.1038/s43588-022-00250-z. PubMed PMID: WOS:000888207600013.

Stanisz, G.J., Szafer, A., Wright, G.A., Henkelman, R.M., 1997. An analytical model of restricted diffusion in bovine optic nerve. Magn. Reson. Med. 37 (1), 103–111. https://doi.org/10.1002/mrm.1910370115. PubMed PMID: WOS:A1997VZ62900014.

Stejskal, E.O., Tanner, J.E., 1965. Spin diffusion measurements: spin echoes in the presence of a time-dependent field gradient. J. Chem. Phys. 42 (1), 288. https://doi.org/10.1063/1.1695690. PubMed PMID: WOS:A19656099500051.

Stepisnik, J., 1993. Time-dependent self-diffusion by NMR spin-echo. Physica B 183 (4), 343–350. https://doi.org/10.1016/0921-4526(93)90124-o. PubMed PMID: WOS:A1993LE11400001.

Szafer, A., Zhong, J.H., Gore, J.C., 1995. Theoretical-model for water diffusion in tissues. Magn. Reson. Med. 33 (5), 697–712. https://doi.org/10.1002/mrm.1910330516. PubMed PMID: WOS:A1995QV05800015.

Tariq, M., Schneider, T., Alexander, D.C., GandiniWheeler-Kingshott, C.A., Zhang, H., 2016. Bingham-NODDI: mapping anisotropic orientation dispersion of neurites using diffusion MRI. Neuroimage 133, 207–223. https://doi.org/10.1016/j.neuroimage.2016.01.046. PubMed PMID: WOS:000377048600018.

Tavor, I., Hofstetter, S., Assaf, Y., 2013. Micro-structural assessment of short term plasticity dynamics. Neuroimage 81, 1–7. https://doi.org/10.1016/j.neuroimage.2013.05.050. PubMed PMID: WOS:000322934400001.

Tax, C.M.W., Bastiani, M., Veraart, J., Garyfallidis, E., Irfanoglu, M.O., 2022. What's new and what's next in diffusion MRI preprocessing. Neuroimage, 249. https://doi.org/10.1016/j.neuroimage.2021.118830. PubMed PMID:WOS:000745464600002.

Topgaard, D., Malmborg, C., Soderman, O., 2002. Restricted self-diffusion of water in a highly concentrated W/O emulsion studied using modulated gradient spin-echo NMR. J. Magn. Reson. 156 (2), 195–201. https://doi.org/10.1006/jmre.2002.2556. PubMed PMID: WOS:000177480900005.

Torrey, H.C., 1956. Bloch equations with diffusion terms. Phys. Rev. 104 (3), 563–565. https://doi.org/10.1103/PhysRev.104.563. PubMed PMID: WOS:A1956WK97700002.

Tourbier, S., Rue-Queralt, J., Glomb, K., Aleman-Gomez, Y., Mullier, E., Griffa, A., Schöttner, M., Wirsich, J., Tuncel, M.A., Jancovic, J., 2022. Connectome mapper 3: a flexible and open-source pipeline software for multiscale multimodal human connectome mapping. J. Open Source Softw. 7 (74), 4248.

Tournier, J.D., Smith, R., Raffelt, D., Tabbara, R., Dhollander, T., Pietsch, M., Christiaens, D., Jeurissen, B., Yeh, C.H., Connelly, A., 2019. MRtrix3: a fast, flexible and open software framework for medical image processing and visualisation. Neuroimage, 202. https://doi.org/10.1016/j.neuroimage.2019.116137. PubMed PMID: WOS:000491861000070.

Tustison, N.J., Cook, P.A., Klein, A., Song, G., Das, S.R., Duda, J.T., Kandel, B.M., van Strien, N., Stone, J.R., Gee, J.C., Avants, B.B., 2014. Large-scale evaluation of ANTs and FreeSurfer cortical thickness measurements. Neuroimage 99, 166–179. https://doi.org/10.1016/j.neuroimage.2014.05.044. PubMed PMID: WOS:000339860000018.

Vangelderen, P., Despres, D., Vanzijl, P.C.M., Moonen, C.T.W., 1994. Evaluation of restricted diffusion in cylinders—phosphocreatine in rabbit leg muscle. J. Magn. Reson. B 103 (3), 255–260. https://doi.org/10.1006/jmrb.1994.1038. PubMed PMID: WOS:A1994ND15700007.

Veraart, J., Novikov, D.S., Fieremans, E., 2018. TE dependent diffusion imaging (TEdDI) distinguishes between compartmental T2 relaxation times. Neuroimage 182, 360–369.

Wang, Z.X., Zhang, S., Liu, C.X., Yao, Y.H., Shi, J.J., Zhang, J., Qin, Y.Y., Zhu, W.Z., 2019. A study of neurite orientation dispersion and density imaging in ischemic stroke. Magn. Reson. Imaging 57, 28–33. https://doi.org/10.1016/j.mri.2018.10.018. PubMed PMID: WOS:000458096100004.

Wedeen, V.J., Hagmann, P., Tseng, W.Y.I., Reese, T.G., Weisskoff, R.M., 2005. Mapping complex tissue architecture with diffusion spectrum magnetic resonance imaging. Magn. Reson. Med. 54 (6), 1377–1386. https://doi.org/10.1002/mrm.20642. PubMed PMID: WOS:000233655200008.

Yeh, C.H., Jones, D.K., Liang, X., Descoteaux, M., Connelly, A., 2021. Mapping structural connectivity using diffusion MRI: challenges and opportunities. J. Magn. Reson. Imaging 53 (6), 1666–1682.

Zeng, Q., Shi, F.N., Zhang, J.M., Ling, C.H., Dong, F., Jiang, B.A., 2018. A modified tri-exponential model for multi-b-value diffusion-weighted imaging: a method to detect the strictly diffusion-limited compartment in brain. Front. Neurosci., 12. https://doi.org/10.3389/fnins.2018.00102. PubMed PMID: WOS:000426079800001.

Zhan, L., Zhou, J.Y., Wang, Y.L., Jin, Y., Jahanshad, N., Prasad, G., Nir, T.M., Leonardo, C.D., Ye, J.P., Thompson, P.M., 2015. Alzheimer's dis neuroimaging. I. Comparison of nine tractography algorithms for detecting abnormal structural brain networks in Alzheimer's disease. Front. Aging Neurosci., 7. https://doi.org/10.3389/frogi.2015.00048. PubMed PMID: WOS:000353909300001.

Zhang, Y.Y., Brady, M., Smith, S., 2001. Segmentation of brain MR images through a hidden Markov random field model and the expectation-maximization algorithm. IEEE Trans. Med. Imaging 20 (1), 45–57. https://doi.org/10.1109/42.906424. PubMed PMID: WOS:000167324900005.

Zhang, H., Schneider, T., Wheeler-Kingshott, C.A., Alexander, D.C., 2012. NODDI: practical in vivo neurite orientation dispersion and density imaging of the human brain. Neuroimage 61 (4), 1000–1016. https://doi.org/10.1016/j.neuroimage.2012.03.072. PubMed PMID: WOS:000305920600028.

Glossary

Antipodal Diametrical opposition, relating to the antipodes or situated at opposite sides of a circle or sphere.

Attenuation Weakening

Bias Deviation of results or inferences from the truth, or processes leading to such systematic deviation. Any trend in the collection, analysis, interpretation, publication, or review of data that can lead to conclusions that are systematically different from the truth.

Biomarker A characteristic that is objectively measured and can be viewed as an indicator of a normal biological process, a disease process, or a typical response to a drug or therapy; for example, blood pressure.

Correlation A statistical term to describe the relationship between two variables (e.g., calcium intake and bone growth).

Demyelination A loss of myelin with relative preservation of axons

Deterministic A process or model in which the output is determined solely by the input and initial conditions, thereby always returning the same results (opposed to stochastic).

Deviation Failure to meet a critical limit.

Diffusivity Diffusivity is a rate of diffusion. A measure of the rate at which (water) particles can spread in brain tissue. (Please see Parameters of the compartment model functions section for more details.)

Diffusion-Weighted Imaging (DWI) Diffusion-weighted magnetic resonance imaging (DWI or DW-MRI) is the use of specific MRI sequences as well as software that generates images from the resulting data that use the diffusion of water molecules to generate contrast in MR images.

Distribution The frequency and pattern of health-related characteristics and events in a population. In statistics, the observed or theoretical frequency of values of a variable.

Efficacy How well something works in relation to predefined standards or expectations.

Fractional Anisotropy (FA) It is a scalar value between zero and one that describes the degree of anisotropy of a diffusion process. Please see Fig. 2 for more details.

Free water fraction (Ffw) It is the measure of volume fraction contributed by the isotropic compartment in the NODDI (Zhang et al., 2012) model (v_{iso}). It is used to estimate the extent of CSF contamination (Kraguljac et al., 2023).

Heterogeneity (of data) In statistics, homogeneity and its opposite, heterogeneity, arise in describing the properties of a dataset, or several datasets. They relate to the validity of the often-convenient assumption that the statistical properties of any one part of an overall dataset are the same as any other part. In meta-analysis, which combines the data from several studies, homogeneity measures the differences or similarities between the several studies.

Mean Diffusivity (MD) It a scalar measure of the total diffusion within a voxel. Please see Fig. 2 for more details.

Natural Language Processing (NLP) It is a field in artificial intelligence and is also related to linguistics. On a high level, the goal of NLP is to program computers to automatically understand human languages and also to automatically write/speak in human languages.

Neurite Density Index (NDI) It is a measure that quantifies the packing density of axons or dendrites, calculated using NODDI (Zhang et al., 2012) model. Please see NODDI/Bingham-NODDI section.

Orientation Dispersion Index (ODI) A NODDI (Zhang et al., 2012) model derived measure, which assesses the orientational coherence of neurites. Please see Neurite orientation and dispersion section for more details.

Pathological Indicative of, or caused by, a disease or condition.

Problem Formulation The process of defining the specific problem being addressed. It involves articulating a question and defining how it may be answered (e.g., by identifying the endpoints to be measured).

Range The difference between the largest and smallest values in a distribution. In common use, the span of values from smallest to largest.

Sampling Sampling is the process of selecting a few cases from all the cases in a particular group or universe.

Sensitivity It is the probability of a positive test result, conditioned on the individual truly being positive (true positive rate).

Specificity It is the probability of a negative test result, conditioned on the individual truly being negative (true negative rate)

Validity The degree to which a measurement actually measures or detects what it is supposed to measure.

Variable Any characteristic or attribute that can be measured.

Variance A measure of the dispersion shown by a set of observations, defined by the sum of the squares of deviations from the mean, divided by the number of degrees of freedom in the set of observations.

Dynamic brain network models: How interactions in the structural connectome shape brain dynamics

Joana Cabral[a] and John D. Griffiths[b,c]

[a]*Life and Health Sciences Research Institute (ICVS), Medical School, University of Minho, Braga, Portugal,* [b]*Krembil Centre for Neuroinformatics, Centre for Addiction and Mental Health, Toronto, ON, Canada,* [c]*Department of Psychiatry & Institute of Medical Sciences, University of Toronto, Toronto, ON, Canada*

Introduction to dynamic brain network models

Dynamic brain network models come in many different forms, ranging from highly abstract mathematical models to moderately detailed and biologically realistic simulations of interacting neuronal populations. They can be used to explore a wide range of questions, from the basic principles of macroscale organization such as collective brain rhythms, to the prediction of therapeutic outcomes following clinical interventions such as (invasive or noninvasive) neurostimulation, surgery, and drugs.

The main theoretical context for brain network models consists in the fact that brain areas do not work in isolation but are embedded in a network, and the activity in a given brain area is affected by the activity of the brain regions to which it is coupled. These in turn are also coupled to other brain regions, and so it is crucial to consider the entire brain even when investigating the function of a specific brain region (Sporns and Tononi, 2001). In other words, neuroscientists face a version of the so-called "N-body problem," which lies at the foundations of complexity science and chaos theory, for which exact mathematical solutions cannot be adequately obtained, and henceforth, computer simulations are a powerful tool to investigate the complex behavior of interacting units. Indeed, the simulated dynamics resulting from brain network models link directly with studies describing interesting dynamic phenomena occurring in complex networks, irrespective of their scale and nature (Boccaletti et al., 2006; Newman et al., 2006; Arenas et al., 2006; Yeung and Strogatz, 1999).

Computational and Network Modeling of Neuroimaging Data. https://doi.org/10.1016/B978-0-443-13480-7.00001-6

Generally, a brain network model is a system of interacting dynamical units, each representing a distinct brain subsystem, coupled together according to a specific wiring diagram representing the axonal tracts that physically connect brain subsystems together. Brain network models were introduced in the first decade of the 21st century, first using the wiring diagram obtained from ex vivo tracing studies of one hemisphere of the macaque brain (Kötter, 2004), and rapidly extended to humans following the development of in vivo human tractography studies using diffusion MRI, providing the so-called structural connectome (Hagmann et al., 2008), see (Griffiths et al., 2022).

Over the years, brain network models have evolved in two primary and divergent directions. On the one hand, mechanistic models aim to identify essential properties and fundamental mechanisms that give rise to observed patterns. This is usually achieved by replicating with a minimal number of variables certain qualitative features of brain activity such as long-range correlations, plasticity mechanisms, collective brain rhythms, nonstationary dynamics, or transitions between different brain states. On the other hand, predictive models focus on enhancing the fit with recorded data by refining parameters and increasing biophysical realism, opening the door for personalized planning of therapeutic interventions.

Although brain network models can vary substantially in terms of biological realism, spatial scale, and assumptions, they are generally built by specifying the following three ingredients:

I. The *local dynamics* at the network nodes, which can exhibit fixed point, oscillatory, or chaotic equilibria with varying degrees of biochemical and biophysical realism and heterogeneity.
II. The *coupling function*, which defines how each node responds to the activity of the nodes coupled to it, considers for instance, excitatory or diffusive coupling, and includes or neglects the time delays between nodes.
III. The *connectivity matrix*, which defines the wiring diagram of the network, weighs each link as a function of the structural neuroanatomical connections detected between each pair of brain areas.

In the following sections, we provide an overview of the main distinct models that have been proposed in the field, highlighting the insights they have brought to neuroscience. In addition to exploring the nontrivial link between structural and functional connectivity, these models have been used to better understand the self-organizing principles of brain activity at the macroscale, the differentiating factors across brain states, and ultimately their value for personalized planning of surgical or pharmacological interventions (Vohryzek et al., 2022; Kurtin et al., 2023; Lang et al., 2023) (Fig. 1).

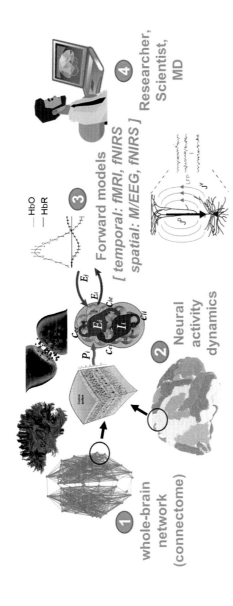

FIG. 1

Brain network modeling schematic. (1) Whole-brain activity is modeled with a large-scale brain network (anatomical connectome), with nodes representing brain regions and edges representing long-range white matter fiber tracts. (2) Neural activity dynamics at each network node is described by a mesoscopic neural model, summarizing local cortical microcircuitry in terms of, for example, interacting excitatory and inhibitory neural populations. (3) Hemodynamic and electromagnetic signals generated by local neural activity are mapped on to channel-level measurements via spatial (M/EEG ≥ Maxwell equations + electromagnetic sensitivity matrix), temporal (fMRI ≥ Balloon-Windkessel model), or spatiotemporal (fNIRS fDOT ≥ balloon + optical sensitivity matrix) measurement models. (4) Modeler/research scientist/clinician end-user uses the above process (technically or nontechnically presented, as needed) to generate personalized brain simulations and to estimate brain state and physiological parameters from neuroimaging and neurophysiology measurements.

Examples of dynamic brain network models
Emergence of correlated brain activity patterns in the resting state

In the advent of brain Connectomics research, one crucial question concerned whether the long-range correlations detected in fMRI signals between brain areas (Biswal et al., 1995)—generally referred to as functional connectivity—could be directly related to neuroanatomical connectivity. Christopher Honey, Olaf Sporns, Patrick Hagmann, and colleagues (Honey et al., 2009) used the by-then recently-available whole-brain structural connectome obtained from applying advanced tractography algorithms to in vivo diffusion-weighted MRI from humans. Honey et al. (2009) demonstrated that, despite the differences between structural and functional connectivity maps, human resting-state functional connectivity could be predicted from structural connectivity using a dynamic brain network model. Interestingly, comparable performance in reconstructing fMRI functional connectivity was observed in models with rich (nonlinear chaotic) local dynamics and models with simpler (linear) local dynamics—an unexpected result that has been repeatedly confirmed in subsequent studies over the past decade (Cabral et al., 2011; Hansen et al., 2015; Deco et al., 2017).

Until then, previous dynamic brain network models had used a connectome obtained from a collection of ex vivo tracing studies on the macaque brain, termed CoCoMac (Stephan et al., 2001; Kötter, 2004), which had served to demonstrate how the specific network topology of the mammalian brain could shape brain activity in space and time, as previously hypothesized and mimicked using synthetic network topologies (Jirsa, 2004). Although not directly comparing with fMRI functional connectivity, the simulations on the CoCoMac network were relevant to put forward the key role of the connectivity structure and the time delays between brain regions in shaping slow and macroscale spatiotemporal patterns in brain dynamics (Honey et al., 2007; Ghosh et al., 2008a,b; Deco et al., 2009).

Dynamic brain network models have served to demonstrate how the transient correlations observed in resting-state fMRI—usually referred to as dynamic functional connectivity—can be linked to metastable synchronization between brain regions, a phenomenon observed in systems of coupled oscillators in nature. Adapting the Kuramoto model of coupled oscillators (Kuramoto, 1984; Acebrón et al., 2005) to incorporate realistic coupling weights and time delays between brain areas, simulations revealed a critical range of coupling and delays, where the system is positioned between incoherence and synchrony, with subsets of nodes transiently synchronizing and desynchronizing even in the absence of noise (Cabral et al., 2011). During periods of transient synchrony, a collective slowdown in the frequency of oscillations generates correlated amplitude fluctuations, which could explain the slow and spatiotemporally organized fluctuations detected in EEG/MEG (Cabral et al., 2014). This mechanism was later demonstrated to be robust in more realistic settings, i.e., by replacing phase oscillators with damped oscillators or even Wilson-Cowan units (Cabral et al., 2022; Castaldo et al., 2023). The switching behavior of functional connectivity over time has also been reproduced using a brain network model without time delays, by placing in each node a variant of the "dynamic mean field" model (Wong and Wang, 2006; Deco et al., 2013), minimally modified to introduce bi-stability between high and low firing rate states (Hansen et al., 2015). This approach enhanced the dynamic repertoire of the whole-brain model, leading to spontaneous switching between functional connectivity states, which shared qualitative features with known resting state networks.

Traveling waves and spiral waves have also been shown to emerge spontaneously from a brain network model of coupled neural masses, demonstrating that metastability and waves are compatible dynamical regimes that can be explained with a unified mechanism (Roberts et al., 2019). For a critical value of coupling weight between brain areas, a rich array of three-dimensional wave patterns could be observed even in the absence of noise, with nonlinear instabilities driving constant reconfigurations of underlying phase flows.

It is also important to note that some features of the signals simulated using brain network models can be solved analytically from the model equations without the need to numerically simulate the network behavior in a computer. Resting-state networks observed in fMRI signals have been shown to be predicted from the brain network structure using established algebraic expressions for the steady-state correlation structure of a discrete- or continuous-time linear system given its connectivity matrix (e.g., Honey et al., 2009; Robinson, 2012; Deco et al., 2013) or its connectivity Laplacian (Atasoy et al., 2016; Abdelnour et al., 2018). Additionally, incorporating time delays in these linear equations improves the detection of canonical functional networks (Xie et al., 2021). These studies provide supporting evidence that the brain network structure plays a key role in shaping the emergence of functional networks observed with fMRI, and furthermore that "slow" features of resting-state dynamics such as static long-term correlation patterns are not as dependent as faster time-varying features on the dynamical richness introduced by nonlinearity, time delays, and metastability.

Perturbation with electrical/magnetic/acoustic/optical stimulation

In both experimental and clinical contexts, electromagnetic stimulation techniques such as transcranial magnetic stimulation (TMS), transcranial electrical stimulation (TES), and deep brain stimulation (DBS) are commonly employed in an attempt to characterize and modulate neural activity.

Therapeutically, neurostimulation paradigms are highly successful in some conditions (e.g., DBS for tremor in Parkinson's disease) but only moderately effective in others (e.g., TMS and DBS for major depressive disorder). Part of the reason limiting the success of brain stimulation therapies can be attributed to a lack of comprehensive mechanistic understanding, compounded by a multitude of potential combinations of stimulation parameters, such as location, frequency, duration, and voltage, that cannot be iteratively tested in vivo due to ethical constraints. Brain network models offer the opportunity to test this extensive array of stimulation parameters and predict its impact on brain signals. However, the effectiveness of planning stimulation in a model depends on its alignment with the biophysical reality, on an adequate description of heterogeneous neural processes, and on how the neuroanatomy and other physiological processes configure the collective dynamics (Kurtin et al., 2023; Vohryzek et al., 2023).

Brain stimulation paradigms, across all invasive and noninvasive modalities, can be broadly classified into two general types: immediate effects and persistent effects. Immediate effects are brain activity changes observed either during a block of sustained repetitive stimulation (steady-state response) or as a transient following a single stimulation pulse (evoked response). Persistent effects are responses that are visible minutes, hours, or days after cessation of the input stimuli and are thus, by definition, instances of stimulation-induced neural plasticity, which at the level of synaptic weights may be positive (long-term potentiation; LTP) or negative (long-term depression; LTD). The majority of brain network modeling work in this area has focused on immediate stimulation effects, leaving the investigation of enduring LTP/LTD effects relatively underexplored.

A notable line of work exploring immediate stimulation effects has been led by Spiegler and colleagues (Fig. 2C), who investigated this using brain network models in the human (Spiegler et al., 2016) and mouse brain (Spiegler et al., 2020). The latter of these is particularly notable due to its use of extremely high-quality and rarely-used experimental data, high-resolution and "ground-truth" anatomical connectivity data (Allen Mouse Brain Atlas) and whole-cortex wide-field calcium imaging of neural activity, both of which allow superior spatial and temporal resolution than the noninvasive methods used in humans. Two key questions of interest in this work were: (a) the balance between local (cortical neural field) vs long-range (white matter) connectivity and (b) whether canonical ("dynamically responsive") resting-state networks could be evoked by stimulation at multiple cortical locations. To study these relationships, brain network models with Fitzhugh-Nagumo equations for local dynamics were developed, and tens of thousands of in silico stimulation experiments were performed, an exercise that would be completely infeasible to undertake experimentally. The conclusions are broadly similar in the human and mouse studies, that a set of "dynamically responsive networks," closely resembling resting state networks seen in resting-state functional connectivity, are observable in stimulation-evoked responses for a wide variety of stimulus locations.

Focusing more specifically on established immediate stimulation effects in humans, Bensaid et al. (2019) used brain network modeling to demonstrate how the modulation of thalamo-cortical connectivity could explain differences in the TMS-evoked potential (TEP) EEG response during wakefulness vs sleep or vs patients with disorders of consciousness. In related work, Momi et al. (2023) studied the role of recurrent feedback in shaping specific aspects of the TEP waveform (Fig. 2A). Using a brain network model with tractography-based anatomical connectivity and Jansen-Rit cortical units, these authors demonstrated that the N100 component within the TEP is heavily dependent on recurrent activity reverberating across the connectome network away from, and then re-converging on, the stimulated region. The magnitude of this canonical brain stimulation response was also found to be strongly influenced by the level of inhibitory feedback within the cortical circuit. This study is also notable as it demonstrated the feasibility of fitting whole-brain network models to subject-specific TEP waveforms, employing a novel machine learning-based parameter estimation approach. Earlier work from the same authors (Griffiths and Lefebvre, 2019; Griffiths et al., 2020) examined immediate steady-state stimulation responses to periodic drive. Using a model of resting-state EEG activity able to reproduce frequency band-specific functional connectivity as well as an alternation between "idling" alpha and "active" gamma states, these studies observed that the alpha state exhibits greater stability and is less susceptible to periodic stimulation effects than the gamma state, consistent with the notion of alpha as an active neurocognitive suppression mechanism.

Work on brain network models of plasticity effects in computational brain stimulation modeling has tended to focus on the mechanisms and dynamics of plasticity processes at the level of meso-scale local dynamics (Shouval et al., 2002; Fung and Robinson, 2013; Wilson et al., 2018, 2021). Notably, Wilson et al. (2018, 2021), extending the calcium-dependent synaptic plasticity formulations of previous authors (Fung and Robinson, 2013; Shouval et al., 2002), developed a model of how multiple minutes of repetitive TMS (rTMS) lead to LTP or LTD depending on waveform characteristics including stimulation frequency, (for theta-burst rTMS specifically) carrier frequency, and number of pulses per burst. A key component of these models is the "'omega function" (Fig. 2A), which implements the experimental observation that lowered intracellular calcium levels lead to synaptic depression, raised levels lead to synaptic enhancement, and baseline levels have no effect. Brain network models have also been extended to integrate Hebbian learning mechanisms at the level of long-range excitatory connections between brain regions, with a view to explain experimental findings of frequency-dependent plasticity in neuroimaging data. These model results support the hypothesis that the intrinsic resonant properties of the brain network may play a role in plasticity processes at the whole-brain level (Lea-Carnall et al., 2017).

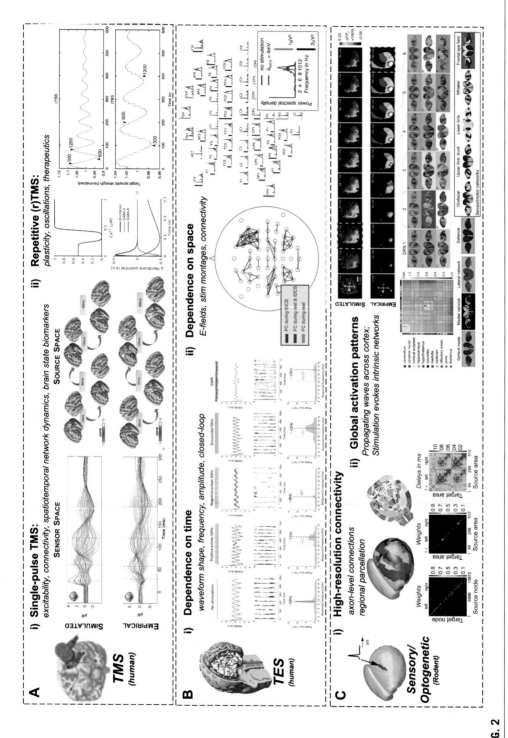

FIG. 2

Examples of brain stimulation modeling studies. (A) Models of transcranial magnetic stimulation (TMS) stimulation; (i) example of recorded TMS evoked potentials in EEG sensors (top) and simulated signals (bottom) (Momi et al., 2023), (ii) same for brain signals reconstructed in source space. (B) Models of transcranial electric stimulation (TES) serving to study the impact of stimulation on either (i) the signal responses over time (Hutt et al., 2018) or (ii) alterations in the spatial configuration of the signals (Kunze et al., 2016). (C) Models of sensory/optogenetic stimulation; (i) high spatial resolution (Spiegler et al., 2020) and (ii) map of empirical vs simulated activity and activation of dynamically responsive networks following optogenetic stimulation in mice (Spiegler et al., 2020).

Psychopharmacological perturbation and the role of the neuromodulatory systems

Brain network models have also been used to explain large-scale effects of molecular neuromodulation observed in neuroimaging data (Kringelbach et al., 2020; Shine et al., 2021). Considering how a given neurotransmitter affects neuronal activity at the node level, it is possible to incorporate those effects in the model equations, simulate the resulting outcome at the network level and evaluate its predictive value by comparing it with neuroimaging data (Fig. 3). For instance, using the receptor density maps of serotonin obtained using PET imaging, Deco and colleagues demonstrated how the alterations observed at the level of dynamic functional connectivity following the infusion of LSD (which binds to specific serotonin receptors) could be predicted by selectively tuning the slope of the sigmoid function converting excitatory currents into neuronal firing rates in each node of the network as a function of receptor density (Deco et al., 2018).

Assumptions of brain network models

Brain network models are based on several key assumptions, including:

- *The brain is a complex network:* Brain network models assume that the brain is a complex network made up of interconnected regions or nodes. These nodes can represent individual neurons, groups of neurons, or larger brain regions.
- *Nodes interact through connections:* Brain network models assume that nodes in the brain are connected through some form of communication or interaction such as the synaptic transmission of electrical or chemical signals via long-range axonal projections.
- *Network structure affects function*: Brain network models assume that the structure of the network, including the pattern of connections between nodes, has a direct impact on brain function. For example, some models suggest that certain network structures are more efficient for processing information, while others may be better suited for generating complex patterns of activity.

Overall, brain network models provide a framework for understanding how different regions of the brain work together to produce complex behaviors and cognitive processes. By simulating the behavior of these networks, researchers can gain insights into how the brain functions under different conditions and develop new strategies for treating neurological and psychiatric disorders.

Despite the success and development of the brain network modeling paradigm over the past decade, it remains unclear whether network models per se are the best representation of brain activity in order to understand its function. An alternative and reemerging scenario is aligned with wave/field theory (Jirsa and Haken, 1996),

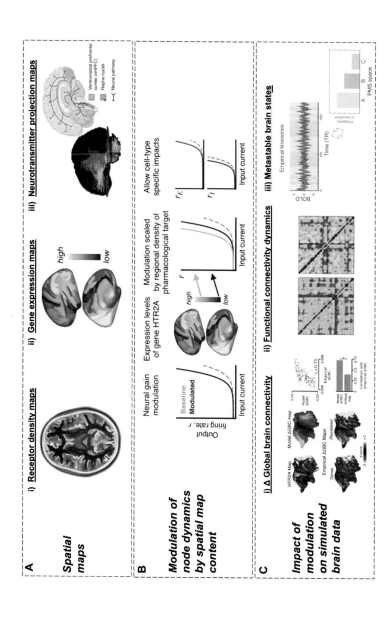

FIG. 3

Modeling brain neuropharmacology interventions. (A) The effect of pharmacological neuromodulation can be heterogeneously modulated by altering the local dynamics in each brain region as a function of a given spatial map, i.e., taking into account; (i) the density of neurotransmitter receptors as captured with PET imaging (Deco et al., 2018), (ii) the expression of specific genes (Burt et al., 2021), or (iii) the projection maps of neurotransmitters (Shine et al., 2021). (B) The impact of neuromodulation on the local node dynamics can be modeled by changing how the firing rate depends on the input current. For instance, a higher concentration of a given gene can be modeled by increasing the slope of the gain function (Burt et al., 2021). (C) The impact of neuromodulation in neuroimaging data can be assessed by comparing the simulation results with empirical data, either in terms of (i) the global brain connectivity (GBC) (Burt et al., 2021), (ii) the functional connectivity dynamics (Deco et al., 2017), or (iii) the relative occupancy of different states of functional connectivity, also known as probabilistic metastable substates (PMS) (Kringelbach et al., 2020).

where the brain is treated as a continuum system and its activity is described by a macroscale wave function reflecting the superposition of fundamental wave patterns extending across the whole brain evolving in time (Pang et al., 2023; Cabral et al., 2023). Although the network and the wave/field approaches are not necessarily mutually exclusive and may reflect different perspectives of a common process, it is important to embrace the insights that field theories may provide in order to improve our models of macroscale brain activity.

Building, testing and interpreting dynamic brain network models

Building brain network models is a complex and interdisciplinary endeavor that requires expertise in neuroscience, mathematics, computational modeling, and often machine learning. Collaboration between researchers with different skill sets is crucial for successful model development. The process involves several steps, which we have enumerated briefly in the following:

1. Define scope and goals of the project

   This consists in determining the specific aspects of brain function one wants to model and the hypothetical mechanistic scenarios. This could range from oscillatory activity, functional connectivity, neurotransmitter modulation, stimulation outcome, and lesion impact, among others.

2. Choose an appropriate modeling approach

   A suitable modeling approach should have an adequate level of biophysical realism to address the specific goals of the project, balancing the right level of complexity to capture the desired emergent behaviors with sufficiently simplicity to allow interpretability. This involves defining the scale of brain parcellation, the node model, the coupling function and the variable parameters that will be used to regulate the dynamics of the system.

3. Consider nodal behavior

   Regarding the node model, the equations used to describe the activity of neural populations at each individual node can vary widely between studies, ranging from more realistic biophysical models considering the properties of active and passive membrane conductance as well as neurotransmitter concentrations and decay times (Honey et al., 2009; Naskar et al., 2021), to reduced models of neural populations such as the Wilson-Cowan model (Wilson and Cowan, 1972; Deco et al., 2009; Castaldo et al., 2023), through to linearized, algebraic model forms for phenomena such as power spectra (Lopes da Silva et al., 1974; Robinson et al., 2003; Abeysuriya et al., 2015) and functional connectivity (Robinson, 2012; Abdelnour et al., 2018; Verma et al., 2022).

4. Consider coupling functions

    The coupling function may consider the excitatory nature of long-range axonal connections, such that the response function of a node has a sigmoidal shape or instead assume a diffusive coupling, where the response of a node is a function of the difference with respect to coupled nodes. In the case of excitatory coupling, the general form is to consider a threshold input current above which the firing rate of a node will increase with increasing input current, up to a point where the firing rate reaches a plateau of high firing rate. However, given that long-range connections are usually assumed to be purely excitatory, this approach tends to lead to a globally excited state, and therefore, to reach an adequate fit with spontaneous brain activity, a local feedback inhibition is added to keep the global network in a low-firing state.

5. Data gathering

    Collect data from various sources such as electrophysiological recordings, functional MRI (fMRI) data, neuroanatomical data, and other experimental observations. This data will serve as the foundation for your model.

6. Model Development

    Based on your chosen approach, develop the model's architecture and equations. This involves defining the properties of neurons, synapses, connectivity patterns, and other relevant components. This step may require adjusting parameters to match the available experimental data.

7. Structural connectivity, equations describing local node dynamics, and coupling term

    Since the connection topology changes when anatomical scales are traversed, so will the corresponding network dynamics. As a consequence, different types of networks are encountered at different levels of neural organization.

8. Simulation and Validation

    Implement the model in a computational framework and run simulations. Compare the model's behavior and output with experimental data to validate its accuracy and performance. This may involve tweaking parameters and model structure to improve the fit.

9. Parameter Tuning and Optimization

    Fine-tune the model's parameters to better match experimental data and produce desired behaviors. This process might involve using optimization algorithms to find the parameter values that minimize the difference between model predictions and empirical observations.

10. Analysis and Interpretation

    Analyze the model's behavior and output to gain insights into the neural mechanisms and processes being simulated. This could involve examining firing rates, network dynamics, information flow, and other relevant measures.

11. Hypothesis Testing and Predictions

    Use the validated model to test hypotheses about the brain's functioning or make predictions about how the brain might respond under different conditions. This step helps refine our understanding of neural processes and guides further experimental research.

12. Model Refinement and Iteration

    Models are rarely perfect on the first attempt. Refine your model based on new data, feedback from the neuroscience community, and improved computational methods. Iterative refinement helps create more accurate and meaningful models over time.

Dynamic network models consist of sets of coupled differential equations that can be solved in any simulation environment, including, among others, C++, Matlab, Python, Julia, etc. The Virtual Brain (www.thevirtualbrain.org) is a software developed with the aim to *"Simulate a human brain, right on your PC!,"* facilitating the access of nonexperts to the endeavors of brain network modeling, including several options for brain parcellations and node models (Sanz-Leon et al., 2015). For researchers with programming skills, it is recommended to develop their own scripts, making use of some toolboxes, such as the Brain Dynamics Toolbox (https://bdtoolbox.org) (Heitmann and Breakspear, 2022), which offers examples for solving coupled differential equations.

Pitfalls of brain network modeling

Brain network modeling has emerged as a powerful tool for understanding the intricate workings of the human brain. However, it is essential to acknowledge that this approach comes with its own set of challenges and pitfalls that must be carefully navigated to ensure accurate and meaningful insights. Here we list some of the key pitfalls associated with brain network modeling.

1. Oversimplification of Complexity

   One of the primary pitfalls in brain network modeling is the risk of oversimplifying the complexity of the brain's intricate neural connections. While simplification is often necessary to depict the most primitive mechanisms and to make models computationally manageable, overly simplistic representations can lead to inaccurate conclusions. Failing to account for the rich diversity of neuron types, synaptic strengths, and dynamic interactions between regions can result in unrealistic predictions.

2. Lack of Ground Truth

   Brain network models heavily rely on empirical data, but the availability of accurate ground-truth data is often limited. This lack of ground truth can lead to models that are not adequately validated or calibrated. Without proper validation, it becomes challenging to ascertain whether the model's predictions truly reflect the underlying biological reality.

3. **Parameter Sensitivity and Overfitting**

     Brain network models often involve a multitude of parameters that control the strength of connections, time constants, and other dynamic properties. Fitting these parameters to experimental data can be prone to overfitting, where the model captures noise rather than meaningful patterns. Moreover, slight changes in parameter values can lead to dramatically different model outcomes, making the models sensitive to initial conditions and potentially less generalizable.

4. **Ignoring Temporal Dynamics**

     The brain's activity is highly dynamic, with neural processes occurring across various timescales. Neglecting temporal dynamics or oversimplifying them can lead to unrealistic network behaviors. For instance, ignoring delays in signal transmission can result in overly synchronous network activity that does not align with empirical observations.

5. **Limited Incorporation of Individual Variability**

     Individual differences in brain anatomy, connectivity, and function are significant factors that shape neural activity. However, many brain network models treat brains as homogeneous entities, ignoring these critical sources of variability. Neglecting individual variability can lead to models that fail to capture the full range of human brain functioning.

6. **Interpretability and Validation**

     Brain network models can produce intricate and complex results, making their interpretation and validation challenging. Ensuring that the model outcomes are biologically plausible and interpretable is crucial. Validating the model's predictions against multiple datasets and experimental designs can help establish the model's credibility.

7. **Incorporating Realistic Stimulation and Intervention**

     Interventions such as brain stimulation techniques are often explored using network models. However, accurately representing these interventions and their effects within the model can be intricate. Failure to capture the biophysics of interventions can lead to misleading conclusions about their potential efficacy.

8. **Computational Complexity**

     Developing and simulating brain network models often requires significant computational resources. As the complexity of models increases, so does the demand for computational power and memory. This can limit the accessibility and practicality of these models for researchers with limited resources.

In conclusion, while brain network modeling offers valuable insights into the brain's functional and structural organization, researchers must be aware of the potential pitfalls associated with this approach. Navigating these pitfalls requires a balanced consideration of model complexity, empirical validation, the incorporation of individual variability, and careful interpretation of results. Addressing these challenges will contribute to the development of more accurate and meaningful brain network models that advance our understanding of the brain's complexities.

Open challenges and future directions

Whole-brain modeling and computational connectomics have raised great expectations for understanding brain disorders at a causal mechanistic level and provide novel, more effective therapeutic interventions, e.g., through drug discovery and new targets for deep brain stimulation (Deco and Kringelbach, 2014; Braun et al., 2018). However, although whole-brain network models have brought remarkable insights into brain organization and dynamics over the last decades, it is important to temper the notion of a "successful brain network model" in the sense that models have served mostly to test the validity of hypothetical mechanistic scenarios or to simulate the impact of perturbation, but their predictive value still needs to be improved and validated in preclinical models before they can serve to design successful interventions in clinical settings (Bassett and Sporns, 2017).

A central goal of the field is to build personalized replicas—or "digital twins"—of the human brain to test different types of interventions to define personalized therapeutic strategies for different brain disorders (Jirsa et al., 2010; Kurtin et al., 2023; Vohryzek et al., 2023; Lang et al., 2023; Luppi et al., 2023). However, whether improvements in computer power or the use of machine-learning algorithms to optimize parameter search will be sufficient, or an entirely new paradigm shift is needed, remains an open question.

In terms of insight for cognitive neuroscience, brain network models have served to support the view of mental function as an emergent property resulting from interactions between distinct brain regions rather than a localized neuronal process (Deco et al., 2011). Describing the brain as a collective dynamical system, composed of multiple units capable of engaging in a multitude of collective activity patterns, links with theories of emergence, which may ultimately open up to a better understanding of mental function and coordinated behavior. Still, the link between models of brain activity and theories of consciousness remains correlational, and more efforts are needed to understand how brain network models can serve to solve still unanswered mechanistic questions regarding the emergence of consciousness and cognitive function (Cabral et al., 2017; Demertzi et al., 2019).

Take-home points

- Brain network models serve to simulate the behavior of interacting brain areas.
- There are a variety of models for node dynamics and coupling functions, combined with choices on free and fixed parameter values.
- Models explain some features of neuroimaging data, but their predictive value remains insufficiently validated to translate into clinics.
- Personalized brain network models are poised to turn into digital twins to optimize pharmacological and/or electromagnetic interventions.

References

Abdelnour, F., Dayan, M., Devinsky, O., Thesen, T., Raj, A., 2018. Functional brain connectivity is predictable from anatomic network's Laplacian eigen-structure. Neuroimage 172, 728–739. Available at: https://doi.org/10.1016/j.neuroimage.2018.02.016.

Abeysuriya, R.G., Rennie, C.J., Robinson, P.A., 2015. Physiologically based arousal state estimation and dynamics. J. Neurosci. Methods 253, 55–69. Available at: https://doi.org/10.1016/j.jneumeth.2015.06.002.

Acebrón, J.A., Bonilla, L.L., Pérez Vicente, C.J., Ritort, F., Spigler, R., 2005. The Kuramoto model: a simple paradigm for synchronization phenomena. Rev. Mod. Phys. 77, 137–185. Available at: https://doi.org/10.1103/RevModPhys.77.137.

Arenas, A., Díaz-Guilera, A., Pérez-Vicente, C.J., 2006. Synchronization reveals topological scales in complex networks. Phys. Rev. Lett. 96, 114102. Available at: https://doi.org/10.1103/PhysRevLett.96.114102.

Atasoy, S., Donnelly, I., Pearson, J., 2016. Human brain networks function in connectome-specific harmonic waves. Nat. Commun. 7, 10340. Available at: https://doi.org/10.1038/ncomms10340.

Bassett, D.S., Sporns, O., 2017. Network neuroscience. Nat. Neurosci. 20, 353–364. Available at: https://doi.org/10.1038/nn.4502.

Bensaid, S., Modolo, J., Merlet, I., Wendling, F., Benquet, P., 2019. COALIA: a computational model of human EEG for consciousness research. Front. Syst. Neurosci. 13, 59. Available at: https://doi.org/10.3389/fnsys.2019.00059.

Biswal, B., Yetkin, F.Z., Haughton, V.M., Hyde, J.S., 1995. Functional connectivity in the motor cortex of resting human brain using echo-planar MRI. Magn. Reson. Med. 34, 537–541. Available at: https://doi.org/10.1002/mrm.1910340409.

Boccaletti, S., Latora, V., Moreno, Y., Chavez, M., Hwang, D., 2006. Complex networks: structure and dynamics. Phys. Rep. 424, 175–308. Available at: https://linkinghub.elsevier.com/retrieve/pii/S037015730500462X.

Braun, U., Schaefer, A., Betzel, R.F., Tost, H., Meyer-Lindenberg, A., Bassett, D.S., 2018. From maps to multi-dimensional network mechanisms of mental disorders. Neuron 97, 14–31. Available at: https://doi.org/10.1016/j.neuron.2017.11.007.

Burt, J.B., Preller, K.H., Demirtas, M., Ji, J.L., Krystal, J.H., Vollenweider, F.X., Anticevic, A., Murray, J.D., 2021. Transcriptomics-informed large-scale cortical model captures topography of pharmacological neuroimaging effects of LSD. Elife, 10. Available at: https://doi.org/10.7554/eLife.69320.

Cabral, J., Hugues, E., Sporns, O., Deco, G., 2011. Role of local network oscillations in resting-state functional connectivity. Neuroimage 57, 130–139. Available at: https://doi.org/10.1016/j.neuroimage.2011.04.010.

Cabral, J., Luckhoo, H., Woolrich, M., Joensson, M., Mohseni, H., Baker, A., Kringelbach, M.L., Deco, G., 2014. Exploring mechanisms of spontaneous functional connectivity in MEG: how delayed network interactions lead to structured amplitude envelopes of band-pass filtered oscillations. Neuroimage 90, 423–435. Available at: https://doi.org/10.1016/j.neuroimage.2013.11.047.

Cabral, J., Vidaurre, D., Marques, P., Magalhães, R., Silva Moreira, P., Miguel Soares, J., Deco, G., Sousa, N., Kringelbach, M.L., 2017. Cognitive performance in healthy older adults relates to spontaneous switching between states of functional connectivity during rest. Sci. Rep. 7, 5135. Available at: https://doi.org/10.1038/s41598-017-05425-7.

Cabral, J., Castaldo, F., Vohryzek, J., Litvak, V., Bick, C., Lambiotte, R., Friston, K., Kringelbach, M.L., Deco, G., 2022. Metastable oscillatory modes emerge from synchronization in the brain spacetime connectome. Commun. Phys. 5, 1–13. Available at: https://www.nature.com/articles/s42005-022-00950-y. (Accessed 19 September 2023).

Cabral, J., Fernandes, F.F., Shemesh, N., 2023. Intrinsic macroscale oscillatory modes driving long range functional connectivity in female rat brains detected by ultrafast fMRI. Nat. Commun. 14, 375. Available at: https://doi.org/10.1038/s41467-023-36025-x.

Castaldo, F., Páscoa Dos Santos, F., Timms, R.C., Cabral, J., Vohryzek, J., Deco, G., Woolrich, M., Friston, K., Verschure, P., Litvak, V., 2023. Multi-modal and multi-model interrogation of large-scale functional brain networks. Neuroimage 277, 120236. Available at: https://doi.org/10.1016/j.neuroimage.2023.120236.

Deco, G., Kringelbach, M.L., 2014. Great expectations: using whole-brain computational connectomics for understanding neuropsychiatric disorders. Neuron 84, 892–905. Available at: https://doi.org/10.1016/j.neuron.2014.08.034.

Deco, G., Jirsa, V., McIntosh, A.R., Sporns, O., Kötter, R., 2009. Key role of coupling, delay, and noise in resting brain fluctuations. Proc. Natl. Acad. Sci. U. S. A. 106, 10302–10307. Available at: https://doi.org/10.1073/pnas.0901831106.

Deco, G., Jirsa, V.K., McIntosh, A.R., 2011. Emerging concepts for the dynamical organization of resting-state activity in the brain. Nat. Rev. Neurosci. 12, 43–56. Available at: https://doi.org/10.1038/nrn2961.

Deco, G., Ponce-Alvarez, A., Mantini, D., Romani, G.L., Hagmann, P., Corbetta, M., 2013. Resting-state functional connectivity emerges from structurally and dynamically shaped slow linear fluctuations. J. Neurosci. 33, 11239–11252. Available at: https://doi.org/10.1523/JNEUROSCI.1091-13.2013.

Deco, G., Kringelbach, M.L., Jirsa, V.K., Ritter, P., 2017. The dynamics of resting fluctuations in the brain: metastability and its dynamical cortical core. Sci. Rep. 7, 3095. Available at: https://doi.org/10.1038/s41598-017-03073-5.

Deco, G., Cruzat, J., Cabral, J., Knudsen, G.M., Carhart-Harris, R.L., Whybrow, P.C., Logothetis, N.K., Kringelbach, M.L., 2018. Whole-brain multimodal neuroimaging model using serotonin receptor maps explains non-linear functional effects of LSD. Curr. Biol. 28, 3065–3074. Available at: https://doi.org/10.1016/j.cub.2018.07.083.

Demertzi, A., Tagliazucchi, E., Dehaene, S., Deco, G., Barttfeld, P., Raimondo, F., Martial, C., Fernández-Espejo, D., Rohaut, B., Voss, H.U., Schiff, N.D., Owen, A.M., Laureys, S., Naccache, L., Sitt, J.D., 2019. Human consciousness is supported by dynamic complex patterns of brain signal coordination. Sci. Adv. 5, eaat7603. Available at: https://doi.org/10.1126/sciadv.aat7603.

Fung, P.K., Robinson, P.A., 2013. Neural field theory of calcium dependent plasticity with applications to transcranial magnetic stimulation. J. Theor. Biol. 324, 72–83. Available at: https://doi.org/10.1016/j.jtbi.2013.01.013.

Ghosh, A., Rho, Y., McIntosh, A.R., Kötter, R., Jirsa, V.K., 2008a. Noise during rest enables the exploration of the brain's dynamic repertoire. PLoS Comput. Biol. 4, e1000196. Available at: https://doi.org/10.1371/journal.pcbi.1000196.

Ghosh, A., Rho, Y., McIntosh, A.R., Kötter, R., Jirsa, V.K., 2008b. Cortical network dynamics with time delays reveals functional connectivity in the resting brain. Cogn. Neurodyn. 2, 115–120. Available at: https://doi.org/10.1007/s11571-008-9044-2.

Griffiths, J.D., Lefebvre, J, 2019. Shaping brain rhythms: dynamic and control-theoretic perspectives on periodic brain stimulation for treatment of neurological disorders.

In: Curtsudis, V (Ed.), Handbook of Multi-Scale Models of Brain Disorders: From Microscopic to Macroscopic Assessment of Brain Dynamics. Springer, London. Available at: https://doi.org/10.1007/978-3-030-18830-6_18.

Griffiths, J.D., Lefebvre, J., McIntosh, A.R., 2020. A connectome-based, corticothalamic model of state- and stimulation-dependent modulation of rhythmic neural activity and connectivity. Front. Comput. Neurosci. 14, 575143. Available at: https://doi.org/10.3389/fncom.2020.575143.

Griffiths, J.D., Bastiaens, S.P., Kaboodvand, N., 2022. Whole-brain modelling: Past, present, and future. In: Giugliano, M., Negrello, M., Linaro, D. (Eds.), Computational Modelling of the Brain: Modelling Approaches to Cells, Circuits and Networks. Springer International Publishing, Cham, pp. 313–355. Available at: https://doi.org/10.1007/978-3-030-89439-9_13.

Hagmann, P., Cammoun, L., Gigandet, X., Meuli, R., Honey, C.J., Wedeen, V.J., Sporns, O., 2008. Mapping the structural core of human cerebral cortex. PLoS Biol. 6, e519. Available at: https://doi.org/10.1371/journal.pbio.0060159.

Hansen, E.C.A., Battaglia, D., Spiegler, A., Deco, G., Jirsa, V.K., 2015. Functional connectivity dynamics: modeling the switching behavior of the resting state. Neuroimage 105, 525–535. Available at: https://doi.org/10.1016/j.neuroimage.2014.11.001.

Heitmann, S., Breakspear, M., 2022. Brain Dynamics Toolbox. Available at: https://zenodo.org/record/7070703.

Honey, C.J., Kötter, R., Breakspear, M., Sporns, O., 2007. Network structure of cerebral cortex shapes functional connectivity on multiple time scales. Proc. Natl. Acad. Sci. U. S. A. 104, 10240–10245. Available at: https://doi.org/10.1073/pnas.0701519104.

Honey, C.J., Sporns, O., Cammoun, L., Gigandet, X., Thiran, J.P., Meuli, R., Hagmann, P., 2009. Predicting human resting-state functional connectivity from structural connectivity. Proc. Natl. Acad. Sci. U. S. A. 106, 2035–2040. Available at: https://doi.org/10.1073/pnas.0811168106.

Hutt, A., Griffiths, J.D., Herrmann, C.S., Lefebvre, J., 2018. Effect of stimulation waveform on the non-linear entrainment of cortical alpha oscillations. Front. Neurosci. 12, 376. Available at: https://doi.org/10.3389/fnins.2018.00376.

Jirsa, V.K., 2004. Connectivity and dynamics of neural information processing. Neuroinformatics 2, 183–204. Available at: https://doi.org/10.1385/NI:2:2:183.

Jirsa, V.K., Haken, H., 1996. Field theory of electromagnetic brain activity. Phys. Rev. Lett. 77, 960–963. Available at: https://doi.org/10.1103/PhysRevLett.77.960.

Jirsa, V.K., Sporns, O., Breakspear, M., Deco, G., McIntosh, A.R., 2010. Towards the virtual brain: network modeling of the intact and the damaged brain. Arch. Ital. Biol. 148, 189–205. Available at: https://www.ncbi.nlm.nih.gov/pubmed/21175008.

Kötter, R., 2004. Online retrieval, processing, and visualization of primate connectivity data from the CoCoMac database. Neuroinformatics 2, 127–144. Available at: https://doi.org/10.1385/NI:2:2:127.

Kringelbach, M.L., Cruzat, J., Cabral, J., Knudsen, G.M., Carhart-Harris, R., Whybrow, P.C., Logothetis, N.K., Deco, G., 2020. Dynamic coupling of whole-brain neuronal and neurotransmitter systems. Proc. Natl. Acad. Sci. U. S. A. 117, 9566–9576. Available at: https://doi.org/10.1073/pnas.1921475117.

Kunze, T., Hunold, A., Haueisen, J., Jirsa, V., Spiegler, A., 2016. Transcranial direct current stimulation changes resting state functional connectivity: a large-scale brain network modeling study. Neuroimage 140, 174–187. Available at: https://doi.org/10.1016/j.neuroimage.2016.02.015.

Kuramoto, Y., 1984. Chemical turbulence. In: Chemical Oscillations, Waves, and Turbulence. Springer series in synergetics, Springer Berlin Heidelberg, Berlin, Heidelberg, pp. 111–140. Available at: https://doi.org/10.1007/978-3-642-69689-3_7.

Kurtin, D.L., Giunchiglia, V., Vohryzek, J., Cabral, J., Skeldon, A.C., Violante, I.R., 2023. Moving from phenomenological to predictive modelling: progress and pitfalls of modelling brain stimulation in-silico. Neuroimage 272, 120042. Available at: https://doi.org/10.1016/j.neuroimage.2023.120042.

Lang, S., Momi, D., Vetkas, A., Santyr, B., Yang, A.Z., Kalia, S.K., Griffiths, J.D., Lozano, A., 2023. Computational modeling of whole-brain dynamics: a review of neurosurgical applications. J. Neurosurg. 140, 218–230. Available at: https://doi.org/10.3171/2023.5.JNS23250.

Lea-Carnall, C.A., Trujillo-Barreto, N.J., Montemurro, M.A., El-Deredy, W., Parkes, L.M., 2017. Evidence for frequency-dependent cortical plasticity in the human brain. Proc. Natl. Acad. Sci. U. S. A. 114, 8871–8876. Available at: https://doi.org/10.1073/pnas.1620988114.

Lopes da Silva, F.H., Hoeks, A., Smits, H., Zetterberg, L.H., 1974. Model of brain rhythmic activity. The alpha-rhythm of the thalamus. Kybernetik 15, 27–37. Available at: https://doi.org/10.1007/BF00270757.

Luppi, A.I., Cabral, J., Cofre, R., Mediano, P.A.M., Rosas, F.E., Qureshi, A.Y., Kuceyeski, A., Tagliazucchi, E., Raimondo, F., Deco, G., Shine, J.M., Kringelbach, M.L., Orio, P., Ching, S., Sanz Perl, Y., Diringer, M.N., Stevens, R.D., Sitt, J.D., 2023. Computational modelling in disorders of consciousness: closing the gap towards personalised models for restoring consciousness. Neuroimage 275, 120162. Available at: https://doi.org/10.1016/j.neuroimage.2023.120162.

Momi, D., Wang, Z., Griffiths, J.D., 2023. TMS-evoked responses are driven by recurrent large-scale network dynamics. Elife 12, e83232. Available at: https://doi.org/10.7554/eLife.83232.

Naskar, A., Vattikonda, A., Deco, G., Roy, D., Banerjee, A., 2021. Multiscale dynamic mean field (MDMF) model relates resting-state brain dynamics with local cortical excitatory-inhibitory neurotransmitter homeostasis. Netw. Neurosci. 5, 757–782. Available at: https://doi.org/10.1162/netn_a_00197.

Newman, M., Barabasi, A.-L., Watts, D.J., 2006. The Structure and Dynamics of Networks. Princeton University Press, Princeton, NJ. Available at: https://doi.org/10.1515/9781400841356/html.

Pang, J.C., Aquino, K.M., Oldehinkel, M., Robinson, P.A., Fulcher, B.D., Breakspear, M., Fornito, A., 2023. Geometric constraints on human brain function. Nature 618, 566–574. Available at: https://doi.org/10.1038/s41586-023-06098-1.

Roberts, J.A., Gollo, L.L., Abeysuriya, R.G., Roberts, G., Mitchell, P.B., Woolrich, M.W., Breakspear, M., 2019. Metastable brain waves. Nat. Commun. 10, 1056. Available at: https://doi.org/10.1038/s41467-019-08999-0.

Robinson, P.A., 2012. Interrelating anatomical, effective, and functional brain connectivity using propagators and neural field theory. Phys. Rev. E Stat. Nonlinear Soft Matter Phys. 85, 011912. Available at: https://doi.org/10.1103/PhysRevE.85.011912.

Robinson, P.A., Rennie, C.J., Rowe, D.L., O'Connor, S.C., Wright, J.J., Gordon, E., Whitehouse, R.W., 2003. Neurophysical modeling of brain dynamics. Neuropsychopharmacology 28 (Suppl 1), S74–S79. Available at: https://doi.org/10.1038/sj.npp.1300143.

Sanz-Leon, P., Knock, S.A., Spiegler, A., Jirsa, V.K., 2015. Mathematical framework for large-scale brain network modeling in the virtual brain. Neuroimage 111, 385–430. Available at: https://www.sciencedirect.com/science/article/pii/S1053811915000051.

Shine, J.M., Müller, E.J., Munn, B., Cabral, J., Moran, R.J., Breakspear, M., 2021. Computational models link cellular mechanisms of neuromodulation to large-scale neural dynamics. Nat. Neurosci. 24, 765–776. Available at: https://doi.org/10.1038/s41593-021-00824-6.

Shouval, H.Z., Bear, M.F., Cooper, L.N., 2002. A unified model of NMDA receptor-dependent bidirectional synaptic plasticity. Proc. Natl. Acad. Sci. U. S. A. 99, 10831–10836. Available at: https://doi.org/10.1073/pnas.152343099.

Spiegler, A., Hansen, E.C.A., Bernard, C., McIntosh, A.R., Jirsa, V.K., 2016. Selective activation of resting-state networks following focal stimulation in a connectome-based network model of the human brain. eNeuro 3. Available at: https://doi.org/10.1523/ENEURO.0068-16.2016.

Spiegler, A., Abadchi, J.K., Mohajerani, M., Jirsa, V.K., 2020. In silico exploration of mouse brain dynamics by focal stimulation reflects the organization of functional networks and sensory processing. Netw. Neurosci. 4, 807–851. Available at: https://doi.org/10.1162/netn_a_00152.

Sporns, O., Tononi, G., 2001. Classes of network connectivity and dynamics. Complexity 7, 28–38. Available at: https://doi.org/10.1002/cplx.10015.

Stephan, K.E., Kamper, L., Bozkurt, A., Burns, G.A., Young, M.P., Kötter, R., 2001. Advanced database methodology for the collation of connectivity data on the macaque brain (CoCoMac). Philos. Trans. R. Soc. Lond. B Biol. Sci. 356, 1159–1186. Available at: https://doi.org/10.1098/rstb.2001.0908.

Verma, P., Nagarajan, S., Raj, A., 2022. Spectral graph theory of brain oscillations—revisited and improved. Neuroimage 249, 118919. Available at: https://doi.org/10.1016/j.neuroimage.2022.118919.

Vohryzek, J., Cabral, J., Vuust, P., Deco, G., Kringelbach, M.L., 2022. Understanding brain states across spacetime informed by whole-brain modelling. Philos. Trans. A Math. Phys. Eng. Sci. 380, 20210247. Available at: https://doi.org/10.1098/rsta.2021.0247.

Vohryzek, J., Cabral, J., Castaldo, F., Sanz-Perl, Y., Lord, L.-D., Fernandes, H.M., Litvak, V., Kringelbach, M.L., Deco, G., 2023. Dynamic sensitivity analysis: defining personalised strategies to drive brain state transitions via whole brain modelling. Comput. Struct. Biotechnol. J. 21, 335–345. Available at: https://doi.org/10.1016/j.csbj.2022.11.060.

Wilson, H.R., Cowan, J.D., 1972. Excitatory and inhibitory interactions in localized populations of model neurons. Biophys. J. 12, 1–24. Available at: https://doi.org/10.1016/S0006-3495(72)86068-5.

Wilson, M.T., Fulcher, B.D., Fung, P.K., Robinson, P.A., Fornito, A., Rogasch, N.C., 2018. Biophysical modeling of neural plasticity induced by transcranial magnetic stimulation. Clin. Neurophysiol. 129, 1230–1241. Available at: https://doi.org/10.1016/j.clinph.2018.03.018.

Wilson, M.T., Moezzi, B., Rogasch, N.C., 2021. Modeling motor-evoked potentials from neural field simulations of transcranial magnetic stimulation. Clin. Neurophysiol. 132, 412–428. Available at: https://doi.org/10.1016/j.clinph.2020.10.032.

Wong, K.-F., Wang, X.-J., 2006. A recurrent network mechanism of time integration in perceptual decisions. J. Neurosci. 26, 1314–1328. Available at: https://doi.org/10.1523/JNEUROSCI.3733-05.2006.

Xie, X., Cai, C., Damasceno, P.F., Nagarajan, S.S., Raj, A., 2021. Emergence of canonical functional networks from the structural connectome. Neuroimage 237, 118190. Available at: https://doi.org/10.1016/j.neuroimage.2021.118190.

Yeung, M.K.S., Strogatz, S.H., 1999. Time delay in the Kuramoto model of coupled oscillators. Phys. Rev. Lett. 82, 648–651. Available at: https://doi.org/10.1103/PhysRevLett.82.648.

Neural graph modeling

Stephen José Hanson and Catherine Hanson
RUBIC, Psychology and CMBN, Newark, NJ, United States

Introduction to neural graph modeling
The need for network neuroscience

In 1976, George Miller and Mike Gazzinaga, while discussing the biological foundations of human cognition over martinis at the Rockefeller University Faculty Club, coined the term *cognitive neuroscience* to describe a research direction that would encompass neuroscience, computational neuroscience, and cognitive psychology (Bruer, 2009; Personal Communication to S. Hanson from George A. Miller, 1989). A decade later, cognitive neuroscience had gained sufficient interest to prompt both the James S. McDonnell Foundation and the Pew Trusts to devote over 10 years and fifty million dollars supporting research aimed at understanding "… *computations supporting cognitive function in brain tissue*" (RFP brochure for the McDonnell-Pew Cognitive Neuroscience Program 1990–2000).

Until very recently, research in cognitive neuroscience was dominated less by computation and more by localization. Researchers seeking to understand the role of the brain in human cognition saw their mission to be one of cataloging individual brain regions and their associated functions. A sluggish BOLD response, limited analysis options (GLM thresholding), and simple insertion logic led to studies focused more on blobology (Hanson, 2022) than on the dynamics characterizing brain function. The focus on blobology has produced innumerable studies identifying face areas, body areas, place areas, theory of mind areas, disgust areas, and so on.

However, 30 years of assigning specific functionality to individual brain regions has revealed some general limitations with this approach. First, the identification of a given brain region depends upon the brain atlas being used, with some atlases (e.g., Talairach) dividing the brain into much broader regions than others (e.g., Harvard-Oxford). Thus, the relation of a given function to a region will vary depending on how the region is defined. Second, a given brain region is not generally dedicated to a single function. Rather, the function of a given brain region is task dependent, with the functionality of a region being expressed differently across potentially related but separate tasks. For example, dorsolateral prefrontal cortex, which is

Computational and Network Modeling of Neuroimaging Data. https://doi.org/10.1016/B978-0-443-13480-7.00011-9

associated with working memory, may control task switching, inhibition, planning, or selective attention depending on whether the task involves problem-solving, language processing, decision-making, or learning. Third, the mapping of regions to function is not unique, as different brain regions may exhibit the same functionality depending on the context. Finally, brain regions do not act independently. The brain is anatomically and functionally a network, not a collection of independent units or modules. So, a focus on cataloging individual regions ignores the inherent dynamism underlying brain function; specifically, how networks recruit brain regions during the performance of cognitive tasks as well as the way in which recruited areas interact (McIntosh, 1998; Hanson, 2022).

What is a neural graph model?

A Neural Graph Model is an analysis of network connectivity and dynamics based on changes in activation of brain regions over time. This approach fuses standard neural network activity learning rules (e.g., backpropagation, recurrent feedback) with Bayesian parameter estimation (Ramsey et al., 2010; Mumford and Ramsey, 2014) to create an architecture in which the connectivity of nodes (brain regions), and the edge weights (strength of influence) among nodes, is assessed. Specifically, data collected through neuroimaging (e.g., fMRI, EEG, ECoG) is used to generate a graph model in which the parameters are determined using Bayes estimators. The optimal Neural Graph Model is determined through the application of goodness-of-fit metrics such as Bayesian Information Criterion (BIC) or Akaike Information Criterion (AIC) to searches performed in parallel over data collected from multiple subjects.

This hybrid between neural networks and graph models offers a number of advantages in research focused on cognitive function. First, parameters of the Neural Graph Model are grounded by empirical data collected from subjects (e.g., fMRI, EEG, ECoG), rather than by theoretical assumptions or simulated data, resulting in a network configuration unique to the subjects, the task performed, and the measurements collected. Second, Neural Graph Models draw on the structural constraints of the graph model and the functional constraints of the neural network model (learning rules) to generate a network model that is applicable to both small and large-scale networks. Third, the NGM results from a model *search* based on goodness-of-fit measures, not from a model *match* to user specification based on hypothetical assumptions (e.g., DCM). Finally, seeding with empirical data makes the NGM model more biologically plausible and ecologically valid than models generated from hypothetical network properties.

Neural networks
The basic neural network

Neural networks have been used for decades to model cognitive function (for an early historical review, see Anderson and Rosenfeld (1998)). Neural networks, over time, have evolved from very simple algorithms (McCulloch and Pitts, 1943;

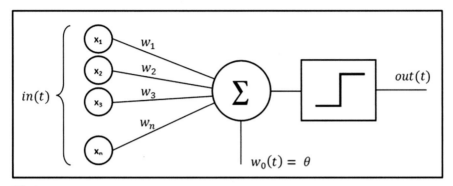

FIG. 1

Perceptron neural network. This was one of the first and simplest neural networks, it involved only input and output(s) units, no intermediate units, which were later called "hidden units" because they were explicitly between input and output and could create novel features.

Rosenblatt, 1961) to more complex algorithms and theories (Rumelhart et al., 1986; Hinton and Sejnowski, 1986; Hanson and Burr, 1990) and most recently, deep learning algorithms (Hinton et al., 2006). In each evolutionary step, the research applications that have resulted demonstrated the viability of using a model composed of only nodes ("neurons") and directed edges ("synapses") to capture cognitive and perceptual function. In engineering fields, these have also been called artificial neural networks (ANNs).

In the most basic form (see Fig. 1) nodes and connections provide one of the simplest algorithms that can converge over learning updates: the perceptron. In this case, we have a known input with a known target that the network attempts to match over presentations of consistent samples of input and target. Each node in the network has a connection from the input to an output that is intended to match the target. The process is simple and involves computing a weighted sum from each input to the output layer. This is also known as a *dot product* and can be represented as a sum over the input (in(t) = $x1, x2, x3 \ldots x4$) in the first sample multiplied by the connection weight on each connection so that:

$$out(t) = x1*w11 + x2*w12 + x3*x13 \ldots xn*w1n$$

Mathematically, we can write this compactly as a dot product: $s_i = \sum_i w_{ij} x_i$.

Finally, there is a threshold that is applied to the sum (s), which produces normalized output (e.g., 0,1 or $-1,1$): $out(t) = thr$.

The perceptron neural network was one of the earliest neural network models to be developed (Rosenblatt, 1961). In the 1980s, Hinton and Sejnowski (1986) added a single layer of intermediate units between input and output called *hidden units*, which generalized the model and dramatically increased its computational power. Learning is achieved most commonly by using a rule based on computing the error gradient between the output predicted by the neural network for a given input sample and the associated target (Rumelhart et al., 1986). These inputs, outputs, and targets can be either binary (or binary vector codes) or real continuous values (e.g., visual coordinates, motor responses, speech). This basic form of the neural network is a feedforward, acyclic type of network where input passes from layer to layer to produce an output contingent on input.

Recurrent neural networks

A more biologically plausible type of network than the basic feedforward network was introduced in the 1980s. By modifying the basic neural network feedforward architecture to allow updating of network activity incrementally, the recurrent neural network (RNN) uses activity changes to provide phased feedback and to evolve over time. RNNs are special since they can alter their internal states, allowing them to model sequential input (motor responses, Jordan, 1986; language, Elman, 1990) and episodic memory (Hanson and Hanson, 1996). This makes RNNs harder to train, but ideal for modeling dynamic changes in actual brain activity, particularly interactions and communication among brain areas. Since these models are a type of instantiation of functional brain networks, albeit at a less detailed neural level, it is not surprising that they have been applied with considerable success in characterizing biological neural networks (e.g., Cadieu et al., 2014; Yamins et al., 2014), as well as perceptual brain function (Hanson et al., 2018b).

Understanding the dynamics in neural networks is key to understanding the behavior of neural networks and what they can effectively model. The basic dynamics of an RNN arise from the connectivity (weights to nodes) and patterns of connections that will define unique temporal activity in the RNN. These patterns of connection can be described with various tools from differential calculus. One such tool is called a "phase plot," which provides a temporal trace between the relevant variables in the neural network over time, in this case, the nodes and their connectivity. Another critical tool is the *degree distribution*. Coming from complexity/information theory degree distributions can be used to characterize the connectivity patterns in the aggregate and help predict its behavior based on the input/output activity of each node and how much each node may be a hub for the connectivity and help organize the flow and structure of neural activity (we define later).

Although connectivity implies some sort of communication, it does not determine the amount of "memory" capacity (e.g., the number of independent patterns that can be stored in a network) for specific neural circuits. In the context of neural network modeling, two immediate conclusions about neural memory capacity can be drawn by considering simple linear models. These type of neural networks simply compute dot product between units without thresholds or nonlinearities (e.g., sigmoids), these are called linear *activation functions*. First, recurrence of a linear network is simply the power of the weight matrix, where the power is the number of time steps taken. This implies there is no improvement by real-time simulation of the neural network. Second, the simple recurrent neural networks (Hopfield, 1982; Elman, 1990; Hanson and Hanson, 1996). Normally in an RNN one would have to simulate the node activity

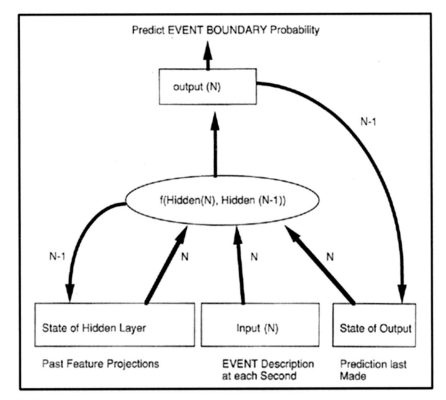

Predict EVENT BOUNDARY Probability

output (N)

N-1

f(Hidden(N), Hidden (N-1))

N-1   N   N   N

State of Hidden Layer   Input (N)   State of Output

Past Feature Projections   EVENT Description   Prediction last
at each Second   Made

FIG. 2

Simple recurrent neural network, showing how sequential episodic memory information may
be modeled. The input has a feedback state value from the hidden layers (at one step prior)
and then is fed back as new input, thus creating a simple input memory (exponentially
weighted in time). The output is also fed back to the input as the last output state providing a
prediction history of the RNN predictions over time.

updates by updating nodes over time, there is no way to compute final values (rather
than a closed-form equation) from the initial weight matrix and node values alone due
to the nonlinearities. Another limitation with linear systems appears with feedback
which will converge to asymptotic values very rapidly. Consequently, these systems
are very limited in memory capacity (e.g., log number of nodes). However, it is known
that a small change in the activation function to be semi-linear or logistic can provide,
with appropriate learning rules (gradient error), a much larger memory capacity and
can represent extremely large neural networks that can learn simple grammars
(>1000s of nodes in a finite state machine) as well as in more complex context-free
grammar (Giles et al., 1995; Hanson and Negishi, 2002).

One example of an RNN simulating episodic memory (Hanson and Hanson,
1996) is shown in Fig. 2. In this case, the model simulates a subject watching a

video of two people playing a game of Monopoly in which the actors, actions, and objects are used as input. In this model, there are two recurrent network elements, one from the input to the hidden layer, which creates a sequential memory (short term) of the input data. The second is a prediction from the output layer to an input state that induces a sequential prediction of the next state, providing expectations (plans) of what is about to occur. Both recurrent sources had sigmoidal activation functions.

More recently, new developments in AI such as deep learning have been highlighting successful applications in biomedical research, transportation, and more recently with large language models (LLMs, i.e., ChatBOTS), which appear to have acquired language and reasoning abilities akin to humans (although this is controversial). The concept of deep learning originally due to Geoff Hinton and his students has created a revolution in artificial intelligence based on both the scale of the architectures (e.g., $>1\,T$ connections) and data (e.g., $>1B$ sentences in text corpora). It should be noted that the nature of the NN representations are actually poorly understood at this point, though we may expect rapid advances in the nature of the computations.

Another kind of neural network model is an autoencoder, one of the simplest kinds of sequential learning machine (networks that use a copy of the input to use for target output to reproduce or predict input; see also Cottrell, 1988; Hanson and Kegl, 1987; Hanson and Hanson, 1996). Autoencoders were precursors to the GPTs (transformer architectures), which use one-step sequence prediction with massive covariance (what they call "attention," but really a metaphor) to encode, weight, and re-encode information producing unique and powerful pattern recognition results (Hinton et al., 2006) and apparently enormous storage capacity or memory *without at the same time overfitting data.*

The brain, of course, can be seen as an extremely large ($100\,T$ connections) set of recurrent networks with multiple reentrant paths for feedback (e.g., visual pathways; attentional pathways) and control of these networks. Let us consider how we might begin to model and understand those dynamics, albeit on a much smaller scale using Bayesian Graph estimation. This direction, we show, can result in highly sensitive biomarkers representing connectivity disturbance especially related to various brain diseases, which we will discuss later in the neural graph modeling applications.

Although highly useful in understanding brain dynamics, network models are limited by their nature in capturing fairly generic aspects (e.g., stable, unstable, chaotic) of network dynamics. Network models do not generally differentiate node types or edges (connections between nodes). However, graph models do provide a means of modeling these important network features.

Graph models

Graph models provide a visualization of network structure as shown in Fig. 3. Different types of nodes and edges provide the means of analyzing and modeling networks to reveal the specific properties of that network. The term *clique* is used to identify a set of nodes that are connected. Connections between nodes are known as edges and can be either undirected or directed. Undirected edges represent simply

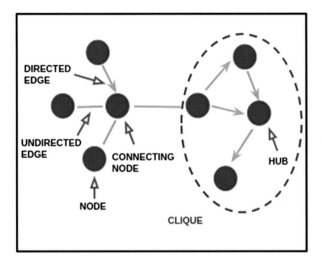

FIG. 3

Graph model structure: Nodes are the blue balls, and the green lines between nodes are called edges or connections. Note there are undirected edges (associations) that have no arrows and directed edges that have a single arrow indicating an influence or potential "cause." Note the box showing a set of nodes that are highly connected called a "clique" or cluster. Note the one edge extending from the clique is a single-cut node isolating the clique from the rest of the network.

that an association between nodes exists, whereas directed edges provide information about the direction of influence between nodes.

In addition to different types of edges, there are a number of different types of nodes, each having distinct characteristics depending on how it functions within the network. Simple nodes represent the activity associated with a given network unit (e.g., a brain region). A connecting node has an edge (connection) to a separate clique; so, removing that edge isolates the two cliques (see Fig. 3). A node within a clique that has maximal degree (in/out connections in the graph) is called a hub.

Metrics used in graph models

Before discussing the various types of graph models that can be used, it is important to understand the metrics used to characterize network behavior and properties. Three basic metrics are defined here, and more examples and graph metrics have been described by Bullmore and Sporns (2009).

Path length

First is the distance between two nodes in the graph. This is called *path length* (see Fig. 4), which defines the shortest or longest length between two nodes. Pairwise measures can provide an average maximal path length or minimal path length throughout an entire graph network, indicating the density of sparsity of the connection overall. This could be useful for determining hyper- or hypo-connectivity in specific applications.

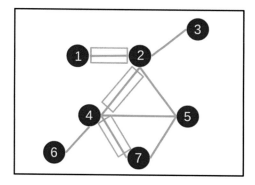

FIG. 4

Path length is the link distance between Node X to Node Y. Green links indicate the connectivity among nodes and the shortest path $(1,7)=3$ is highlighted with orange boxes. This metric is often used in MRI measures such as diffusion to measure distributions over path length indicating that significant path lengths tend to be either sparse (hypo-connected) or dense (hyper-connected).

Clustering coefficient

A second useful network metric, the clustering coefficient, is a measure of the node density in a specific group of nodes. This node density can be used to determine whether the cluster is a "clique." A clique is indicated, with a threshold, that the number of connections between nodes is maximal as in the first case shown in Fig. 5, where $C=1$. In this case, every/node is connected to every other nearby neighbor. In the third case where the $C=0$ the connections are minimal indicating no clustering present, as any node in this set could participate in another nearby clique. If the path length

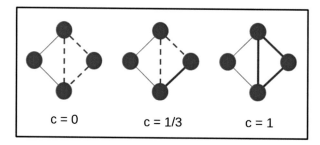

FIG. 5

Clustering coefficient: It is measured with the following equation within any graph structure: $(2L)/(k-l)*k$, where $L=$ sum of neighbor links and $k=$ potential candidate neighbor link. This metric can be used to identify hubs and hub structures within the larger network.

increases the locality of the clique is compromised and this can be thresholded as well. So both the locality and the clustering coefficient can vary to create cluster maps over the estimated node connectivity space. Clustering coefficients are therefore metrics that could be used to identify a hub, in effect a node with the greatest connections in undirected or directed edges with the maximal input/output connections.

Degree distribution

A third metric, *degree distribution* (see Fig. 6), is very useful for understanding the topography (hierarchical or random) of a network. First note that *degree* is a property of a specific node. It is a measurement of the number of links to other nodes in the network and much like the clustering coefficient metric, describes the density of connections in a graph. However, the degree metric provides more information than the clustering coefficient by characterizing the entire network structure in terms of hierarchical structure that is typical in feedforward structures.

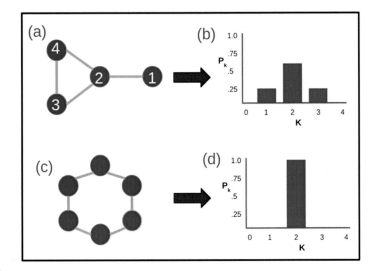

FIG. 6

To the right of each graph is the degree distribution (DD) showing the number of nodes with degree values (e.g., 1, 2, 3). The graph shown in (A) has a heterogeneity of connections, some with one connection and some with two or three connections. In (B) the distribution shows that the network has more than one kind of connection. In contrast, (C) shows a network where each node has only two connections. Consequently, in (D) the degree distribution has a single bar at degree 2 with all the number of the connections shown as a frequency on the y-axis as a single bar.

More complex graph structures that involve extensive feedback connections, hub structures, or any deviation from strict hierarchical structure (scale-free structures), will result in a more general degree distribution. These distributions would deviate from power functions depending on that lack of hub structures and strict hierarchy. Fig. 7 displays an example of a scale-free network, in this case, airline transportation routes.

Because biological systems tend to be scale-free, the degree distribution metric is particularly useful. Degree distributions tend to be Gaussian for networks in which the structure of the nodes and edges are not critical for functionality (i.e., variations do not affect performance). However, when the structure of the nodes and edges are based on optimal performance (e.g., when airlines try to minimize fuel costs) the degree distribution tends to be based on a power function (bottom in Fig. 7).

The degree distribution exponent (in the case of scale-free, a power function exponent) has been used successfully to model excursions from scale-free networks and identify biomarkers based on connectivity (Hanson et al., 2016). These degree distribution models involve the underlying dynamics of a large-scale recurrent (feedback) network based on a resting state. It can also be used to focus on distinct change points predicated on the evolution of the disease and network communication breakdown apparent in the development of progressive diseases such as Alzheimer's and schizophrenia.

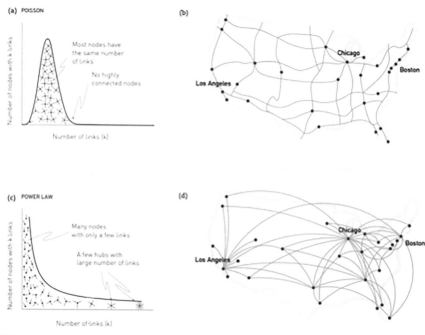

FIG. 7

Examples of degree distributions (DD). The top row on the left shows a Poisson DD based on city incoming and outgoing roads. The bottom row on the left shows power scale-free DD for incoming and outgoing airline paths.

Modelling network connectivity

To model network connectivity, the relevant time series from the set of ROIs identified for the task must be obtained. Once the ROI time series data is obtained from the neuroimaging data (e.g., fMRI or EEG), the graph and its significant edges are then estimated. Initially, networks are formed by connections between regions of interest (ROIs). The selection of ROIs for network analysis should have a theoretical or empirical basis and not be arbitrary or contrived. Next, the time series should be measured during a resting state, for example, when the subject is asked to do nothing but rest and relax. In a brilliantly simple experiment, Biswal et al. (2010) found that seeding a resting-state condition with one ROI time series, obtained from a previously executed motor task, would produce activity in all the other ROIs. In effect, the same motor network would pop out in the resting condition as if the networks were latently present in the background and waiting to be summoned forth. The resting state, therefore, became a robust benchmark for functional connectivity studies.

Graph models of network connectivity characterize the relation among the nodes. The most simple form of connectivity (functional) is able to show the associations that exist in a network, but not the influence that nodes may have on one another or the directionality of that influence (effective connectivity). Not surprisingly, the graph models used to specify what is called functional or effective connectivity in a network employ very different methods.

Functional connectivity: Basic association through correlation

There are many kinds of approaches that might be used to measure network connectivity, the weakest (in a statistical sense) is functional connectivity, which measures associations between regions of interest (ROIs), but without direction or influence. The time series of active ROIs, once properly conditioned to remove drift, spikes, and irrelevant periodicity, could be correlated. But how much of the correlation is noise? Pearson correlation coefficient is the measure that has been used to determine the association between ROIs. Simulations show that due to oversimplifying assumptions, Pearson "r" correlations can be terribly misleading. For example, white or Gaussian noise is often assumed to be an additive factor (which is not accurate), leading to benchmarks that overinflate goodness of fit making differences small to statistically nonexistent. To correct these assumptions, researchers use raw (or slightly conditioned or corrected for trends) correlation matrices for analyzing brain connectivity. There are many problems with this strategy, not the least of which is the arbitrary thresholding of the correlation matrix, inducing as many or as few clusters as one would like to find. This results in the correlation matrix being examined directly, as if the raw correlation matrix was itself a type of analysis rather than a way station to some valid testable inferential model. In addition to the thresholding issue, there are a number of problems with relying on a similarity measure based on correlation or association.

1. *Correlations are associations, not influences. Consequently, they cannot orient edges.*
2. *Correlations fail to detect structure. Most recent work in brain networks involving "functional connectivity," often estimated using correlations of bold time series, has been shown to miss actual causal structure in known time series simulation tests* (Hanson and Glymour, 2010; Ramsey et al., 2010).
3. *Correlations are unstable. Test–retest of resting-state data in individuals reveals that correlation graph dissimilarity increases with sample size whereas Bayes network estimated graph dissimilarity decreases with sample size as it should* (Hanson et al., 2018a,b).
4. *Correlations across individual differences are statistically heterogeneous. Resting-state correlation matrices based on 235 ROIs over 70 subjects (pairwise) all tests fail a null hypothesis chi-square test, indicating that the correlations are rife with averaging artifacts* (Hanson et al., 2018a,b).

Some of these issues could be avoided by using latent analysis methods (e.g., principal component analysis, factor analysis, MDS) as most statisticians would recommend, which are unfortunately not commonly used. The additional precaution of applying latent analysis would not affect the detection of patterns based on correlation, but any underlying assumptions would be properly met and rationally modeled even if there were millions of ROIs instead of just a handful (see Reid et al., 2019). Despite this oversight, using the Pearson "r" correlation to model functional connectivity has proved to be effective as a basic analysis of network structure (Cole et al., 2014). However, only through the use of stronger statistical approaches can effective connectivity patterns in network behavior be discovered.

Effective connectivity: "Causality" through Bayes graph estimators

Associative measures can not estimate the direction or influence of edges (connections between nodes), but this can be achieved if the search for edges is conditioned on the time series data from multiple ROIs. Activity for ROIs can be sequenced and used by the search algorithm to determine which direction of influence best fits the time series data.

Methods for doing graph searches that are robust and accurate include IMaGES (Ramsey et al., 2010) and Gimme (Gates et al., 2014). The IMaGES algorithm is shown in Fig. 8. Methods such as IMaGES are based on conditional independence and constraint satisfaction to find DAGs (directed acyclic graphs) from the ROI time series.

Beyond brain imaging signals, Bayesian search methods have been used in a broad range of research areas (Glymour, 1980). By the 1990s there was growing consensus in modeling domains that real or hypothetical causal relations are better represented by directed graphs (DIGRAPHS); that is, graphs in which a node represents the neural activity of a localized region and a directed edge between two nodes indicates the influence that one node has (relative to the set of represented

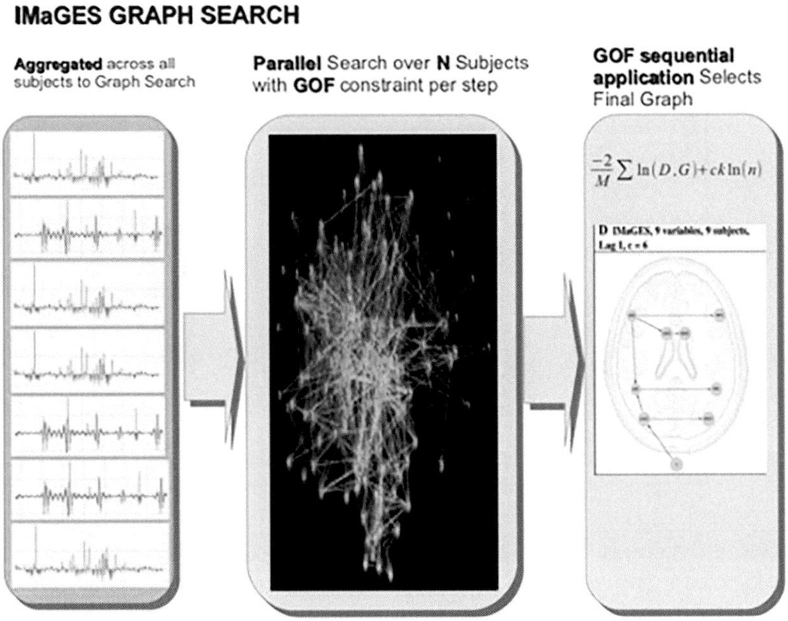

IMaGES GRAPH SEARCH

Aggregated across all subjects to Graph Search

Parallel Search over **N** Subjects with **GOF** constraint per step

GOF sequential application Selects Final Graph

$$\frac{-2}{M}\sum \ln(D.G) + ck\ln(n)$$

D IMaGES, 9 variables, 9 subjects, Lag L,c = 6

FIG. 8

IMaGES algorithm. On the left neuroimaging signals (fMRI) time series are input into multiple (subjects) parallel search. As edges are proposed an aggregate goodness of fit is applied across all existing subject graphs at each step, thus creating a set of individual difference constraints. These constraints require a majority resolution that at the same time doesn't result in destructive averaging (Simpson's paradox). The resultant graph is the best compromise among all individual graphs that could be identified.

regions) on the other. This type of directed graph, including cyclic graphs representing feedback relations, implies a fundamental influence (effective connectivity) within statistical and dynamical models such as structural equation models (SEM; McIntosh, 1998), Granger causality models (GCM; Granger, 1969), and Bayesian graph models such as IMaGES (Ramsey et al., 2010). Bayesian graph modeling uses discovery (search) methods (e.g., GES, GIMME, IMaGES, FASK; cyclic estimator) to search the possible model space defined by time series data, and use goodness-of-fit measures, such as BIC or AIC (Vrieze, 2012), to determine the model that best fits the data. For example, IMaGES implements a forward-backward search in which edges are added and removed incrementally based on group-wise log-likelihood per-step parallel search. This approach was found in independent benchmarking (Smith et al., 2011) to generate highly accurate models (recall/precision of edges/orientation 99/95%). Finally, IMaGES can be implemented for both small-scale and large-scale (YEO-1000 node, 80–150k edges; see Fig. 9) networks.

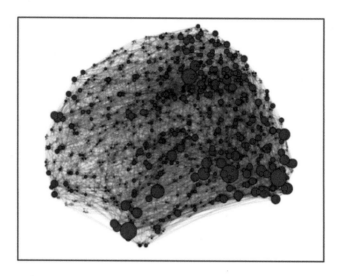

FIG. 9

A large-scale example of fitting IMaGES to the 1000 node YEO parcellation. The black lines are the directed arrows (edges ~200k) and the nodes are the red balls where the size of the node indicates the degree (in/out connections).

Modeling feedforward (acyclic) activity in networks: IMaGES (independent multiple sample greedy equivalence search)

Because IMaGES was developed specifically for fMRI and individual differences, it can create a highly robust estimator (Ramsey et al., 2010). Specifically, IMaGES incrementally fits fMRI time series data (empirical data that is input into IMaGES) until a better model is discovered over all N subjects simultaneously. Starting with an empty graph, each edge across all N subjects is tested with a goodness-of-fit metric based on common constraints arising from each subject graph. Forward edges are only added if they pass a threshold indicating they are common to the majority of the subjects. The final step tests edges in a backward direction, removing the ones that in the full model configuration fail significance, now based on the discovered forward model. This procedure (Meek, 1997) specifies a theorem that proves this method is asymptotically correct given some minimal conditions. These Bayesian search methods have been extremely effective for fMRI analysis and other signals as well (Ramsey et al., 2011; Hanson et al., 2013).

Modeling feedback in networks: FASK (fast adjacency skewness)

Just as network models can be either acyclic (basic linear feedforward structure) or cyclic (structured recurrent neural networks), graph models also fall into acyclic and cyclic forms. Acyclic graph models provide a snapshot of the pattern of activation present in a given network (node influence and directionality), capturing the stable properties of that network. Most of the graph models used to model effective connectivity are acyclic and therefore unable to assess feedback influence in networks, but in recent years our group has developed a cyclic graph model, FASK (Sanchez-Romero et al., 2018), that is able to identify feedback effects.

The FASK algorithm is basically a universal edge orientation procedure. As with IMaGES, the ROI time series is used to condition the fit of the edge orientation. In this case, we adapt a rule based on information flow (Hyvärinen and Smith, 2013) to compute the direction of influence between two nodes (i.e., random variables) $\mathbf{X}$ and $\mathbf{Y}$. We construct a likelihood ratio to test the direction of influence as a conditional information flow toward either $\mathbf{X}$ or $\mathbf{Y}$. This procedure is repeated for all pairs of nodes in the graph, which provides a complete oriented graph structure. FASK is a very general approach that can, in principle, detect very complex feedback structures since the orientations are only locally determined in the graph. Thus the global structure could, in theory, take any complex feedback form, including a completely recurrent network where nodes connect to all other nodes in the graph.

Tests of FASK show (Sanchez-Romero et al., 2018) that over a large set of feedback orientation graph benchmarks providing more than 80% correct orientation over 30+ cases. Although, in principle, FASK may be able to identify any complex graph structure, in reality, due to the limited BOLD dynamic range, and the temporal resolution, long-distance (memory) feedback transmission will prevent the identification of true graph structure derived from time series signals. In general, we will require more consistent time series and higher resolution (e.g., OPM-MEG) signals to achieve the true potential of this very powerful procedure.

Examples of successful neural graph models

There are two forms of the Neural Graph Model: (1) Small-scale models involve mapping a given task onto a specific neural graph. These models can be static or dynamic and generally model 20 nodes or fewer. (2) Large-scale models are used to explore the underlying aggregate behavior of the network (graph) in terms of complexity measures such as hub in/out connectivity, correlation length, and degree distributions as described earlier. These large-scale models are designed to model networks having a large number of nodes (20+ to 1000).

Example 1: Differentiating models of autistic spectrum disorder

Three competing hypotheses have been proposed to account for the cognitive impairment associated with autism spectrum disorder (ASD): (1) Theory of Mind (Baron-Cohen et al., 1985), (2) face processing (Grelotti et al., 2002), or (3) action understanding (Gallese, 2009). Networks associated with each type of cognitive function are small, having fewer than 10 ROIs. Hanson et al. (2013) used Neural Graph Models to compare the three hypotheses and examined both behavioral and fMRI data collected during various tasks involving simple sequential judgments based on eye gaze, directional cueing, and face processing. The tasks were designed to systematically target each of the three networks (TOM, FACE, and ACTION) posited as the source of impaired social cognition.

Using Neural Graph Modeling allowed a direct comparison of the three different hypotheses about ASD impairment based on network activity rather than localized ROI functionality. Graph analysis of the fMRI data using IMaGES revealed striking differences in the number of edges, edge direction, and edge strength between ASD and NT subjects. Whereas neurotypicals (NT) appear to have fairly consistent effective connectivity across both subjects and tasks for any given network, high-functioning individuals with ASD demonstrate considerable variance both across subjects and across tasks. The graph analyses revealed that the greatest difference between ASD and neurotypical individuals appears in the network associated with the Theory of Mind (TOM), shown in Fig. 10, and the least difference in the network associated with face processing (FACE).

The results obtained in this study offer support for existing hypotheses about network connectivity differences between neurotypical and ASD populations with a

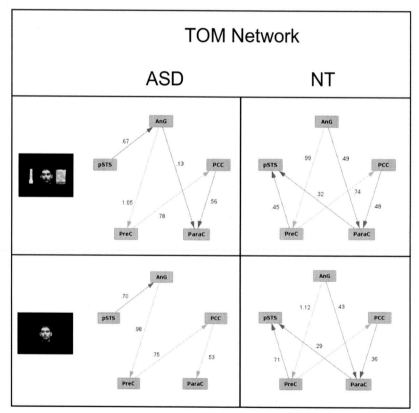

FIG. 10

Effective connectivity for Theory of Mind network in individuals with autism spectrum disorder (ASD) and neurotypicals (NT). Graphs were generated with IMaGES for the eyes only (EO) and eyes left–right (ELR) task.

level of detail that was not previously possible. Moreover, the approach taken, using NGMs to compare network properties across subject populations and tasks, provides a template for resolving competitive hypotheses and clarifying network functions in any research involving cognitive function.

Example 2: Modeling attention and memory

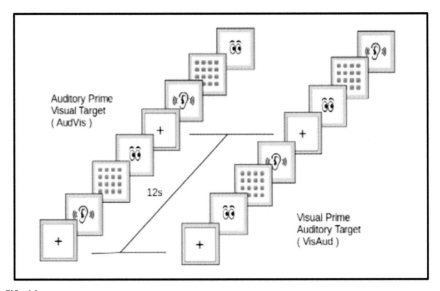

FIG. 11

Depiction of trial runs for the two tasks used in the crossmodal memory study.

Example 2 demonstrates another implementation of using small-scale Neural Graph Models to examine cognition function (Hanson and Hanson, 2023). In this case, a novel feedforward direct modeling approach was used to examine the role of top-down and bottom-up processing in a task involving attention and memory. Specifically, Neural Graph Modeling of fMRI data, collected while subjects performed a task involving priming and memory retrieval, was used to assess network dynamics underlying attention and memory and how those dynamics might change over time.

The task was relatively simple (see Fig. 11). There were two conditions that were performed by all subjects: (1) one in which a visual prime was used to cue an auditory target (VisAud), and (2) the other in which an auditory prime was used to cue a visual target (AudVis). Subjects were cued and asked to select a cell from a 4 × 4 matrix that held an associated image or sound. If an image was used as the cue (e.g., *cat*), a sound was the target (e.g., *meow*) and vice versa. The cells of the matrix were masked so that the content was unknown until a selection was made. Once a selection was made, the cell's content would be revealed. This required subjects to initially guess when selecting a cell, but with time they became familiar with the location of different targets. Different stimuli were used in the two conditions and the order of the conditions was randomized across subjects.

The inclusion of a crossmodal component provided a means of validating the causal connectivity detected with IMaGES. We hypothesized that the directionality of the edges between brain regions that process different sensory inputs would differ as a function of the condition, but that other edges that were not directly involved in sensory processing would be directed similarly across conditions. In Fig. 12A, the hypothetical connectivity of cognitive processes and their associated brain regions underlying the task are depicted. A number of 2-cycle feedback loops were expected as was an oriented edge from Heschl's gyrus to lateral occipital lobule. Fig. 12B displays the actual graph analysis using IMaGES (blue arrows) and FASK (green arrows) for both the VisAud and AudVis conditions. The red arrow represents the directed edge between visual (lateral occipital lobe) and auditory (Heschl's gyrus) sensory processing regions and reverses between the two conditions as hypothesized.

A secondary goal for this study was to monitor how feedback loops were affected by practice. Fig. 12B reveals that feedback edges (shown in green) decreased by 60% from the first to the second run (indicating the emergence of automaticity). "Snapshots" of the learning process for the AudVis condition are shown in Fig. 13. The snapshots were created by running IMaGES on the time series data to produce a "base" acyclic graph, breaking the ROI times into four separate blocks, and then freezing those edges. The time series data were then fit again to the remaining variance using FASK to identify those edges associated with the feedback structure. The resultant "feedback edges" linearly decrease per block over time with a collapse to zero in the fourth quarter snapshot. The priming edge that appears in the second quarter from audio to visual indicates a shift from controlled processing to automatic processing (Schneider and Shiffrin, 1977), essentially a practice effect.

Example 3: Identifying biomarkers for Alzheimer's

Large-scale Neural Graph Models are extremely effective when examining network dynamics in progressive diseases such as schizophrenia and Alzheimer's disease (e.g., Mastrovito et al., 2018). In this example (Hanson et al., 2023) the application of large-scale Neural Graph Models, using resting-state fMRI data, is used to identify biomarkers for Alzheimer's disease, a polygenetic neurodegenerative brain disorder resulting in neocortical atrophy developing over decades and characterized by an increasing loss of synapses and neurons.

Alzheimer's progressively kills neurons in the brain by creating neural plaques and tangles that disrupt local connectivity and eventually systemic and global

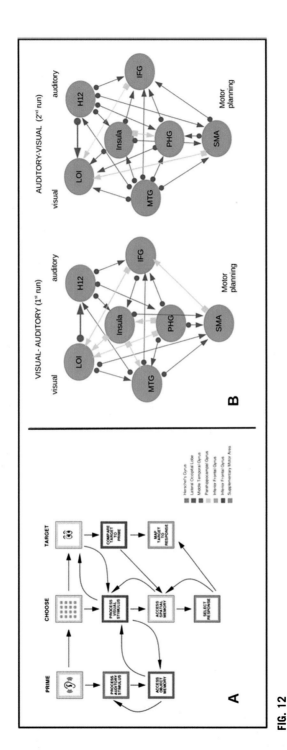

FIG. 12

Panel A shows the hypothetical connectivity for cognitive processes and associated brain regions in the crossmodal task. In Panel B cyclic and acyclic graphs for the two conditions are shown. The red arrow highlights the edge reversal between visual and auditory areas consistent with the condition. Blue arrows indicate acyclic edges (IMaGES) and green arrows indicate cyclic edges (FASK).

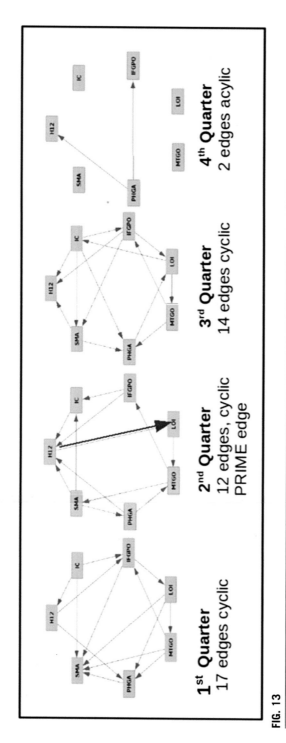

FIG. 13

Shown are four FASK graphs generated from data obtained while subjects performed the AudVis task. The graphs represent the feedback dynamics as the task progresses, providing "snapshots" of changes as subjects become more familiar with the task.

network communication. This progressive loss can increase rapidly even before significant plaques are detected. Consequently, existing biomarkers are lagged predictors and often precede (by as much as a decade) the eventual decline observed in behavior (Sharma, 2016), making it difficult to detect the change points marking cognitive decline during the progression of the disease.

Often the underlying dynamics of a physical or biological system can be very sensitive to local variation or small progressive loss that can be seen in the resulting phase space (the signature of the dynamics in terms of stability and evolution of the sets of variables). The brain activity associated with the resting state is known to have complex dynamics (Biswal et al., 2010; Mastrovito et al., 2018) that are known to be correlated with many mental illnesses, especially age-related dementia and Alzheimer's disease (Badhwar et al., 2017).

The YEO-1000 graph (Fig. 9), a model of default mode network (DMN) dynamics during resting state, has 150–200k edges estimated with FASK. The edge weights vary over a wide range (-200 to 200) with moderate variance as estimated from fitting the BOLD time series with the model. Brain activity can be simulated by seeding all 1000 ROIs with the actual initial fMRI time series values (we often use two or three time steps to enter empirical data). Node activity can then be updated by using the estimated FASK weights. Each node is updated asynchronously (weighted activation from all other connected nodes) over all time steps. The dynamics were simulated over 100 time steps shown in Fig. 14A and B. These graphs actually show a random sample of 100 ROIs given that all 1000 would be impossible to visualize. In this case, we see trajectories that tend asymptotically toward 1, -1, or 0 (the activation function was normalized from 1 to -1 instead of the usual 0, 1) and many that oscillate around zero and some that oscillate in the positive or negative quadrant. Hypothetical differences between neurotypicals and those subjects diagnosed with Alzheimer's disease can then be tested.

Projecting the ROI time series into a lower dimensional manifold using nMDS (non-metric multidimensional scaling), on a distance (Euclidian) matrix of all 1000 time series, accounted for more than 70% of the total variance. In Fig. 14C, we show the comparison of the two subject groups by rotating the 3D phase space of a neurotypical subject and that of a patient with Alzheimer's disease into the same 3D space. Note that although the trajectories end up in a similar region of the space (left corner), the initial trajectory (beginning of the simulated resting stage bottom left) is highly chaotic for the patient with Alzheimer's disease (red line). This result cannot be seen by using actual BOLD data because it has such a low signal-to-noise ratio. However, the RNN/FASK model provides temporal filtering that enhances the signal-to-noise ratio to reveal a complex but distinct *phase space* trajectory (the time trace of the ROI time series) shown in Fig. 14.

The scale-free DD was applied to data obtained from the ADNI Alzheimer's disease database (http://adni.loni.usc.edu) to provide convergence across multiple models. Using the method described earlier, the scale-free DD was fit to data collected from 11 male subjects with Alzheimer's disease and 11 age-matched controls. The scale-free DD (*power* distribution associated with scale-free networks) for

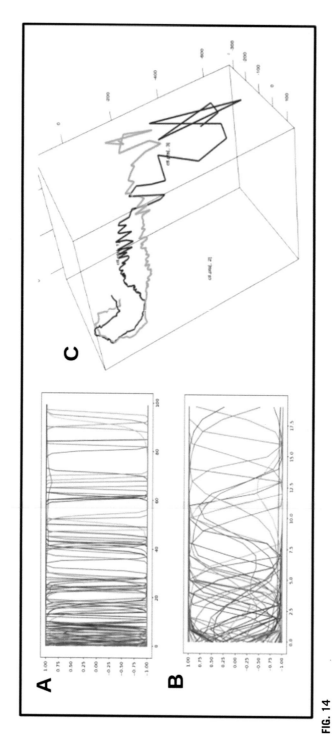

FIG. 14

RNN simulations for large-scale parcellations of neurotypicals (A) and Alzheimer's (B); from the LONI archive. The time series (oscillations) graphs show the simulated time series seeded with actual fMRI data during resting-state tasks for both subject groups. The graphs were fit with FASK and weights were estimated from the 1000 time series. With the architecture and the weights fMRI was used to seed all 1000 nodes and then standard RNN update rules were used to estimate the change in each node. These are plotted in panels A and B. Panel C shows the phase plot of the oscillations projected in three dimensions (using MDS).

Alzheimer's-diagnosed patients and age-matched controls is shown in Fig. 15. These degree (or "order") distributions are derived from IMaGES fit graphs to the 1000 ROI time series. There is a 10% increase in the exponent for the subjects diagnosed with Alzheimer's disease compared to that for age-matched controls. Although this is a small change in a global exponent, it represents a significant local change in network structure.

Cluster cliques based on node weights derived from IMaGES were input as a distance matrix to a WARD's agglomerative cluster analysis (Euclidean distance on estimated weights on each node; with entropy clustering). Seven to eight clusters were identified and the color-coded clusters from the YEO-1000 model are displayed in Fig. 15. Note that the red (temporal lobe) and purple (prefrontal, orbital frontal) areas are significantly disrupted in the subject with Alzheimer's disease (right bottom), with most hub redistributions (decreases in posterior areas and increases in anterior areas) occurring near the anterior temporal lobe.

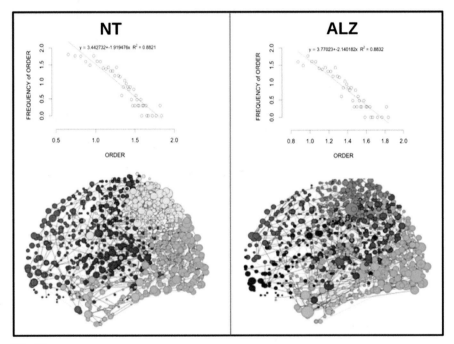

FIG. 15

The same data from LONI used in Fig. 9 is shown here. Note the loss of structure (region clustering) based on degree distribution deviations due to the Alzheimer's group (ALZ), which showed a change in exponent near 10%. Nonetheless, the resultant cluster region graphs based on weights estimated earlier showed massive cluster loss through prefrontal to MTL in ALZ subjects.

Based on unpublisehd figure from R. Sanchez Romeo and S.J. Hanosn.

This result indicates that even relatively small changes in the DD exponent can represent a significant disruption in cortical networks. Examination of the DD also suggests that there may be systematic errors in the fits (which are good >88% over three orders of magnitude), which may lead to a different DD that may provide better sensitivity than the scale-free DD used in this example. The approach described here reveals local network disruptions and provides the means of monitoring the progression of Alzheimer's disease by tracking significant change point trajectories over time.

Assumptions of neural graph models

The Neural Graph Model framework has a number of assumptions. Because the structure of the graph is estimated with Bayesian graph estimation, it assumes the nodes are linear and Gaussian over time. Once the estimated graph network is incorporated into a neural network model, the node variables could be used in a nonlinear update (sigmoid, threshold) with no specific distributional assumptions. This mismatch between graph estimator and use could produce misspecified structure and reduced goodness of fit. We will discuss how to improve this in the section later on future directions.

Another assumption of the Neural Graph Model framework is that the variables are complete and consistent. If, for example, the brain ROIs that are identified are only a subset of potential ROIs, or they are spatially inaccurate, or if other unidentified ROIs exist that may be correlated with the identified ROIs, you may have a *latent* variable. Latent variables are unobserved variables that influence measured or observed variables. There are other methods that are available that can model both observed and latent variables, but that is beyond the scope of this chapter. We provide resources to further research this type of latent variable modeling (in Tetrad).

Finally, the modeling of a task depends on the requisite time resolution that is relevant to the phenomena that are being modeled. Many models using slower time signals (e.g., fMRI) assume a resampling assumption, where the use of many trials will allow the model to sample processes and model them over time as an aggregate effect. This can fail in a number of ways, leaving your model without sufficient information and thus will not likely be the true model. We will discuss how to approach this issue in the next section and attempt to provide ways to create valid models, even in this particular case.

Building, testing, interpreting neural graph models

This is a moment when a strong interest in brain networks, coupled with the availability of novel computational approaches (Reid et al., 2019; Ramsey et al., 2010; Sanchez-Romero et al., 2018; Smith et al., 2011; Farahani et al., 2019), can be leveraged to produce real advances in understanding brain network structure and dynamics. The Neural Graph Model framework adapts the neural network model to a graph with nodes and edges (i.e., weights) where there potentially exists nonlinear activity between nodes (also see Cole, this volume). In this section, we describe considerations when building, testing, and interpreting Neural Graph Models.

How to build a neural graph model

The key to building an effective Neural Graph Model is to determine the goal of the model. Three different applications of how Neural Graph Models can be used in research on cognitive function have been described earlier. Each application used models with slightly different properties based on the questions being addressed. The key properties to consider when building a Neural Graph Model are network size, node selection, and model type.

Network size

Before choosing a Neural Graph Model, the size of the network under investigation must be considered. Small-scale networks are designed for networks with 20 or fewer nodes. These networks are best when modeling influence between brain areas (ROIs) recruited when cognitive or perceptual processing is engaged. A very different approach must be taken for large-scale networks. For networks having more than 20, and up to 1000 nodes, it is the relation between the aggregate statistical properties of connectivity and probability distributions, rather than individual nodes, that must be modeled.

Node selection

For any size network, it is necessary to determine how many and which nodes will be included in the Neural Graph Model. When building a small-scale neural graph it is often advisable to consult the extant literature for hypothetical networks associated with the data being modeled. Networks identified through meta-analysis are also useful when choosing nodes to be used in a Neural Graph Model. For fMRI data, the Neurosynth site (neurosynth.org) provides an extensive collection of studies on all types of cognitive performance. Given your task, it is advisable to do GLM or MVPA (Haxby, 2010) to get the best estimate. Further you should validate these ROIS with your hypotheses (based on edge direction and ROI activity consistent with your activation function) and knowledge of the relevant brain areas.

Type of model

Although there are many types of graph estimators, we focus on two here: IMaGES, an acyclic graph estimator, and FASK, a cyclic graph estimator. Both kinds of estimators have similar kinds of assumptions as discussed earlier, but they are very different in terms of their algorithms and the way they use neural signal time series. In terms of neural networks, we are also focused in this chapter on feedforward and recurrent neural networks, both of which can be more complex than their associated graph estimators, in that neural networks can be non-Gaussian and nonlinear. In terms of software resources, there are a number of possible tools for graph estimators. IMaGES (available as an R CRAN library or from Tetrad) and FASK (available from Tetrad). The Tetrad modeling tools were developed by the Causal Discovery group (https://sites.google.com/view/tetradcausal) using Glymour's original tools for causal modeling. The CRAN R model was developed by the Hanson lab.

Testing a neural graph model

Ideally, a model should be evaluated by how well it accounts for variance in the data. An easy way to account for data variance is to increase the number of parameters that are used to model that data. However, increasing the number of parameters used to fit the data can weaken the model even though the amount of variance accounted for is increased. Without constraints placed on the number of parameters used, it is possible to simply increase the number of parameters until no variance exists. But, as the number of parameters approaches the number of data points, the value of the model as a flexible predictive tool is lost.

A good model optimizes the number of parameters used to account for the variance in the data so that the greatest amount of variance can be accounted for by the fewest number of parameters. Neural Graph Models are tested using standard goodness-of-fit measures (e.g., AIC and BIC), which assess the trade-off between the number of parameters used to model data and the amount of variance accounted for. Goodness-of-fit scores drop as the number of parameters increases, thereby providing a robust and useful means of assessing the value of a given model and comparing across models.

Interpreting a neural graph model

The Neural Graph Model (NGM) is a visualization of the network under investigation. The NGM represents the relation among regions of interest (ROIs) and provides insight into the network dynamics underlying cognitive performance. The NGM is a powerful modeling framework that can be scaled to address different levels of network analysis. NGMs can be used to model individual ROIs in small-scale networks and to characterize organization and complexity in large-scale networks.

A small-scale NGM is used when research is focused on the role a particular set of ROIs may play in a given task. Small-scale NGMs are useful in detecting which ROIs are recruited and how they influence one another. By focusing on specific ROIs and their influence, small-scale NGMs can reveal the flow of information through a network, both feedforward and feedback, that underlies learning and practice effects associated with a given task.

Whereas small-scale models can be used to reveal hypothesized influence between ROIs during cognitive or perceptual processing, it is difficult, if not impossible, to model cognitive functions in large-scale networks due to the extremely large number of nodes and connections. However, cognitive function can be assessed by using probability distributions that characterize the connectivity, clustering, and hub structure of large-scale networks. These emergent properties will determine the dynamics and temporal behavior of the network relative to a population, experimental conditions, or interventions.

Large-scale models are necessary to assess the systemic or global properties of a network because smaller networks are less likely to produce stable degree distribution parameters. For example, to get valid estimates of node degree, at least three orders of magnitude (1000) are required for stable and reliable estimates of the distributional shape (Broido and Clauset, 2019). All characteristic degree distributions are based on developmental growth models, ones that provide some preferential attachment based on existing properties of the network at a given point in time. Limiting cases are random distributions (Erdős and Rényi, 1959), which result in binomial (or with large N, Gaussian/Poisson) distributions over degree. Strictly hierarchical models, where attachment preference is a function of the degree already present, allow the "rich to get richer" and produce scale-free distributions that are power functions (Barabási and Albert, 1999) over degree.

Pitfalls of neural graph models

Potential vulnerabilities should be considered when using Neural Graph Models. First, ROI selection is crucial when developing a small-scale NGM. Detection of interesting effects depends on selecting ROIs that will account for a substantial amount of the variance related to the functional mapping. When possible, the selection of ROIs should be based on brain areas known to be critical to the cognitive function being modeled. Another way to address this vulnerability is to incorporate specific expectations about edges (presence and/or influence) in the model. The selection of ROIs is probably less of an issue in large-scale modeling as long as there is good sampling coverage over the cortical regions of interest.

Second, the time series measured from each node should have a significant variance to avoid errors from centroid estimates of the ROIs that are misestimated based on a particular parcellation algorithm. Choosing ROIs in small-scale cases based on GLM or other contrastive methods could minimize this problem, since these estimates are made manually using specific known locations.

Third, in large-scale applications, the resolution of the parcellation must be large enough to provide statistically significant results. In general, 300–500 ROIs are sufficient for degree distribution estimates; however, increasing the parcellation resolution by three orders of magnitude ($n > 1000$) produces more statistically stable exponents.

Open challenges and future directions

One challenge for the NGM framework is the characterization of early and late learning in small-scale networks. Learning tends to evolve over time from controlled processing to more implicit or automatic processing. In the NGM this is evidenced by feedback links gradually dropping out and being replaced with feedforward links (see Fig. 13). Currently, IMaGES is used to fit the data and then FASK is applied manually to that fit. A factorization method that automates the fit using both IMaGES and FASK would increase accuracy by providing a better estimate of feedforward and feedback edges.

Currently, the NGM relies on the scale-free (power function) degree distribution to fit large-scale networks. However, the scale-free degree distribution assumes that the network is strictly hierarchical, which, of course, is not true of the brain. Although the human brain may have some hierarchical structure, recurrent connections characterize all functional pathways. Expanding the standard scale-free model with other candidates that might better fit the network structure will allow a more sensitive categorization of the degree distributions and network hub structure associated with progressive diseases such as Alzheimer's and schizophrenia.

Another challenge for Neural Graph Models is the time resolution of the neural signal being modeled. Although the sampling rate for fMRI data can be increased by using multiband acquisition, the BOLD signal itself is still relatively sluggish. New methods for collecting brain data, such as OPMEG (optically pumped magnetoencephalography), hold the possibility of millisecond resolution with complete cortical coverage. Increasing the time resolution of the time series data used in the NGM model will undoubtedly prove transformative.

Take-home points

- The transition from blobology to network modeling is a critical step in the evolution of cognitive neuroscience.
- The Neural Graph Model (NGM) is a biologically plausible network modeling framework that merges statistical graph analysis and neural network learning rules to model and simulate neuroimaging data.
- NGMs provide a means of testing hypotheses in small-scale networks by modeling specific ROIs involved in cognitive function.
- In large-scale networks, NGMs can be used to identify network properties (e.g., centrality, clusters, node degree, and potential hubs) associated with systemic change over time.
- The NGM framework has been used successfully to evaluate competing hypotheses about cognitive performance, simulate learning processes, and identify biomarkers sensitive to change points in progressive diseases.

References

Anderson, J.A., Rosenfeld, E., 1998. Talking Nets: An Oral History of Neural Networks. MIT Press.

Badhwar, A., Tam, A., Dansereau, C., Orban, P., Hoffstaedter, F., Bellec, P., 2017. Resting-state network dysfunction in Alzheimer's disease: a systematic review and meta-analysis. Alzheimer's Dement. 8, 73–85. https://doi.org/10.1016/j.dadm.2017.03.007.

Barabási, A.-L., Albert, R., 1999. Emergence of scaling in random networks. Science 286 (5439), 509–512. https://doi.org/10.1126/science.286.5439.509.

Baron-Cohen, S., Leslie, A.M., Frith, U., 1985. Does the autistic child have a "theory of mind"? Cognition 21 (1), 37–46. https://doi.org/10.1016/0010-0277(85)90022-8.

Biswal, B.B., Mennes, M., Zuo, X.N., Gohel, S., Kelly, C., Smith, S.M., Beckmann, C.F., Adelstein, J.S., Buckner, R.L., Colcombe, S., Dogonowski, A.M., Ernst, M., Fair, D., Hampson, M., Hoptman, M.J., Hyde, J.S., Kiviniemi, V.J., Kötter, R., Li, S.J., Lin, C. P., Lowe, M.J., Mackay, C., Madden, D.J., Madsen, K.H., Margulies, D.S., Mayberg, H.S., McMahon, K., Monk, C.S., Mostofsky, S.H., Nagel, B.J., Pekar, J.J., Peltier, S.J., Petersen, S.E., Riedl, V., Rombouts, S.A., Rypma, B., Schlaggar, B.L., Schmidt, S., Seidler, R.D., Siegle, G.J., Sorg, C., Teng, G.J., Veijola, J., Villringer, A., Walter, M., Wang, L., Weng, X.C., Whitfield-Gabrieli, S., Williamson, P., Windischberger, C., Zang, Y.F., Zhang, H.Y., Castellanos, F.X., Milham, M.P., 2010. Toward discovery science of human brain function. Proc. Natl. Acad. Sci. U. S. A. 107 (10), 4734–4739. https://doi.org/10.1073/pnas.0911855107. Epub 2010 Feb 22. PMID: 20176931; PMCID: PMC2842060.

Broido, A.D., Clauset, A., 2019. Scale-free networks are rare. Nat. Commun. *10* (1). https://doi.org/10.1038/s41467-019-08746-5. Article 1.

Bruer, J.T., 2009. Mapping cognitive neuroscience: two-dimensional perspectives on twenty years of cognitive neuroscience research. In: Gazzaniga, M.S. (Ed.), The Cognitive Neurosciences, fourth ed. The MIT Press, pp. 1221–1234, https://doi.org/10.7551/mitpress/8029.003.0108.

Bullmore, E., Sporns, O., 2009. Complex brain networks: graph theoretical analysis of structural and functional systems. Nat. Rev. Neurosci. 10 (3), 186–198. https://doi.org/10.1038/nrn2575.

Cadieu, C.F., Hong, H., Yamins, D.L.K., Pinto, N., Ardila, D., Solomon, E.A., Majaj, N.J., DiCarlo, J.J., 2014. Deep neural networks rival the representation of primate it cortex for core visual object recognition. PLoS Comput. Biol. 10 (12). https://doi.org/10.1371/journal.pcbi.1003963.

Cole, M.W., Bassett, D.S., Power, J.D., Braver, T.S., Petersen, S.E., 2014. Intrinsic and task-evoked network architectures of the human brain. Neuron 83 (1), 238–251. https://doi.org/10.1016/j.neuron.2014.05.014.

Cottrell, G., 1988. Image Compression by Back-Propagation: An Example of Extensional Programming. https://api.semanticscholar.org/CorpusID:58450507.

Elman, J.L., 1990. Finding structure in time. Cognit. Sci. 14 (2), 179–211. https://doi.org/10.1016/0364-0213(90)90002-E.

Erdős, P., Rényi, A., 1959. On the evolution of random graphs. Publ. Math. Debrecen 6, 45.

Farahani, F.V., Karwowski, W., Lighthall, N.R., 2019. Application of graph theory for identifying connectivity patterns in human brain networks: a systematic review. Front. Neurosci. 13. https://doi.org/10.3389/fnins.2019.00585.

Gallese, V., 2009. Motor abstraction: a neuroscientific account of how action goals and intentions are mapped and understood. Psychol. Res. 73 (4), 486–498. https://doi.org/10.1007/s00426-009-0232-4.

Gates, K.M., Molenaar, P.C.M., Iyer, S.P., Nigg, J.T., Fair, D.A., 2014. Organizing heterogeneous samples using community detection of GIMME-Derived resting state functional networks. PLoS One 9 (3). https://doi.org/10.1371/journal.pone.0091322.

Giles, L.C., Horne, B.G., Lin, T., 1995. Learning a class of large finite state machines with a recurrent neural network. Neural Netw. 8 (9), 1359–1365. https://doi.org/10.1016/0893-6080(95)00041-0.

Glymour, C.N., 1980. Theory and Evidence. Princeton University Press. https://books.google.com/books?id=HQmLQgAACAAJ.

Granger, C.W.J., 1969. Investigating causal relations by econometric models and cross-spectral methods. Econometrica 37 (3), 424–438. https://doi.org/10.2307/1912791.

Grelotti, D.J., Gauthier, I., Schultz, R.T., 2002. Social interest and the development of cortical face specialization: what autism teaches us about face processing. Dev. Psychobiol. 40 (3), 213–225. https://doi.org/10.1002/dev.10028.

Hanson, S.J., 2022. The failure of blobology: fMRI misinterpretation, maleficience and muddle. Front. Hum. Neurosci. 16. https://doi.org/10.3389/fnhum.2022.870091.

Hanson, S.J., Burr, D.J., 1990. What connectionist models learn: learning and representation in connectionist networks. Behav. Brain Sci. 13 (3), 471–489.

Hanson, S.J., Glymour, C., 2010. 11 Discovering how brains do things. In: Hanson, S.J., Bunzl, M. (Eds.), Foundational Issues in Human Brain Mapping, p. 115.

Hanson, C., Hanson, S.J., 1996. Development of schemata during event parsing: Neisser's perceptual cycle as a recurrent connectionist network. J. Cogn. Neurosci. 8 (2), 119–134. https://doi.org/10.1162/jocn.1996.8.2.119.

Hanson, C., Hanson, S.J., 2023. Network Dynamics Underlying a Cross Modal Priming Task. *Manuscript in Preparation.*

Hanson, S., Kegl, J., 1987. PARSNIP: A connectionist network that learns natural language grammar from exposure to natural language sentences. In: Proceedings of the Ninth Annual Conference of the Cognitive Science Society. Erlbaum, Hillsdale, NJ, pp. 106–119.

Hanson, S.J., Negishi, M., 2002. On the emergence of rules in neural networks. Neural Comput. 14 (9), 2245–2268. https://doi.org/10.1162/089976602320264079.

Hanson, C., Hanson, S.J., Ramsey, J., Glymour, C., 2013. Atypical effective connectivity of social brain networks in individuals with Autism. Brain Connect. 3 (6), 578–589. https://doi.org/10.1089/brain.2013.0161.

Hanson, S.J., Mastrovito, D., Hanson, C., Ramsey, J., Glymour, C., 2016. *Scale-free exponents of resting state are biomarkers of neuro-typical and atypical brain activity* [preprint]. Bioinformatics. https://doi.org/10.1101/068841.

Hanson, S.J., Hanson, C., Sanchez-Romero, R., Mastrovito, D., 2018a. NIH Brain Theories Workshop. Washington DC.

Hanson, C., Caglar, L.R., Hanson, S.J., 2018b. Attentional bias in human category learning: the case of deep learning. Front. Psychol. 9. https://doi.org/10.3389/fpsyg.2018.00374.

Hanson, S.J., Yardiv, V., & Hanson, C. (2023). Network Disruptions Provide Biomarkers for Alzheimer's Disease. Manuscript in Preparation.

Hinton, G.E., Sejnowski, T.J., 1986. Learning and relearning in Boltzmann machines. In: Parallel Distributed Processing Explorations in the Microstructure of Cognition: *Vol. 1: Foundations.* MIT Press, Cambridge, MA, pp. 282–317. https://papers.cnl.salk.edu/PDFs/Learning%20and%20Relearning%20in%20Boltzmann%20Machines%201986-3239.pdf.

Haxby, J.V., 2010. Multivariate pattern analysis of fMRI data: high-dimensional spaces for neural and cognitive representations. In: Hanson, S.J., Bunzl, M. (Eds.), Foundational Issues in Human Brain Mapping. MIT Press.

Hinton, G.E., Osindero, S., Teh, Y.-W., 2006. A fast learning algorithm for deep belief nets. Neural Comput. 18 (7), 1527–1554. https://doi.org/10.1162/neco.2006.18.7.1527.

Hopfield, J.J., 1982. Neural networks and physical systems with emergent collective computational abilities. Proc. Natl. Acad. Sci. U. S. A. 79 (8), 2554–2558. https://doi.org/10.1073/pnas.79.8.2554.

Hyvärinen, A., Smith, S.M., 2013. Pairwise likelihood ratios for estimation of non-gaussian structural equation models. J. Mach. Learn. Res. 14 (1), 111–152.

Jordan, M.I., 1986. Serial Order: A Parallel Distributed Processing Approach. ICS Report 8604. Institute for Cognitive Science, UCSD, La Jolla.

Mastrovito, D., Hanson, C., Hanson, S.J., 2018. Differences in atypical resting-state effective connectivity distinguish autism from schizophrenia. NeuroImage : Clin. 18, 367–376. https://doi.org/10.1016/j.nicl.2018.01.014.

McCulloch, W.S., Pitts, W., 1943. A logical calculus of the ideas immanent in nervous activity. Bull. Math. Biophys. 5, 115–133. http://www.cse.chalmers.se/~coquand/AUTOMATA/mcp.pdf.

McIntosh, A., 1998. Understanding neural interactions in learning and memory using functional neuroimaging. Ann. N. Y. Acad. Sci. 855 (1), 556–571. https://doi.org/10.1111/j.1749-6632.1998.tb10625.x.

Meek, C., 1997. Graphical Models: Selecting Causal and Statistical Models. [PhD Thesis]. Carnegie Mellon University.

Mumford, J.A., Ramsey, J.D., 2014. Bayesian networks for fMRI: a primer. Neuroimage 86, 573–582. https://doi.org/10.1016/j.neuroimage.2013.10.020.

Ramsey, J.D., Hanson, S.J., Hanson, C., Halchenko, Y.O., Poldrack, R.A., Glymour, C., 2010. Six problems for causal inference from fMRI. Neuroimage 49 (2), 1545–1558.

Ramsey, J.D., Spirtes, P., Glymour, C., 2011. On meta-analyses of imaging data and the mixture of records. Neuroimage 57 (2), 323–330. https://doi.org/10.1016/j.neuroimage.2010.07.065.

Reid, A.T., Headley, D.B., Mill, R.D., Sanchez-Romero, R., Uddin, L.Q., Marinazzo, D., Lurie, D.J., Valdés-Sosa, P.A., Hanson, S.J., Biswal, B.B., Calhoun, V., Poldrack, R.A., Cole, M.W., 2019. Advancing functional connectivity research from association to causation. Nat. Neurosci. 22 (11), 1751–1760. https://doi.org/10.1038/s41593-019-0510-4.

Rosenblatt, F., 1961. Principles of Neurodynamics: Perceptrons and the Theory of Brain Mechanisms. Spartan Books.

Rumelhart, D.E., Hinton, G.E., Williams, R.J., 1986. Learning internal representations by error propagation. In: Parallel Distributed Processing: Explorations in the Microstructure of Cognition, vol. 1: Foundations, p. 318.

Sanchez-Romero, R., Ramsey, J.D., Zhang, K., Glymour, M.R.K., Huang, B., Glymour, C., 2018. Estimating feedforward and feedback effective connections from fMRI time series: assessments of statistical methods. Netw. Neurosci. 3, 274–306. https://doi.org/10.1162/netn_a_00061.

Schneider, W., Shiffrin, R.M., 1977. Controlled and automatic human information processing: I. Detection, search, and attention. Psychol. Rev. 84 (1), 1–66. https://doi.org/10.1037/0033-295X.84.1.1.

Sharma, N., 2016. Exploring biomarkers for Alzheimer's disease. J. Clin. Diagn. Res. https://doi.org/10.7860/JCDR/2016/18828.8166.

Smith, S.M., Miller, K.L., Salimi-Khorshidi, G., Webster, M., Beckmann, C.F., Nichols, T.E., Ramsey, J.D., Woolrich, M.W., 2011. Network modelling methods for FMRI. Neuroimage 54 (2), 875–891. https://doi.org/10.1016/j.neuroimage.2010.08.063.

Vrieze, S.I., 2012. Model selection and psychological theory: a discussion of the differences between the Akaike information criterion (AIC) and the Bayesian information criterion (BIC). Psychol. Methods 17 (2), 228–243. https://doi.org/10.1037/a0027127.

Yamins, D.L.K., Hong, H., Cadieu, C.F., Solomon, E.A., Seibert, D., DiCarlo, J.J., 2014. Performance-optimized hierarchical models predict neural responses in higher visual cortex. Proc. Natl. Acad. Sci. U. S. A. 111 (23), 8619–8624. https://doi.org/10.1073/pnas.1403112111.

Machine learning and neuroimaging: Understanding the human brain in health and disease

Zijin Gu[a], Keith W. Jamison[b,c], Mert R. Sabuncu[a,c], and Amy Kuceyeski[b,c]

[a]*School of Electrical and Computer Engineering, Cornell University and Cornell Tech, New York, NY, United States,* [b]*Department of Computational Biology, Cornell University, Ithaca, NY, United States,* [c]*Department of Radiology, Weill Cornell Medicine, New York, NY, United States*

Introduction to machine learning in neuroimaging

Machine learning (ML), lying at the intersection of computer science and statistics, is a subfield of artificial intelligence and one of the most dramatically progressing technical fields today (Jordan and Mitchell, 2015). The historical roots of ML can be traced back to the 1950s when pioneers like Alan Turing and Arthur Samuel began exploring the concept of machines that could perform tasks of human intelligence level and learn from data (Turing, 2009; Samuel, 1959). However, it was not until the development of powerful computational hardware and the availability of vast amounts of data that ML models started gaining prominence. An ML model, different from traditional programming where explicit instructions are needed for a particular task, is a mathematical or computational representation of a system that makes predictions or decisions automatically. Typically, ML models are created through training, wherein the model learns patterns or relationships within a large amount of data by iteratively optimizing its parameters to minimize some predefined loss function. Once trained, the model can make predictions on new, unseen data based on the patterns it learned during training. ML research involves addressing both fundamental scientific and engineering questions as well as practical applications across the empirical sciences, including medicine.

ML models began to appear in medical applications as early as the 1960s, where early attempts were made at computer-aided diagnosis with simple rule-based algorithms to assist radiologists in detecting abnormalities. More recently, with the development and widespread adoption of noninvasive neuroimaging techniques like magnetic resonance imaging (MRI), positron emission tomography (PET), and

Computational and Network Modeling of Neuroimaging Data. https://doi.org/10.1016/B978-0-443-13480-7.00010-7

electroencephalography (EEG), data within the neuroscientific community have undergone exponential growth (Kernbach et al., 2022). Consequently, there is an increasing need for advanced and efficient tools to extract meaningful insights from these large datasets, and thus ML models have emerged as a natural fit for analyzing and interpreting neuroimaging data. The application of ML models to neuroimaging data allows identification of complex patterns and relationships that exist within the data, providing a deeper understanding of brain structure, function, and pathology, and how these metrics map to behavior or impairments. One notable example of this is the application of deep neural networks in analyzing functional MRI (fMRI) data to isolate potential biomarkers associated with specific neurological conditions (Talo et al., 2019; Kamnitsas et al., 2017).

The theoretical context for ML applied to neuroimaging encompasses different fields. Statistical and mathematical frameworks, that is, Bayesian statistics and graph theory, have been instrumental in the development and application of ML models of uncertainty in brain data (Friston et al., 2002a, b; Woolrich et al., 2009; Friston, 2012) and characterizing brain networks and their properties (Bullmore and Sporns, 2009; Rubinov and Sporns, 2010; Sporns, 2012). Theoretical and computational neuroscience provide a quantitative framework for the underlying mechanisms and general principles of nervous systems, thus providing insights into developing biologically plausible ML models (Dayan and Abbott, 2005). Advancements in deep learning (DL) have enabled more accurate and efficient identification and interpretation of patterns in brain data. Convolutional neural networks (CNNs), for example, U-Net (Ronneberger et al., 2015) and its variants (Çiçek et al., 2016; Oktay et al., 2018; Zhou et al., 2018), have demonstrated exceptional performance in tasks such as brain image classification and segmentation. Generative models, for example, generative adversarial networks (GANs) or variational autoencoders (VAEs), have shown promising results in generating synthetic brain images that resemble the characteristics and patterns observed in real data (Han et al., 2018), as well as accurately decoding and reconstructing external stimuli from brain signals recorded during the stimulus presentation (Gu et al., 2022a). Synthetic images generated with these models can be useful for creating or augmenting training datasets, particularly when data is limited.

ML, with its ability to extract meaningful information from vast amounts of data, is playing an increasingly indispensable role in analyzing neuroimaging data with the ultimate goal of understanding the most complex object in the known universe—the human brain. ML models can be used to detect and capture complicated patterns within complex, multimodal data, including nonlinear relationships that may not be apparent through traditional statistical methods. For instance, CNNs have demonstrated remarkable performance in image classification tasks, and have been applied to segment specific brain regions or abnormalities (Esteva et al., 2017). ML models have also been used to identify reliable biomarkers for neurological and psychiatric disorders, thus facilitating more accurate disease diagnosis, monitoring, and early intervention. For example, researchers have been analyzing the large-scale Alzheimer's Disease Neuroimaging Initiative (ADNI) dataset (Jack et al., 2008) to develop ML models that extract features and patterns associated with Alzheimer's disease (AD) progression in order for earlier diagnosis and thus intervention (Arbabshirani et al., 2017; Salvatore et al., 2015). ML models even offer opportunities for personalized disease treatment by leveraging individual characteristics, neuroimaging biomarkers, and clinical information to tailor interventions based on specific patient profiles and predicted treatment responses (Myszczynska et al., 2020; Shah et al., 2019). In the exploration of brain connectivity and functional networks, ML models, combined with graph theory, have shown success in characterizing the relationship between brain network properties and cognitive processes as well as clinical neurodevelopmental outcomes (Bassett and Sporns, 2017; Kawahara et al., 2017). Finally, ML models, with their data-driven nature, contribute to advancing basic neuroscientific knowledge of how the brain works. For instance, ML models have shed light on the neural correlates of specific cognitive tasks or mental states (Pereira et al., 2009) and interindividual and interregional differences in brain region responses to visual stimuli (Gu et al., 2022c). ML models provide valuable insights on many fundamental questions about brain structure, function, and cognitive processes, thus advancing our understanding of brain-behavior relationships.

Examples of successful machine learning in neuroimaging

The field of neuroscience has witnessed remarkable advances in understanding brain-behavior relationships through the combination of ML techniques and big neuroimaging data. In this section, we delve into some successful applications of ML for brain behavior mapping, where the goal is to uncover patterns in neuroimaging data that are associated with specific behavioral or cognitive outcomes. According to different neuroimaging data types and applications, we start with classic ML models, for example, support vector machines (SVMs), and move on to more complicated models such as CNNs, graph neural networks (GNNs), and generative models. In each part, we first explore the general theoretical foundations and technical aspects, and then focus on one representative work, with a detailed description of its approach and potential impacts.

Classic ML models like SVM can classify brain states

Classic ML algorithms have played a significant role in the analysis of neuroimaging data. They are usually simple to implement, efficient to compute, and easy to interpret. SVM (Boser et al., 1992) is one of the most widely used ML algorithms in neuroimaging, with applications expanding from classification of clinical populations, prediction of cognitive states, to the identification of biomarkers associated with specific brain disorders.

SVM is a supervised ML algorithm that aims to find a hyperplane that maximally separates data samples from different classes, as shown in Fig. 1. In cases where the data are not linearly separable in the original feature space, a common issue in neuroimaging data, SVM employs a technique called the kernel trick. The kernel trick allows SVM to implicitly transform the data into a higher-dimensional feature space where it becomes linearly separable. Common kernel functions include linear, polynomial, radial basis function (RBF), and sigmoid. SVM, along with other classic ML models, can be easily implemented with Matlab or Python scikit-learn package for various neuroimaging applications (Abraham et al., 2014).

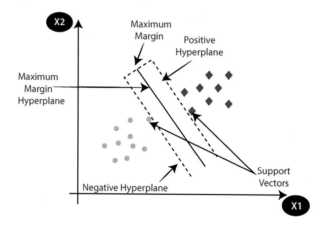

FIG. 1

SVM works by finding a hyperplane that achieves maximum margin.

From https://www.analyticsvidhya.com/blog/2021/10/support-vector-machinessvm-a-complete-guide-for-beginners/.

SVMs have shown promising results in classifying brain states. For example, Mourao-Miranda et al. (2005) used SVM to discriminate fMRI activation patterns from two different visual attention tasks: face matching and location matching. The schematic illustration is shown in Fig. 2. After data preprocessing, PCA is first applied to project the high-dimensional fMRI volumes onto the principal components obtained from all training volumes. Then, in the training phase, SVM with a linear kernel is employed to find the most discriminating voxels between the two brain states. During testing, fMRI volumes from new subjects are projected onto the principal components and fed into the model, which predicts the brain state for the new subjects. The model was evaluated using leave-one-out cross-validation and error rate, sensitivity, and specificity as metrics. When comparing the results from

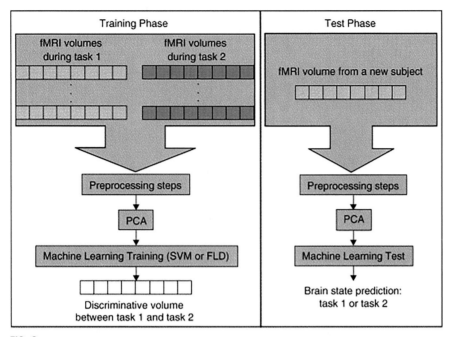

FIG. 2

SVM for brain state classification. During training, the ML algorithm finds the discriminating volume to best distinguish two brain states. During testing, the classifier predicts the brain state for a given new subject. The data are preprocessed and dimensionality reduced using PCA (Mourao-Miranda et al., 2005).

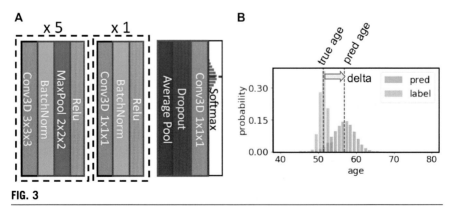

FIG. 3

Simple fully convolutional neural network (SFCN) for brain age prediction using 3D neuroimaging data (Peng et al., 2021). (A) The model contains seven convolutional blocks which map the input brain image to the age probability distribution. (B) An example of soft labels and output probabilities.

SVM with the results from Fisher linear discriminant (FLD), SVM clearly outperforms FLD in both classification performance and robustness of obtained spatial maps.

CNNs can help infer brain age from neuroimages

The CNN architecture was originally inspired by the principles of how the feedforward cortical network processes visual information (DiCarlo et al., 2012). With the development of DL, many variants of CNN have emerged and shown great success in natural image classification and pattern recognition (He et al., 2016; Krizhevsky et al., 2017). This drew people's attention to "transfer learning" where adopting such methods in neuroimaging contexts opens the door for neuroimaging in DL. CNNs have shown great promise in various predictive and diagnostic tasks such as tumor/lesion segmentation, disease classification, and prediction of individual characteristics (i.e., age and sex) (Kamnitsas et al., 2017; Thomas et al., 2020; Peng et al., 2021).

Two-dimensional (2D) CNNs are the most widely used neural network; their key component is the convolution kernel that learns to detect regular features within various parts of an image. By applying multiple different kernels in each convolutional layer, various features can be extracted from the image to make downstream predictions. For three-dimensional (3D) brain data, 3D CNN is a natural extension of conventional 2D CNN for analyzing such type of data by adding an additional depth dimension to its convolutional kernels. 3D CNNs are able to capture both spatial and volumetric dependencies within the input data.

Brain age prediction using T1-weighted MRI scans of brain anatomy is one of the most successful use cases of 3D CNNs (Cole et al., 2017; Jónsson et al., 2019; Peng et al., 2021). Here, we introduce one representative work that proposed a simple fully convolutional neural network (SFCN) model (Peng et al., 2021) for brain age prediction. The architecture of SFCN is based on VGGNet (Simonyan and Zisserman, 2015) with only seven convolutional blocks as shown in Fig. 3A. There is one $3 \times 3 \times 3$ 3D convolutional layer, one batch normalization later, one max pooling layer, and one activation layer, for example, rectified linear unit (ReLU), in each of the first five blocks. This is the first stage of the model and extracts feature maps from the input images. The sixth block contains one $1 \times 1 \times 1$ 3D convolutional layer, one batch normalization later, and one ReLU layer, further increasing the nonlinearity of the model. The last block contains one average pooling layer, one dropout layer, one $1 \times 1 \times 1$ 3D convolutional layer, and one softmax layer; it serves as a classifier to map the extracted features to the age probability distribution. The model is trained and evaluated on UK Biobank dataset (Miller et al., 2016) to minimize a Kullback-Leibler divergence loss between the predicted probability and a Gaussian distribution with true age as mean and 1 year as variance, as shown in Fig. 3B. SFCN demonstrated the best performance in mean absolute error compared to other popular 3D ResNet-based models and simpler regression models.

Graph neural networks identify disease-related biomarkers from brain networks

Brain connectivity networks, that is, brain connectomes, contain information about connectivity patterns between brain regions and are represented by graphs, wherein each brain region corresponds to a node, and the connections between regions corresponds to the edges. This data type is in contrast to the 3D volumes or 2D images normally associated with neuroimaging datasets, and thus require alterations to classic ML approaches developed for regular image data (Sporns et al., 2005).

BrainNetCNN is the first deep CNN architecture designed for the special structure of brain network data (Kawahara et al., 2017). Inspired by image-based CNNs, BrainNetCNN leverages network topology information and generalizes the traditional convolutional filters to accommodate graph data. Specifically, BrainNetCNN has three layer types: edge-to-edge (E2E), edge-to-node (E2N), and node-to-graph (N2G), shown in Fig. 4, to extract meaningful features from brain networks. Given a brain network represented by $G = (A, \Omega)$, where $A \in \mathcal{R}^{|\Omega| \times |\Omega|}$ is a weighted adjacency matrix of the edges connecting the brain regions and Ω is a set of nodes representing brain regions, E2E layer filters the adjacency matrix by taking a weighted combination of the weights of edges that share nodes together, E2N layer then acts similarly but produces one value for each node and outputs a vector of size $|\Omega|$, and finally the N2G layer continues to reduce the dimensionality of the feature map and outputs a single scalar. BrainNetCNN was validated using diffusion MRI-derived structural brain networks of preterm infants to perform postmenstrual age and neurodevelopmental scores prediction.

BrainNetCNN has not only advanced neurodevelopmental research with improved predictive performance and new insights into brain connectivity, but also opened up new possibilities in advancing the methodology for analyzing brain network data. Many specialized CNN-based approaches for network neuroimaging data are inspired by it and have shown success in various types of neuroimaging data and tasks (Meszlényi et al., 2017; Phang et al., 2019; Kam et al., 2019).

In recent years, GNNs, a subfield of geometric DL, have gradually become mainstream in brain network modeling due to the inherent graph structure of the brain networks. Among different types of GNNs, graph convolutional networks (GCNs) are the most widely used model for network neuroscience (Bessadok et al., 2022). In individual level modeling shown in Fig. 5A, each node can be viewed as a brain region and each edge can be viewed as the connection strength between two regions. GNNs can also be utilized on a population level, in which each node represents a particular subject and each edge between two nodes represents the similarity between two subjects, as shown in Fig. 5B.

Applications of GNNs in neuroimaging can be generally categorized into brain graph classification, generation, and integration (Bessadok et al., 2022). The most widely used case in classification is to identify brain state and biomarkers related to certain diseases, such as autism spectrum disorder (ASD) and AD. For graph-based modeling, there was a Siamese-GCN adopted for similarity estimation

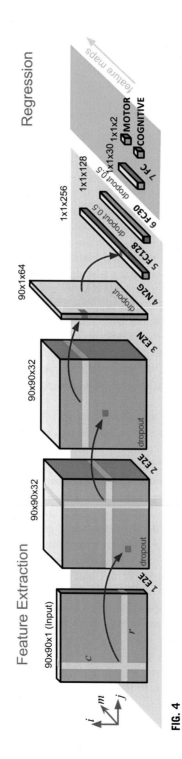

FIG. 4

BrainNetCNN architecture for graph CNN-based brain-behavior mapping. BrainNetCNN takes in the brain network adjacency matrix and passes it through the E2E, E2N, and N2G filters, and uses fully connected (FC) layers to map the extracted features to behavioral/demographic variables (Kawahara et al., 2017).

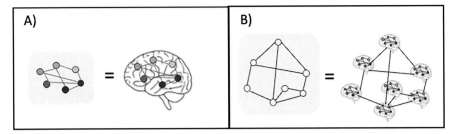

FIG. 5

GNNs can be used in graph-based modeling and population-based modeling. (A) Individual level graph model and (B) population level graph model.

between brain functional connectivities (Ktena et al., 2018), and an interpretable BrainGNN designed to detect salient brain regions having high correspondence with neuroimaging-derived evidence of biomarkers (Li et al., 2021). For population-based modeling, GCNs leveraging nonimaging information were shown to boost prediction performance (Parisot et al., 2018). Other applications of GNNs in neuroimaging include brain network prediction and integration. In brain graph prediction, GNNs are used to synthesize brain graphs for patients who have missing data acquisitions, which in turn helped boost the accuracy of models designed to perform early disease diagnosis (Bessadok et al., 2021). And for brain graph integration, GNNs are used to integrate multimodal brain connectivities into a single brain graph that represents the shared patterns across subjects (Gurbuz and Rekik, 2020).

Synthesizing images with generative models

Generative models are a type of ML tool that learns the underlying distribution of the data in order to generate new samples that closely resemble the original data. Generative models, particularly in computer vision, have achieved excellent performance in high-quality image generation, image-to-image translation, image denoising, etc. (Van Den Oord et al., 2017; Brock et al., 2019; Karras et al., 2019). They have also gained significant attention in neuroimaging with their potentials in many neuroimaging applications.

GANs (Goodfellow et al., 2020) and VAEs (Kingma and Welling, 2014) are two of the most widely used types of generative models. GANs have a generator and a discriminator, where the generator creates synthetic data samples and the discriminator tries to distinguish between real and generated data. The two components are trained in tandem in an adversarial manner such that the quality of synthetic data gets

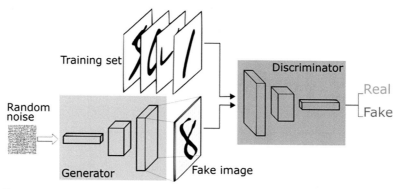

FIG. 6

Illustration of a GAN trained on the MNIST dataset consisting of handwritten digits. The generator creates synthetic digit samples, while the discriminator tries to distinguish between real and generated digits.

From https://www.freecodecamp.org/news/an-intuitive-introduction-to-generative-adversarial-networks-gans-7a2264a81394.

improved over time, see Fig. 6. VAEs combine autoencoders with probabilistic modeling to learn the latent distribution of the data such that new data can be generated by sampling from the latent space, see Fig. 7.

In neuroimaging settings, GANs and VAEs have been mostly used in data augmentation for disease diagnosis (Gao et al., 2021; Lin et al., 2021), anomaly detection of tumors and lesions (Nguyen et al., 2021; Bengs et al., 2021), modeling brain aging and disease progression (Xia et al., 2021; Ravi et al., 2022), and neural response modulation (Ponce et al., 2019; Gu et al., 2023). For example, Lin et al. (2021) use a 3D reversible GAN to synthesize high-quality PET images for missing data imputation. They found that using both the synthesized PET images and existing MRI was beneficial for the diagnosis and prediction of AD. A recent work by Gu et al. (2023) combined a GAN with a neural encoding model to generate region-preferred images for individual brains, and demonstrated that it was possible to modulate macroscale human brain responses with the generated images.

Recently, diffusion models (Ho et al., 2020) have shown superior performance than GANs and VAEs in image synthesis tasks (Dhariwal and Nichol, 2021; Rombach et al., 2022), and thus have gained a growing interest from neuroimaging field. The diffusion probabilistic model is a Markov chain that tries to retrieve the noise-free data from noisy data obtained from a diffusion process, where the input data are gradually perturbed over several steps by adding Gaussian noise, see Fig. 8. Applications of diffusion models in neuroimaging include brain image generation (Güngör et al., 2023), segmentation (Wolleb et al., 2022), anomaly detection (Wolleb et al., 2022), etc.

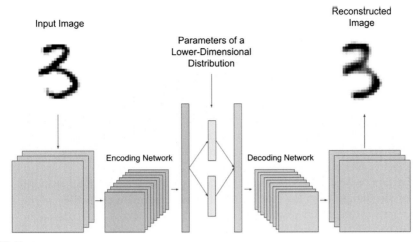

FIG. 7

Illustration of a VAE trained with MNIST dataset of handwritten digits. The encoder encodes the input digit image into a lower-dimensional latent space with learned distribution. The decoder samples from the distribution and decodes it back into the original image space.

From https://www.assemblyai.com/blog/introduction-to-variational-autoencoders-using-keras/.

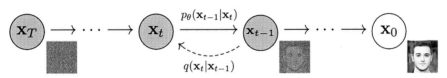

FIG. 8

A directed graphical illustration of a diffusion model (Ho et al., 2020).

Assumptions of machine learning models

ML models make certain assumptions to effectively analyze and capture the structure of neuroimaging data. These assumptions may vary depending on the specific ML algorithm or application, but here are some common assumptions made by ML models applied to neuroimaging data:

- *Data distribution.* ML modeling often needs data samples to be independent and identically distributed (IID). When splitting the data into training and test sets, IID data help to ensure that the learned patterns and relationships from the training set generalize well to unseen test sets during the deployment phase. IID assumptions also help in deriving learning methods, proving consistency, and providing error bounds on estimates. In neuroimaging, this assumption can imply that each voxel, region, or time point in the data is statistically independent and follows the same underlying distribution. However, in practice, these assumptions can be too strong. For example, neuroimaging data often exhibit spatial or temporal dependencies that are best accounted for in the analysis.

- *Relationship between input and output.* When choosing or designing ML algorithms, one often needs to predefine the relationships between the input features and the output variables to be either linear or nonlinear. Linear models, such as linear regression and logistic regression, assume that linear combinations of the input features are enough to represent the target variables. Nonlinear models, on the other hand, make the assumption that more complex nonlinear relationships exist between the features and the target variable(s). Kernel methods and neural networks are popular ML algorithms that have shown effectiveness in modeling nonlinear dependencies in data.

- *Feature relevance.* ML models aim to extract meaningful patterns or representations from the input features to make accurate predictions of the target variable. A key assumption here is that the input features contain relevant information and contribute to the estimation of the target variable. In neuroimaging, datasets are usually high dimensional, wherein some features may be redundant or irrelevant for the task at hand. Therefore, data preprocessing, such as feature selection or dimensionality reduction techniques, is often employed to identify the most informative features and discard irrelevant or redundant ones to improve the accuracy and efficiency of the final model and allow easier model interpretation.

- *Sample size.* ML models in general benefit from larger sample sizes to learn meaningful patterns in the data and avoid overfitting. Insufficient sample sizes may lead to poorly fitted models that cannot generalize to new data effectively, producing unreliable results. In neuroimaging studies, one can choose ML models of different capacity based on the sample size available to obtain reliable and robust ML models. For instance, small data might not be suitable for training a deep neural network.

Building, testing, interpreting machine learning models

In this section, we outline the steps that can be involved in building, testing, and interpreting ML models applied to neuroimaging data.

- *Data preprocessing.* Raw neuroimaging data usually need to follow certain preprocessing pipelines to extract clean signals before being used in ML model building. For example, a typical preprocessing pipeline for fMRI data may include steps such as skull stripping, motion correction, slice-timing correction, temporal/spatial smoothing, registration, and normalization to a template. Various software packages and libraries, such as FSL (Smith et al., 2004), SPM (Penny et al., 2011), or ANTs (Avants et al., 2009), provide tools for neuroimaging preprocessing.

- *Feature extraction.* Feature extraction and selection methods vary depending on the specific neuroimaging data types and research questions being asked. For example, cortical density maps/thicknesses and volumes of brain regions from predefined atlases are common features when assessing structural variation while connectivity analysis and graph measures are two examples of techniques for analyzing fMRI (Rathore et al., 2017).

- *Feature selection and reduction.* The next step following feature extraction is feature selection/reduction, which is essential especially for *individual* level prediction and inference due to the "large p, small n" issue, where the number of predictor variables, for example, voxels, can be much larger than the number of observations, for example, subjects (Fort and Lambert-Lacroix, 2005). Broadly speaking, there are two types of feature reduction techniques, supervised and unsupervised. Supervised techniques can be further divided into "filter" (e.g., Pearson correlation coefficient [PCC]; Guyon and Elisseeff, 2003), "wrapper" (e.g., recursive feature elimination [RFE]; Guyon et al., 2002, searchlight; Kriegeskorte et al., 2006), and "embedded" methods (e.g., LASSO; Tibshirani, 1996), while principal component analysis (PCA) (Abdi and Williams, 2010) and independent component analysis (ICA) (Stone, 2004) are two representative methods for unsupervised approaches (Mwangi et al., 2014). Efficient feature reduction can help avoid overfitting in ML models and therefore improve prediction accuracy and generalization to unseen data.
- *Model selection and design.* The selection and design of ML models is usually informed by the research question and the nature of the data. For some cases where the task is relatively simple, or interpretability and transparency are crucial, classic ML algorithms, including SVMs (Boser et al., 1992) and random forests (RFs) (Breiman, 2001), might be used. Scikit-learn (Pedregosa et al., 2011; Buitinck et al., 2013) is a popular Python toolbox that offers a wide range of implementations of classic ML algorithms, including classification, regression, clustering, and dimensionality reduction algorithms. Nilearn (Abraham et al., 2014) is another Python library specifically designed for neuroimaging data analysis, which leverages scikit-learn for multivariate statistics with applications such as predictive modelling, classification, decoding, or connectivity analysis. When dealing with complex and high-dimensional data, neural networks such as CNNs (Krizhevsky et al., 2017; He et al., 2016) and recurrent neural networks (RNNs) (Hochreiter and Schmidhuber, 1997), usually have better performance. The specific architecture of the ML model may be modified to accommodate the requirements of the problem. Popular DL frameworks include PyTorch (Paszke et al., 2019), TensorFlow (Abadi et al., 2016), and Keras (Chollet et al., 2015).
- *Model training, validation, and evaluation.* The preprocessed data are split into training, validation, and test sets, and this process should be random and each split should be representative of the underlying distribution of the entire dataset. The training set is used to train the model, which typically occupies around 70%–80% of the data. Optimization techniques, such as stochastic gradient descent or Adam (Kingma and Ba, 2015), may be applied to learn the model parameters during training. The validation set (which might be around 10%–15% of the data) is typically used to tune hyperparameters based on some evaluation metrics; this step helps in preventing overfitting and provides an estimate of the model's generalization performance. The test set (which, again, might be around

10%–15% of the data) is used to assess the final performance of the trained model after all the model adjustments and hyperparameter tuning. Commonly used evaluation metrics for ML models are accuracy, precision, recall, F1 score, and area under the receiver operating characteristic curve (AUC-ROC).

- *Interpretation.* Interpretability and explainability in ML become more and more important (Doshi-Velez and Kim, 2017), particularly so in the context of medical applications. It is crucial to understand the learned patterns or relationships captured by the model to make fully informed and transparent clinical decisions. Feature importance analysis, saliency maps, or model-agnostic approaches (e.g., local interpretable model-agnostic explanations [LIME]; Ribeiro et al., 2016 or Shapley additive explanations [SHAP]; Lundberg and Lee, 2017) can be utilized to identify the features or brain regions driving the model's predictions and provide insights into the neurobiological mechanisms.

Pitfalls/considerations of machine learning models

There are some common pitfalls of ML modeling in neuroimaging. Here, we categorize them by their sequence with respect to training: before training, during training, or after training.

Before training

They are two possible biases before training a ML model: one is sample selection bias and the other is feature selection bias, and both may affect the performance and generalizability of the ML model. Sample selection bias refers to a situation where the training, validation, or test sets do not adequately represent the target population or the distribution of the data in the real world. For example, individuals' familial relatedness is an important factor to account for in data splits; this issue exists in a common dataset used by the neuroimaging community called the Human Connectome Project, which enrolled twins and siblings as participants (Van Essen et al., 2012). If the research question is to predict a phenotype that is genetic on an individual level, related subjects, for example, twins and siblings, relatives should be removed from the sampled data, or, alternatively, related individuals should be included in the same split. If twins are split across training and test sets, then the trained model will exhibit inflated accuracy on the test set as it saw a very similar subject during training. Another common pitfall in neuroimaging is that of imbalanced class distribution in splits, wherein the number of samples in different categories of the target variable is significantly unequal. This often occurs when trying to distinguish rare conditions or disorders from a large population of normal subjects, and can lead to biased models that are better at predicting the majority class but much worse at predicting the minority class. Stratified sampling techniques can help mitigate this problem by ensuring a proportional representation of different classes, allowing the model to learn from both majority and minority classes.

Feature selection bias refers to spurious or unreliable feature selection results due to the omission or overemphasis of certain features, and can occur in either subjective or objective feature selection. In subjective selection, features are usually selected based on people's prior knowledge or assumptions, and influenced by preconceived notions or specific hypotheses. In objective selection, even though data-driven statistical tests are conducted to select the relevant features, there are some common mistakes such as the inclusion of test samples in the feature selection, improper validation or correction for multiple testing. More rigorous methods, for example, RFE (Guyon et al., 2002) and minimum-redundancy maximum relevancy (Brown et al., 2012), can be used to avoid introducing such biases into the ML model.

During training

During ML model training, good practice is to monitor both the training loss and the validation loss to avoid potential model overfitting. Overfitting describes a phenomenon when a model starts to memorize the noise or idiosyncrasies present in the training set, rather than learning the underlying patterns or relationships that also exist in unseen data. There are several factors that may cause overfitting, that is, the model is too complex/over-parameterized for the number of samples that exist. Many neuroimaging datasets have this property, where the number of features (voxels) per sample is usually much larger than the total number of samples (subjects). Therefore, overfitting is one major pitfall that should be paid close attention when training ML models on neuroimaging data. If overfitting is observed during training, that is, training loss continues to decrease while validation loss starts to increase, common approaches can be used to overcome this issue. From the model side, simpler models with less learnable parameters or number of layers/neurons (if the model is a neural network) may be used. From the data side, proper feature selection and data augmentation may be used to reduce overfitting. Otherwise, strategies such as cross-validation, regularization, and early stopping can be implemented during training to minimize overfitting.

After training

An essential step after model training is model evaluation; improper assessment of model performance can lead to overconfidence in the model which is particularly dangerous in clinical settings. For classification problems, reporting only the overall accuracy is usually uninformative when the numbers of samples in different classes are unequal, as the model can still perform well by just predicting the majority class (Alberg et al., 2004). Best practice is to summarize classification results in a confusion matrix, where the model's sensitivity, specificity, precision, and F1 score can all be easily assessed. Another way to visualize the performance of the model is to plot the AUC-ROC. The ROC shows the sensitivity against 1-specificity for varied discrimination thresholds, and is usually summarized by AUC where higher AUC value indicates better performance.

For regression tasks, standard evaluation metrics such as mean squared error (MSE) and PCC are often adopted to quantify overall model performance. For individual-level modeling, an easily overlooked aspect is interindividual differences. When using MSE loss during training, the model may learn to predict the population mean in order to minimize the loss, in which case there are small loss values but the loss of interindividual variability renders the prediction meaningless. For example, predicting functional connectomes from structural connectomes is a problem where this issue may occur as predicting the group mean FC or noise around the mean FC results in low MSE. High correlation can still be obtained between predicted FCs and observed FCs across test subjects, but the individual predictions are not accurate. Identifiability can be used as a supplementary metric to assist the assessment of such models, where giving each query subject observed FC, we try to identify that subject from the predicted FCs in the test group based on some distance functions, for example, PCC. By doing so for every subject we get a $n \times n$ matrix where n is the number of subjects in the test set, and ideally we would like the matrix to be diagonal.

Besides choosing proper evaluation metrics, it is also important to report uncertainty estimation, for instance, error bars, alongside the reported average metric values. Training multiple models and testing in multiple test sets via different dataset splits, can give ways of quantifying uncertainty, where sampling based techniques, for example, jackknife (Efron and Stein, 1981) and bootstrapping (Efron and Tibshirani, 1994), can be applied.

Open challenges and future directions

While ML has enabled many promising findings in the area of neuroscience and neuroimaging, it is important to acknowledge the existing challenges and explore the emerging solutions that aim to overcome these hurdles.

Sample size

Comparing with other fields where ML is extensively used, for example, computer vision that has datasets like ImageNet with millions of samples (Russakovsky et al., 2015), neuroimaging studies often face the challenge of limited sample sizes due to the cost and complexity of data collection. It was shown that brain-wide association studies usually require samples with thousands of individuals to achieve reproducible results (Marek et al., 2022), and insufficient sample sizes can lead to low performing, overfitted, unreliable, and/or poorly generalizable ML models (Pereira et al., 2009; Franke et al., 2010; Nieuwenhuis et al., 2012; Button et al., 2013).

To address the "small N" problem in neuroimaging studies, there have been collaborative efforts to advancing open sharing of data across institutions, and, relatedly, establishing standardized protocols for data collection. In recent years, several data sharing platforms and initiatives have begun to make a positive impact. These include platforms like OpenNeuro (https://openneuro.org), Brain Imaging Data Structure (BIDS) (Gorgolewski et al., 2016), International Neuroimaging Datasharing Initiative (INDI) (https://fcon_1000.projects.nitrc.org/), National Institute of Mental Health Data Archive (NDA) (https://nda.nih.gov/) for data dissemination, and open-source datasets like ADNI (Petersen et al., 2010), Autism Brain Imaging Data Exchange (ABIDE) (Di Martino et al., 2014), Adolescent Brain and Childhood Development (ABCD) (Jernigan et al., 2018), Natural Scenes Dataset (NSD) (Allen et al., 2022), and the Human Connectome Project (HCP) (Van Essen et al., 2013).

Besides the widespread adoption of data sharing, efficient data augmentation can also aid the creation of accurate ML models. Commonly employed data augmentation techniques include geometric, elastic and intensity transformations, slice augmentation, and patch extraction (Shin et al., 2016; Esteva et al., 2017). Moreover, taking advantages of existing relevant pretrained models (transfer learning) and ensembling approaches may also lead to more accurate and efficient models than solely relying on small data to train a full model (Wen et al., 2018; Gu et al., 2022b).

Ethical considerations

ML models trained on neuroimaging data collected from human subjects can raise ethical concerns related to data privacy and potential biases. For example, reidentification, which refers to the risk of identifying specific individuals from their neuroimaging data even after deidentification measures have been applied, is a significant concern in human-involved studies (Sarwate et al., 2014). To address this issue and protect individuals' privacy, more advanced data anonymization techniques and rigorous ethical guidelines need to be employed and followed. Additionally, data access policies and safeguards can be implemented to restrict the availability of sensitive information and ensure that only authorized individuals have access to such data. This involves establishing secure data storage and transfer protocols, obtaining informed consent from participants, and complying with relevant data protection regulations.

Biases, such as sample bias and selection bias, also widely exist in neuroimaging studies. One of the most significant biases is the inequitable inclusion of minority ethnic, racial, and other social groups as historically many neuroimaging datasets have been predominantly composed of participants from white or European ancestry. This imbalance in representation has raised concerns about the generalizability of findings to more diverse populations (Poldrack et al., 2017). It is crucial to increasingly emphasize the need for more inclusive study populations by promoting diversity in participant recruitment and ensuring equitable access to research opportunities (Goldfarb and Brown, 2022).

Computational and interpretability challenges from deep learning

Deep learning (DL), an important ML approach, has become more and more popular in dealing with neuroimaging tasks. DL has two main advantages over traditional ML methods, the first of which is automatic feature learning. DL models can do end-to-end learning, that is, they use raw data to learn hierarchical feature representations in a data-driven manner, eliminating manual feature engineering that increases subjectivity in model learning. The other advantage of DL is its flexibility and capability of learning complex and hierarchical features by adjusting the number of neurons in each layer and the number of layers in the model (Plis et al., 2014; Arbabshirani et al., 2017). With these advantages, DL methods can outperform classic ML methods, especially when data are very large, the relationships are complex, and there are no established or known features.

DL models trained on 2D image data have proven quite successful on various computer vision tasks (Simonyan and Zisserman, 2015; Krizhevsky et al., 2017). Training DL models on neuroimaging data that are usually 3D, however, requires much more memory. Directly expanding 2D models to 3D ones will be computationally expensive and possible infeasible, limiting their practical utility in real-time applications (Peng et al., 2021). Solutions such as downsampling image resolution, computing image patches, or taking only 2D slices of the 3D input image may help ease the GPU memory restrictions (Korolev et al., 2017; Kamnitsas et al., 2017; Bashyam et al., 2020). Future directions may benefit from developing efficient algorithms, leveraging hardware accelerators, and exploring model compression techniques to improve computational efficiency for 3D neuroimaging data.

Another challenge associated with DL models is their inherent lack of interpretability; it is often unclear how DL models arrive at their predictions. However, interpretability is crucial in neuroimaging, where explainable models can help gain insights into the underlying biological mechanisms and serve to build trust in the predictions made by the models. Besides the traditional visualization methods (e.g., gradient-weighted class activation mapping) and model-agnostic interpretability methods (LIME and SHAP), researchers need to explore the development of model architectures that are inherently more interpretable.

Take-home points

- ML offers a rich and growing computational toolkit for analyzing neuroimaging data. It can uncover complex patterns within data and produce remarkable performance in many prediction tasks.
- When applying ML techniques, it is important to consider the different assumptions that underpin the model and algorithm.
- We outline a general pipeline for building, testing, and applying ML models, which can be altered according to different data structures, features, and overall goals.
- We present a brief summary of different ML modeling approaches, for example, classic algorithms, CNNs, and GNNs, and how they have been and might be applied to analyze different types of neuroimaging data.
- We underscore potential pitfalls that may occur before, during, and after the training of ML models, such as sample/feature selection bias, overfitting, and improper evaluations.
- We present important challenges for the use of ML models applied to neuroimaging data, including sample size and ethical considerations.

References

Abadi, M., 2016. TensorFlow: Large-Scale Machine Learning on Heterogeneous Systems. Software available from: https://www.tensorflow.org/.

Abdi, H., Williams, L.J., 2010. Principal component analysis. Wiley Interdiscip. Rev. Comput. Stat. 2 (4), 433–459.

Abraham, A., Pedregosa, F., Eickenberg, M., Gervais, P., Mueller, A., Kossaifi, J., Gramfort, A., Thirion, B., Varoquaux, G., 2014. Machine learning for neuroimaging with scikit-learn. Front. Neuroinform. 8, 14.

Alberg, A.J., Park, J.W., Hager, B.W., Brock, M.V., Diener-West, M., 2004. The use of "overall accuracy" to evaluate the validity of screening or diagnostic tests. J. Gen. Intern. Med. 19 (5P1), 460–465.

Allen, E.J., St-Yves, G., Wu, Y., Breedlove, J.L., Prince, J.S., Dowdle, L.T., Nau, M., Caron, B., Pestilli, F., Charest, I., et al., 2022. A massive 7T fMRI dataset to bridge cognitive neuroscience and artificial intelligence. Nat. Neurosci. 25 (1), 116–126.

Arbabshirani, M.R., Plis, S., Sui, J., Calhoun, V.D., 2017. Single subject prediction of brain disorders in neuroimaging: promises and pitfalls. NeuroImage 145, 137–165.

Avants, B.B., Tustison, N., Song, G., et al., 2009. Advanced normalization tools (ants). Insight J. 2 (365), 1–35.

Bashyam, V.M., Erus, G., Doshi, J., Habes, M., Nasrallah, I.M., Truelove-Hill, M., Srinivasan, D., Mamourian, L., Pomponio, R., Fan, Y., et al., 2020. MRI signatures of brain age and disease over the lifespan based on a deep brain network and 14 468 individuals worldwide. Brain 143 (7), 2312–2324.

Bassett, D.S., Sporns, O., 2017. Network neuroscience. Nat. Neurosci. 20 (3), 353–364.

Bengs, M., Behrendt, F., Krüger, J., Opfer, R., Schlaefer, A., 2021. Three-dimensional deep learning with spatial erasing for unsupervised anomaly segmentation in brain MRI. Int. J. Comput. Assist. Radiol. Surg. 16, 1413–1423.

Bessadok, A., Mahjoub, M.A., Rekik, I., 2021. Brain graph synthesis by dual adversarial domain alignment and target graph prediction from a source graph. Med. Image Anal. 68, 101902.

Bessadok, A., Mahjoub, M.A., Rekik, I., 2022. Graph neural networks in network neuroscience. IEEE Trans. Pattern Anal. Mach. Intell. 45 (5), 5833–5848.

Boser, B.E., Guyon, I.M., Vapnik, V.N., 1992. A training algorithm for optimal margin classifiers. In: Proceedings of the Fifth Annual Workshop on Computational Learning Theory, pp. 144–152.

Breiman, L., 2001. Random forests. Mach. Learn. 45, 5–32.

Brock, A., Donahue, J., Simonyan, K., 2019. Large scale GAN training for high fidelity natural image synthesis. International Conference on Learning Representations.

Brown, G., Pocock, A., Zhao, M.J., Luján, M., 2012. Conditional likelihood maximisation: a unifying framework for information theoretic feature selection. J. Mach. Learn. Res. 13, 27–66.

Buitinck, L., Louppe, G., Blondel, M., Pedregosa, F., Mueller, A., Grisel, O., Niculae, V., Prettenhofer, P., Gramfort, A., Grobler, J., Layton, R., VanderPlas, J., Joly, A., Holt, B., Varoquaux, G., 2013. API design for machine learning software: experiences from the scikit-learn project. In: ECML PKDD Workshop: Languages for Data Mining and Machine Learning, pp. 108–122.

Bullmore, E., Sporns, O., 2009. Complex brain networks: graph theoretical analysis of structural and functional systems. Nat. Rev. Neurosci. 10 (3), 186–198.

Button, K.S., Ioannidis, J.P.A., Mokrysz, C., Nosek, B.A., Flint, J., Robinson, E.S.J., Munafò, M.R., 2013. Power failure: why small sample size undermines the reliability of neuroscience. Nat. Rev. Neurosci. 14 (5), 365–376.

Chollet, F., et al., 2015. Keras. Available from: https://keras.io.

Çiçek, Ö., Abdulkadir, A., Lienkamp, S.S., Brox, T., Ronneberger, O., 2016. 3D U-Net: learning dense volumetric segmentation from sparse annotation. In: Medical Image Computing and Computer-Assisted Intervention—MICCAI 2016: 19th International Conference, Athens, Greece, October 17–21, 2016, Proceedings, Part II, 19, pp. 424–432.

Cole, J.H., Poudel, R.P.K., Tsagkrasoulis, D., Caan, M.W.A., Steves, C., Spector, T.D., Montana, G., 2017. Predicting brain age with deep learning from raw imaging data results in a reliable and heritable biomarker. NeuroImage 163, 115–124.

Dayan, P., Abbott, L.F., 2005. Theoretical Neuroscience: Computational and Mathematical Modeling of Neural Systems. MIT Press.

Dhariwal, P., Nichol, A., 2021. Diffusion models beat GANs on image synthesis. Adv. Neural Inf. Process. Syst. 34, 8780–8794.

Di Martino, A., Yan, C.G., Li, Q., Denio, E., Castellanos, F.X., Alaerts, K., Anderson, J.S., Assaf, M., Bookheimer, S.Y., Dapretto, M., et al., 2014. The autism brain imaging data exchange: towards a large-scale evaluation of the intrinsic brain architecture in autism. Mol. Psychiatry 19 (6), 659–667.

DiCarlo, J.J., Zoccolan, D., Rust, N.C., 2012. How does the brain solve visual object recognition? Neuron 73 (3), 415–434.

Doshi-Velez, F., Kim, B., 2017. Towards a rigorous science of interpretable machine learning. arXiv preprint arXiv:1702.08608.

Efron, B., Stein, C., 1981. The jackknife estimate of variance. Ann. Stat. 9, 586–596.

Efron, B., Tibshirani, R.J., 1994. An Introduction to the Bootstrap. CRC Press.

Esteva, A., Kuprel, B., Novoa, R.A., Ko, J., Swetter, S.M., Blau, H.M., Thrun, S., 2017. Dermatologist-level classification of skin cancer with deep neural networks. Nature 542 (7639), 115–118.

Fort, G., Lambert-Lacroix, S., 2005. Classification using partial least squares with penalized logistic regression. Bioinformatics 21 (7), 1104–1111.

Franke, K., Ziegler, G., Klöppel, S., Gaser, C., Alzheimer's Disease Neuroimaging Initiative, et al., 2010. Estimating the age of healthy subjects from T1-weighted MRI scans using kernel methods: exploring the influence of various parameters. NeuroImage 50 (3), 883–892.

Friston, K., 2012. The history of the future of the Bayesian brain. NeuroImage 62 (2), 1230–1233.

Friston, K.J., Glaser, D.E., Henson, R.N.A., Kiebel, S., Phillips, C., Ashburner, J., 2002a. Classical and Bayesian inference in neuroimaging: applications. NeuroImage 16 (2), 484–512.

Friston, K.J., Penny, W., Phillips, C., Kiebel, S., Hinton, G., Ashburner, J., 2002b. Classical and Bayesian inference in neuroimaging: theory. NeuroImage 16 (2), 465–483.

Gao, X., Shi, F., Shen, D., Liu, M., 2021. Task-induced pyramid and attention GAN for multimodal brain image imputation and classification in Alzheimer's disease. IEEE J. Biomed. Health Inf. 26 (1), 36–43.

Goldfarb, M.G., Brown, D.R., 2022. Diversifying participation: the rarity of reporting racial demographics in neuroimaging research. NeuroImage 254, 119122.

Goodfellow, I., Pouget-Abadie, J., Mirza, M., Xu, B., Warde-Farley, D., Ozair, S., Courville, A., Bengio, Y., 2020. Generative adversarial networks. Commun. ACM 63 (11), 139–144.

Gorgolewski, K.J., Auer, T., Calhoun, V.D., Craddock, R.C., Das, S., Duff, E.P., Flandin, G., Ghosh, S.S., Glatard, T., Halchenko, Y.O., et al., 2016. The brain imaging data structure, a format for organizing and describing outputs of neuroimaging experiments. Sci. Data 3 (1), 1–9.

Gu, Z., Jamison, K., Kuceyeski, A., Sabuncu, M., 2022a. Decoding natural image stimuli from fMRI data with a surface-based convolutional network. arXiv preprint arXiv:2212.02409.

Gu, Z., Jamison, K., Sabuncu, M., Kuceyeski, A., 2022b. Personalized visual encoding model construction with small data. Commun. Biol. 5 (1), 1382.

Gu, Z., Jamison, K.W., Khosla, M., Allen, E.J., Wu, Y., St-Yves, G., Naselaris, T., Kay, K., Sabuncu, M.R., Kuceyeski, A., 2022c. Neurogen: activation optimized image synthesis for discovery neuroscience. NeuroImage 247, 118812.

Gu, Z., Jamison, K., Sabuncu, M.R., Kuceyeski, A., 2023. Human brain responses are modulated when exposed to optimized natural images or synthetically generated images. Commun. Biol. 6.

Güngör, A., Dar, S.U.H., Öztürk, Ş., Korkmaz, Y., Bedel, H.A., Elmas, G., Ozbey, M., Çukur, T., 2023. Adaptive diffusion priors for accelerated MRI reconstruction. Med. Image Anal. 88, 102872.

Gurbuz, M.B., Rekik, I., 2020. Deep graph normalizer: a geometric deep learning approach for estimating connectional brain templates. In: Medical Image Computing and Computer Assisted Intervention—MICCAI 2020: 23rd International Conference, Lima, Peru, October 4–8, 2020, Proceedings, Part VII, 23, pp. 155–165.

Guyon, I., Elisseeff, A., 2003. An introduction to variable and feature selection. J. Mach. Learn. Res. 3 (Mar), 1157–1182.

Guyon, I., Weston, J., Barnhill, S., Vapnik, V., 2002. Gene selection for cancer classification using support vector machines. Mach. Learn. 46, 389–422.

Han, C., Hayashi, H., Rundo, L., Araki, R., Shimoda, W., Muramatsu, S., Furukawa, Y., Mauri, G., Nakayama, H., 2018. GAN-based synthetic brain MR image generation. In: 2018 IEEE 15th International Symposium on Biomedical Imaging (ISBI 2018), pp. 734–738.

He, K., Zhang, X., Ren, S., Sun, J., 2016. Deep residual learning for image recognition. In: Proceedings of the IEEE Conference on Computer Vision and Pattern Recognition, pp. 770–778.

Ho, J., Jain, A., Abbeel, P., 2020. Denoising diffusion probabilistic models. Adv. Neural Inf. Process. Syst. 33, 6840–6851.

Hochreiter, S., Schmidhuber, J., 1997. Long short-term memory. Neural Comput. 9 (8), 1735–1780.

Jack Jr., C.R., Bernstein, M.A., Fox, N.C., Thompson, P., Alexander, G., Harvey, D., Borowski, B., Britson, P.J., Whitwell, J.L., Ward, C., et al., 2008. The Alzheimer's disease neuroimaging initiative (ADNI): MRI methods. J. Magn. Reson. Imaging 27 (4), 685–691.

Jernigan, T.L., Brown, S.A., Dowling, G.J., 2018. The adolescent brain cognitive development study. J. Res. Adolesc. 28 (1), 154.

Jónsson, B.A., Bjornsdottir, G., Thorgeirsson, T.E., Ellingsen, L.M., Walters, G.B., Gudbjartsson, D.F., Stefansson, H., Stefansson, K., Ulfarsson, M.O., 2019. Brain age prediction using deep learning uncovers associated sequence variants. Nat. Commun. 10 (1), 5409.

Jordan, M.I., Mitchell, T.M., 2015. Machine learning: trends, perspectives, and prospects. Science 349 (6245), 255–260.

Kam, T.E., Zhang, H., Jiao, Z., Shen, D., 2019. Deep learning of static and dynamic brain functional networks for early MCI detection. IEEE Trans. Med. Imaging 39 (2), 478–487.

Kamnitsas, K., Ledig, C., Newcombe, V.F.J., Simpson, J.P., Kane, A.D., Menon, D.K., Rueckert, D., Glocker, B., 2017. Efficient multi-scale 3D CNN with fully connected CRF for accurate brain lesion segmentation. Med. Image Anal. 36, 61–78.

Karras, T., Laine, S., Aila, T., 2019. A style-based generator architecture for generative adversarial networks. In: Proceedings of the IEEE/CVF Conference on Computer Vision and Pattern Recognition, pp. 4401–4410.

Kawahara, J., Brown, C.J., Miller, S.P., Booth, B.G., Chau, V., Grunau, R.E., Zwicker, J.G., Hamarneh, G., 2017. BrainNetCNN: convolutional neural networks for brain networks; towards predicting neurodevelopment. NeuroImage 146, 1038–1049.

Kernbach, J.M., Ort, J., Hakvoort, K., Clusmann, H., Neuloh, G., Delev, D., 2022. Introduction to machine learning in neuroimaging. In: Machine Learning in Clinical Neuroscience: Foundations and Applications, pp. 121–124.

Kingma, D.P., Ba, J., 2015. Adam: a method for stochastic optimization. International Conference on Learning Representations.

Kingma, D.P., Welling, M., 2014. Auto-encoding variational bayes. International Conference on Learning Representations.

Korolev, S., Safiullin, A., Belyaev, M., Dodonova, Y., 2017. Residual and plain convolutional neural networks for 3D brain MRI classification. In: 2017 IEEE 14th International Symposium on Biomedical Imaging (ISBI 2017), pp. 835–838.

Kriegeskorte, N., Goebel, R., Bandettini, P., 2006. Information-based functional brain mapping. Proc. Natl Acad. Sci. 103 (10), 3863–3868.

Krizhevsky, A., Sutskever, I., Hinton, G.E., 2017. ImageNet classification with deep convolutional neural networks. Commun. ACM 60 (6), 84–90.

Ktena, S.I., Parisot, S., Ferrante, E., Rajchl, M., Lee, M., Glocker, B., Rueckert, D., 2018. Metric learning with spectral graph convolutions on brain connectivity networks. NeuroImage 169, 431–442.

Li, X., Zhou, Y., Dvornek, N., Zhang, M., Gao, S., Zhuang, J., Scheinost, D., Staib, L.H., Ventola, P., Duncan, J.S., 2021. BrainGNN: interpretable brain graph neural network for fMRI analysis. Med. Image Anal. 74, 102233.

Lin, W., Lin, W., Chen, G., Zhang, H., Gao, Q., Huang, Y., Tong, T., Du, M., Alzheimer's Disease Neuroimaging Initiative, 2021. Bidirectional mapping of brain MRI and pet with 3D reversible GAN for the diagnosis of Alzheimer's disease. Front. Neurosci. 15, 646013.

Lundberg, S.M., Lee, S.I., 2017. A unified approach to interpreting model predictions. Adv. Neural Inf. Process. Syst. 30, 4768–4777.

Marek, S., Tervo-Clemmens, B., Calabro, F.J., Montez, D.F., Kay, B.P., Hatoum, A.S., Donohue, M.R., Foran, W., Miller, R.L., Hendrickson, T.J., et al., 2022. Reproducible brain-wide association studies require thousands of individuals. Nature 603 (7902), 654–660.

Meszlényi, R.J., Buza, K., Vidnyánszky, Z., 2017. Resting state fMRI functional connectivity-based classification using a convolutional neural network architecture. Front. Neuroinf. 11, 61.

Miller, K.L., Alfaro-Almagro, F., Bangerter, N.K., Thomas, D.L., Yacoub, E., Xu, J., Bartsch, A.J., Jbabdi, S., Sotiropoulos, S.N., Andersson, J.L.R., et al., 2016. Multimodal population brain imaging in the UK Biobank prospective epidemiological study. Nat. Neurosci. 19 (11), 1523–1536.

Mourao-Miranda, J., Bokde, A.L.W., Born, C., Hampel, H., Stetter, M., 2005. Classifying brain states and determining the discriminating activation patterns: support vector machine on functional MRI data. NeuroImage 28 (4), 980–995.

Mwangi, B., Tian, T.S., Soares, J.C., 2014. A review of feature reduction techniques in neuroimaging. Neuroinformatics 12, 229–244.

Myszczynska, M.A., Ojamies, P.N., Lacoste, A.M.B., Neil, D., Saffari, A., Mead, R., Hautbergue, G.M., Holbrook, J.D., Ferraiuolo, L., 2020. Applications of machine learning to diagnosis and treatment of neurodegenerative diseases. Nat. Rev. Neurol. 16 (8), 440–456.

Nguyen, B., Feldman, A., Bethapudi, S., Jennings, A., Willcocks, C.G., 2021. Unsupervised region-based anomaly detection in brain MRI with adversarial image inpainting. In: 2021 IEEE 18th International Symposium on Biomedical Imaging (ISBI), pp. 1127–1131.

Nieuwenhuis, M., van Haren, N.E.M., Pol, H.E.H., Cahn, W., Kahn, R.S., Schnack, H.G., 2012. Classification of schizophrenia patients and healthy controls from structural MRI scans in two large independent samples. NeuroImage 61 (3), 606–612.

Oktay, O., Schlemper, J., Folgoc, L.L., Lee, M., Heinrich, M., Misawa, K., Mori, K., McDonagh, S., Hammerla, N.Y., Kainz, B., et al., 2018. Attention U-Net: learning where to look for the pancreas. Medical Imaging with Deep Learning.

Parisot, S., Ktena, S.I., Ferrante, E., Lee, M., Guerrero, R., Glocker, B., Rueckert, D., 2018. Disease prediction using graph convolutional networks: application to autism spectrum disorder and Alzheimer's disease. Med. Image Anal. 48, 117–130.

Paszke, A., Gross, S., Massa, F., Lerer, A., Bradbury, J., Chanan, G., Killeen, T., Lin, Z., Gimelshein, N., Antiga, L., et al., 2019. Pytorch: an imperative style, high-performance deep learning library. Adv. Neural Inf. Process. Syst. 32, 8024–8035.

Pedregosa, F., Varoquaux, G., Gramfort, A., Michel, V., Thirion, B., Grisel, O., Blondel, M., Prettenhofer, P., Weiss, R., Dubourg, V., Vanderplas, J., Passos, A., Cournapeau, D., Brucher, M., Perrot, M., Duchesnay, E., 2011. Scikit-learn: machine learning in Python. J. Mach. Learn. Res. 12, 2825–2830.

Peng, H., Gong, W., Beckmann, C.F., Vedaldi, A., Smith, S.M., 2021. Accurate brain age prediction with lightweight deep neural networks. Med. Image Anal. 68, 101871.

Penny, W.D., Friston, K.J., Ashburner, J.T., Kiebel, S.J., Nichols, T.E., 2011. Statistical Parametric Mapping: The Analysis of Functional Brain Images. Elsevier.

Pereira, F., Mitchell, T., Botvinick, M., 2009. Machine learning classifiers and fMRI: a tutorial overview. NeuroImage 45 (1), S199–S209.

Petersen, R.C., Aisen, P.S., Beckett, L.A., Donohue, M.C., Gamst, A.C., Harvey, D.J., Jack, C.-R., Jagust, W.J., Shaw, L.M., Toga, A.W., et al., 2010. Alzheimer's disease neuroimaging initiative (ADNI): clinical characterization. Neurology 74 (3), 201–209.

Phang, C.R., Noman, F., Hussain, H., Ting, C.M., Ombao, H., 2019. A multi-domain connectome convolutional neural network for identifying schizophrenia from EEG connectivity patterns. IEEE J. Biomed. Health Inf. 24 (5), 1333–1343.

Plis, S.M., Hjelm, D.R., Salakhutdinov, R., Allen, E.A., Bockholt, H.J., Long, J.D., Johnson, H.J., Paulsen, J.S., Turner, J.A., Calhoun, V.D., 2014. Deep learning for neuroimaging: a validation study. Front. Neurosci. 8, 229.

Poldrack, R.A., Baker, C.I., Durnez, J., Gorgolewski, K.J., Matthews, P.M., Munafò, M.R., Nichols, T.E., Poline, J.B., Vul, E., Yarkoni, T., 2017. Scanning the horizon: towards transparent and reproducible neuroimaging research. Nat. Rev. Neurosci. 18 (2), 115–126.

Ponce, C.R., Xiao, W., Schade, P.F., Hartmann, T.S., Kreiman, G., Livingstone, M.S., 2019. Evolving images for visual neurons using a deep generative network reveals coding principles and neuronal preferences. Cell 177 (4), 999–1009.

Rathore, S., Habes, M., Iftikhar, M.A., Shacklett, A., Davatzikos, C., 2017. A review on neuroimaging-based classification studies and associated feature extraction methods for Alzheimer's disease and its prodromal stages. NeuroImage 155, 530–548.

Ravi, D., Blumberg, S.B., Ingala, S., Barkhof, F., Alexander, D.C., Oxtoby, N.P., Alzheimer's Disease Neuroimaging Initiative, et al., 2022. Degenerative adversarial neuroimage nets for brain scan simulations: application in ageing and dementia. Med. Image Anal. 75, 102257.

Ribeiro, M.T., Singh, S., Guestrin, C., 2016. "Why should I trust you?" Explaining the predictions of any classifier. In: Proceedings of the 22nd ACM SIGKDD International Conference on Knowledge Discovery and Data Mining, pp. 1135–1144.

Rombach, R., Blattmann, A., Lorenz, D., Esser, P., Ommer, B., 2022. High-resolution image synthesis with latent diffusion models. In: Proceedings of the IEEE/CVF Conference on Computer Vision and Pattern Recognition, pp. 10684–10695.

Ronneberger, O., Fischer, P., Brox, T., 2015. U-Net: convolutional networks for biomedical image segmentation. In: Medical Image Computing and Computer-Assisted Intervention—MICCAI 2015: 18th International Conference, Munich, Germany, October 5–9, 2015, Proceedings, Part III, 18, pp. 234–241.

Rubinov, M., Sporns, O., 2010. Complex network measures of brain connectivity: uses and interpretations. NeuroImage 52 (3), 1059–1069.

Russakovsky, O., Deng, J., Su, H., Krause, J., Satheesh, S., Ma, S., Huang, Z., Karpathy, A., Khosla, A., Bernstein, M., et al., 2015. ImageNet large scale visual recognition challenge. Int. J. Comput. Vis. 115, 211–252.

Salvatore, C., Cerasa, A., Battista, P., Gilardi, M.C., Quattrone, A., Castiglioni, I., Alzheimer's Disease Neuroimaging Initiative, 2015. Magnetic resonance imaging biomarkers for the early diagnosis of Alzheimer's disease: a machine learning approach. Front. Neurosci. 9, 307.

Samuel, A.L., 1959. Some studies in machine learning using the game of checkers. IBM J. Res. Dev. 3 (3), 210–229.

Sarwate, A.D., Plis, S.M., Turner, J.A., Arbabshirani, M.R., Calhoun, V.D., 2014. Sharing privacy-sensitive access to neuroimaging and genetics data: a review and preliminary validation. Front. Neuroinf. 8, 35.

Shah, P., Kendall, F., Khozin, S., Goosen, R., Hu, J., Laramie, J., Ringel, M., Schork, N., 2019. Artificial intelligence and machine learning in clinical development: a translational perspective. NPJ Dig. Med. 2 (1), 69.

Shin, H.C., Roth, H.R., Gao, M., Lu, L., Xu, Z., Nogues, I., Yao, J., Mollura, D., Summers, R.-M., 2016. Deep convolutional neural networks for computer-aided detection: CNN architectures, dataset characteristics and transfer learning. IEEE Trans. Med. Imaging 35 (5), 1285–1298.

Simonyan, K., Zisserman, A., 2015. Very deep convolutional networks for large-scale image recognition. International Conference on Learning Representations.

Smith, S.M., Jenkinson, M., Woolrich, M.W., Beckmann, C.F., Behrens, T.E.J., Johansen-Berg, H., Bannister, P.R., De Luca, M., Drobnjak, I., Flitney, D.E., et al., 2004. Advances in functional and structural MR image analysis and implementation as FSL. NeuroImage 23, S208–S219.

Sporns, O., 2012. From simple graphs to the connectome: networks in neuroimaging. NeuroImage 62 (2), 881–886.

Sporns, O., Tononi, G., Kötter, R., 2005. The human connectome: a structural description of the human brain. PLoS Comput. Biol. 1 (4), e42.

Stone, J.V., 2004. Independent Component Analysis. A Bradford Book. The MIT Press, Cambridge, MA.

Talo, M., Yildirim, O., Baloglu, U.B., Aydin, G., Acharya, U.R., 2019. Convolutional neural networks for multi-class brain disease detection using MRI images. Comput. Med. Imaging Graph. 78, 101673.

Thomas, R.M., Gallo, S., Cerliani, L., Zhutovsky, P., El-Gazzar, A., Van Wingen, G., 2020. Classifying autism spectrum disorder using the temporal statistics of resting-state functional MRI data with 3D convolutional neural networks. Front. Psychiatry 11, 440.

Tibshirani, R., 1996. Regression shrinkage and selection via the lasso. J. R. Stat. Soc. B (Methodol.) 58 (1), 267–288.

Turing, A.M., 2009. Computing Machinery and Intelligence. Springer.

Van Den Oord, A., Vinyals, O., et al., 2017. Neural discrete representation learning. Adv. Neural Inf. Process. Syst. 30.

Van Essen, D.C., Ugurbil, K., Auerbach, E., Barch, D., Behrens, T.E.J., Bucholz, R., Chang, A., Chen, L., Corbetta, M., Curtiss, S.W., et al., 2012. The human connectome project: a data acquisition perspective. NeuroImage 62 (4), 2222–2231.

Van Essen, D.C., Smith, S.M., Barch, D.M., Behrens, T.E.J., Yacoub, E., Ugurbil, K., Wu-Minn HCP Consortium, et al., 2013. The Wu-Minn human connectome project: an overview. NeuroImage 80, 62–79.

Wen, H., Shi, J., Chen, W., Liu, Z., 2018. Transferring and generalizing deep-learning-based neural encoding models across subjects. NeuroImage 176, 152–163.

Wolleb, J., Sandkühler, R., Bieder, F., Valmaggia, P., Cattin, P.C., 2022. Diffusion models for implicit image segmentation ensembles. In: International Conference on Medical Imaging With Deep Learning, pp. 1336–1348.

Woolrich, M.W., Jbabdi, S., Patenaude, B., Chappell, M., Makni, S., Behrens, T., Beckmann, C., Jenkinson, M., Smith, S.M., 2009. Bayesian analysis of neuroimaging data in FSL. NeuroImage 45 (1), S173–S186.

Xia, T., Chartsias, A., Wang, C., Tsaftaris, S.A., Alzheimer's Disease Neuroimaging Initiative, et al., 2021. Learning to synthesise the ageing brain without longitudinal data. Med. Image Anal. 73, 102169.

Zhou, Z., Siddiquee, M.M.R., Tajbakhsh, N., Liang, J., 2018. UNet++: a nested u-net architecture for medical image segmentation. In: Deep Learning in Medical Image Analysis and Multimodal Learning for Clinical Decision Support: 4th International Workshop, DLMIA 2018, and 8th International Workshop, ML-CDS 2018, Held in Conjunction With MICCAI 2018, Granada, Spain, September 20, 2018, Proceedings 4, pp. 3–11.

Decoding models: From brain representation to machine interfaces

Yu Takagi[a,b,c] and Shinji Nishimoto[b,c,d]

[a]*Research and Development Center for Large Language Models, National Institute of Informatics, Tokyo, Japan,* [b]*Graduate School of Frontier Biosciences, Osaka University, Osaka, Japan,* [c]*Center for Information and Neural Networks (CiNet), National Institute of Information and Communications Technology, Osaka, Japan,* [d]*Graduate School of Medicine, Osaka University, Osaka, Japan*

Introduction to decoding models

Decoding refers to the process of translating patterns of neural activity into representations of the outside world or internal cognitive states (Fig. 1). It draws upon a diverse set of disciplines including neuroscience, machine learning, statistics, and computer science.

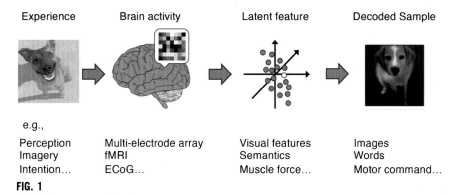

Experience	Brain activity	Latent feature	Decoded Sample

e.g.,

Perception	Multi-electrode array	Visual features	Images
Imagery	fMRI	Semantics	Words
Intention…	ECoG…	Muscle force…	Motor command…

FIG. 1

A schematic diagram of decoding models. Decoding models aim to decipher experiences (such as perception, imagination, or intention) from brain activity. This is typically achieved by recording brain activity, translating it into latent features, and then mapping these features to a sample of interest.

Modern decoding models have their roots in the 1990s and marked significant advances with the advent of multielectrode recording, which allowed for simultaneous recording of the activity of multiple neurons. This development underscored the importance of neural correlations and patterns across populations of neurons

Computational and Network Modeling of Neuroimaging Data. https://doi.org/10.1016/B978-0-443-13480-7.00013-2

rather than single-neuron activity. For example, researchers began to decode motor intentions from the neural activity in the motor cortex, leading to the modern BMI research (Chapin et al., 1999; Wessberg et al., 2000). Later in the 2000s, advancement in noninvasive neuroimaging methods contributed to the emergence in decoding research with human subjects (Blankertz et al., 2003; Cox and Savoy, 2003; Haxby et al., 2001; Kamitani and Tong, 2005; Wolpaw and McFarland, 2004).

Since around 2010, the capabilities of decoding have expanded significantly thanks to the utilization of machine learning, enabling the decoding of sensory stimuli, particularly natural stimuli (Kay et al., 2008; Miyawaki et al., 2008; Nishimoto et al., 2011). Building on this momentum, the latter half of the 2010s witnessed a surge in the application of deep learning to decoding (Horikawa and Kamitani, 2017, among others). This powerful technology has been employed for a myriad of purposes, including feature extraction, and has been harnessed for its generative capabilities of the variables of interest, such as visual images, sounds, and languages (Denk et al., 2023; Shen et al., 2019; Takagi and Nishimoto, 2023; Tang et al., 2023, among other leading examples).

In terms of theoretical context, decoding models generally come from machine learning and statistics. For example, linear regression, support vector machines, and more complex methods such as deep learning have been used for decoding tasks. These methods can be used to find patterns in neural activity data that predict or correlate with external variables, such as the movement of a limb or the perception of a stimulus. These models work by learning a mapping between patterns of neural activity and the variable of interest. This usually involves training the model on a set of data where both the neural activity and the variable of interest are known, and then using the model to predict the variable of interest from new neural activity data. This mapping can reveal how information about the variable of interest is represented in the brain.

Besides "decoding models", there is a different model known as "encoding models" (Fig. 2). It's important to understand how they are similar and how they

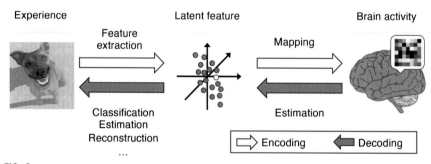

FIG. 2

A schematic illustrating the relationship between encoding and decoding models. Encoding models aim to understand how specific features of experiences are encoded and mapped into brain activity. Decoding models strive to decipher the various attributes of characteristics of experience a subject holds, based on a given pattern of brain activity. Both encoding and decoding models generally assume the existence of latent features. Identifying these relevant features serves as a common approach for understanding the underlying representations in both types of models.

are different (Kriegeskorte and Douglas, 2019; Naselaris et al., 2011). From the standpoint of comprehending the representation of information in the brain, **decoding models** play an essential role in investigating what is being represented within a population of neurons or voxels, such as specific sensory stimuli or cognitive states. By examining the collective activity patterns, decoding models can shed light on how well these neural representations correlate with, or might even drive, particular behaviors or experiences in the subject. This approach focuses on a broader understanding of how groups of neurons function together to create complex representations.

In contrast, **encoding models** concentrate on a more granular level, typically examining representations of individual cells or voxels as well as how they are mapped across certain areas. These are tailored for analyses that seek to understand how specific neurons or small groups respond to particular inputs or conditions, allowing researchers to map precise relationships between given stimuli and neural activity. These complementary methods of encoding and decoding contribute to a multifaceted view of neural function, each suited to different research questions and scales of investigation.

Neural decoding is not limited to unraveling low-level sensory information but has emerged as a vital field in neuroscience for deciphering higher cognitive functions such as contents of working memory (Harrison and Tong, 2009; Stokes et al., 2009), emotional states (Horikawa et al., 2020; Kragel et al., 2016; Saarimäki et al., 2016), and physical pain (Wager et al., 2013). Moreover, decoding models facilitate the creation of diagnostic and treatment tools for neurological or psychiatric disorders by identifying abnormal brain activity patterns (Arbabshirani et al., 2017; Etkin and Wager, 2007). This advancement has further evolved to include the decoding of general cognitive states as well (Nakai and Nishimoto, 2020).

As a bridge between neuroscience and artificial intelligence (AI) (Hassabis et al., 2017; Zador et al., 2023), decoding models drive advancements in machine learning and AI by providing valuable insights into how the brain processes and represents information, thus inspiring the development of more sophisticated models and efficient AI systems. Beyond their scientific and technological applications, decoding models also delve into the philosophical realm by providing insight into consciousness and perception. Through decoding neural correlates of consciousness (Haynes and Rees, 2005), we can investigate how subjective experiences are engendered from brain activity, enriching our understanding of perception and consciousness. The ethical implications of these models also spark crucial discussions about brain privacy and the nature of consciousness, underscoring their broader societal relevance (Yuste et al., 2017).

Decoding models can be used to answer the following types of questions:

- How is information represented in the brain? Decoding models can help us understand how different types of information (e.g., sensory inputs, cognitive states, motor outputs) are encoded in patterns of brain activity. This could include understanding how visual scenes are represented in the visual cortex, how intentions to move are represented in the motor cortex, or how emotions are represented in the limbic system.

- How does the brain process information? By building models that can decode brain activity at different stages of a sensory or cognitive process, we can gain insights into how information is processed and transformed in the brain. This could involve, for example, decoding the neural representations of sensory inputs at different stages of the visual hierarchy to understand the processing steps involved in visual perception.
- Can we predict behavior from brain activity? Decoding models can be used to predict behavior based on brain activity. This could involve predicting an individual's decisions, reactions to stimuli, or even future actions based on their current brain activity.
- How can we interface the brain with machines? Decoding models are essential for developing BMIs, which can translate brain activity into commands for a machine or computer. This has significant implications for assistive technologies and neuroprosthetics, potentially allowing individuals with motor impairments to control devices or prosthetic limbs with their minds.
- Can we identify and monitor neurological disorders? Decoding models can potentially be used to identify abnormal patterns of brain activity associated with specific neurological or psychiatric disorders. They could also be used to monitor disease progression or the effects of treatment over time.
- What is the relationship between AI and human brain function? By aligning and comparing AI and human brain, researchers can identify and analyze the specific areas where the correspondences occur (Güçlü and van Gerven, 2015; Kell et al., 2018; Takagi and Nishimoto, 2023; Yamins et al., 2014, among others). In the case of decoding models, it is possible to establish a correspondence between human brain activity and the internal representations within AI systems. Whether they are reflecting sensory experiences or higher-level cognitive states, the decoding models allow for an understanding of how these two diverse systems—one biological and one artificial—can manifest similar functionalities.

Examples of successful decoding models

Decoding can be categorized into discrete and continuous value decoding (Fig. 3). In the case of the decoding of discrete samples, examples include distinguishing between right and left stimuli or broader tasks such as decoding the keys being typed on a keyboard. On the other hand, continuous value decoding deals with more complex variables, such as the speed of hand movement or predicting the fine pixel values in an image, and can be utilized in the reconstruction of more intricate stimuli. In essence, the decoding of discrete samples often concerns clear, defined choices or actions, providing distinct categorizations. Continuous value decoding, however, dives into more nuanced aspects, allowing for the exploration and interpretation of varying degrees or gradations in a given stimulus.

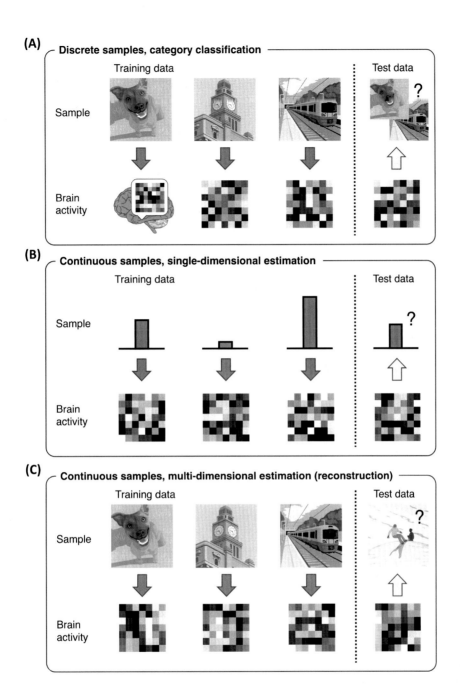

FIG. 3

Schematics of various types of decoding. (A) In the case of discrete samples, decoding models are trained to learn the relationship between brain activity and samples from a fixed set of categories (e.g., visual categories such as animals, buildings, or trains; or arm movements toward the left or right). These models are then tested to determine whether they can infer the specific category represented or intended based on new brain activity. (B) For continuous samples in a single dimension, decoding models are trained to learn the relationship between a specific attribute of experiences (e.g., motion speed or angles) and brain activity. They are then tested to see if the model can predict that attribute based on new brain activity. (C) For continuous samples in multiple dimensions, decoding models aim to infer multidimensional attributes such as images.

Decoding studies have significantly advanced the performance of BMIs, empowering individuals with paralysis or limb loss to control devices such as computer cursors or prosthetic limbs with their thoughts (Birbaumer et al., 1999; Hochberg et al., 2006; Taylor et al., 2002). Andrew Schwartz and his colleagues have made substantial contributions, including the development of a brain-controlled robotic arm (Velliste et al., 2008). In this work, Schwartz and his colleagues set out to see if a monkey could feed itself using a robotic arm, controlled solely by its brain signals. They captured the monkey's neural activity from its motor cortex through a multielectrode array. These data were then transformed in real time to guide the movement and grip of the robotic limb. Even with its natural arms restrained, the monkey adeptly used the brain-directed robotic arm to grasp food and bring it to its mouth. Krishna Shenoy and his colleagues also conducted several pioneering works in BMIs with human patients with motor impairments (e.g., Gilja et al., 2015; Willett et al., 2021). In Willett et al. (2021), by implanting two microelectrode arrays into the subject's motor cortex, they were able to achieve real-time decoding of attempted handwriting movements (Fig. 4). The subject imagined writing individual letters, and the system was able to decode and transcribe these imagined movements into text. Although many of these studies have been conducted with human patients or animals using multielectrode recordings rather than neuroimaging, some researchers achieved to decode motor execution and imagery using electroencephalogram (EEG) (Blankertz et al., 2003, 2007).

Decoding visual perception is another research topic that has been very well investigated, often leveraging fMRI data (Fig. 5). By utilizing fMRI data to identify and reconstruct images or video clips from brain activity, researchers have significantly expanded our understanding of visual perception and consciousness. In 2008, Jack Gallant and his colleagues performed a visual identification of what subjects were viewing, pioneering the field in a substantial way (Kay et al., 2008). Following that, a paper published by Gallant's team in 2011 demonstrated a method for reconstructing natural videos from fMRI data (Nishimoto et al., 2011). In 2012, they also shed light on how various objects and actions visually observed are represented in the human brain, providing a detailed map of semantic information (Huth et al., 2012). Meanwhile, Yukiyasu Kamitani and his colleagues in 2008 presented an approach that involved decomposing visual perception into modules and combining their decoding results to perform visual reconstruction (Miyawaki et al., 2008). Following the work, Kamitani and his colleagues attempted to decode dream content during the early stages of sleep (Horikawa et al., 2013). They used machine learning algorithms trained on waking fMRI data, i.e., when subjects were viewing images, to predict the content of dreams.

The application of cutting-edge technologies has also come into play in the field of decoding visual experience, with deep learning and generative AI trained with large datasets. Such large-scale models can be excellent feature extractors for solving real-world problems that humans also solve, and can therefore be good models of the brain. Furthermore, visualization from predicted latent features is generally a nontrivial and difficult problem, but modern generative models have been optimized to generate realistic images from latent representations. For example, a work from Kamitani's group utilized generative adversarial networks (GANs) to achieve this (Shen et al., 2019), while Takagi and Nishimoto's (2023) study marked another milestone, introducing diffusion models for visual reconstruction (Rombach et al., 2022; Takagi and

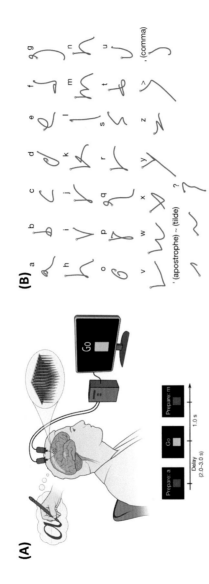

(A)

FIG. 4

(A) To obtain neural activity while attempting to write letters by hand, subjects imagine handwriting one letter at a time according to the instructions given. The figure depicts what was displayed on the screen according to the time axis. (B) The pen trajectories decoded by the decoder are shown for all 31 letters tested. The two-dimensional pen-tip velocities recalled by the subjects were decoded for the letters not used in training using a cross-validation method. Orange circles indicate the start of the trajectory.

Reproduced from Willett, F.R., Avansino, D.T., Hochberg, L.R., Henderson, J.M., Shenoy, K.V., 2021. High-performance brain-to-text communication via handwriting. Nature. 593 (7858), 249–254, with permission.

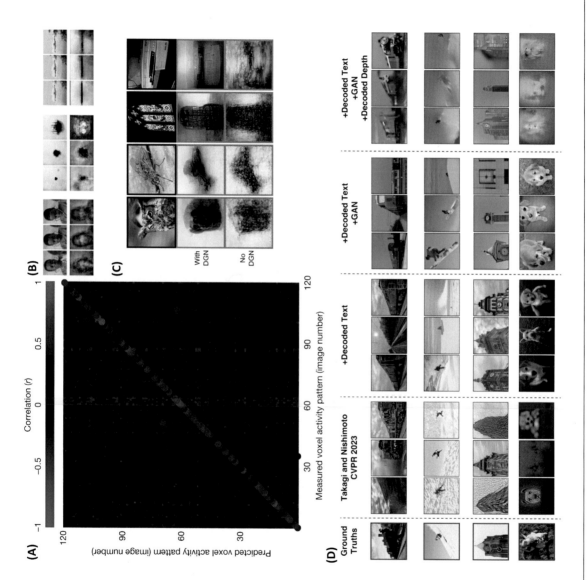

(A)

Correlation (*r*)

Predicted voxel activity pattern (image number)

Measured voxel activity pattern (image number)

(B)

(C)

With DGN

No DGN

(D)

Ground Truths

Takagi and Nishimoto CVPR 2023

+Decoded Text

+Decoded Text +GAN

+Decoded Text +GAN +Decoded Depth

FIG. 5

See figure legend on opposite page

Fig. 5 (A) In Kay et al. (2008), each subject performed an fMRI experiment while viewing novel natural images that had not been used to train the decoding model. This figure shows the performance of one subject using voxel patterns to identify which image they were looking at: the colors in the mth column and nth row represent the correlation between the voxel activity pattern measured for the mth image and the voxel activity pattern predicted for the nth image. The highest correlation in each row is indicated by a color-magnified dot. In this subject, almost all images have the highest correlation with the viewed image itself, i.e., successful identification. (B) The top row shows six frames from a natural movie used in the experiment by Nishimoto et al. (2011). The bottom row shows averaged high posterior (AHP) reconstruction, by averaging the 100 clips in the sampled natural movie prior that had the highest posterior probability. (C) The first, second, and third rows show the images presented to the subject during the fMRI experiment, and reconstructions with and without the deep generative model, respectively (reconstructed from visual cortex activity). (D) Perceived images (leftmost, red box) and examples of reconstructed images from the brain activity of a single subject using diffusion model and several additional techniques (separated by dashed lines). For each method, three generated images from different stochastic noises were chosen randomly.

(A) Reproduced from Kay, K.N., Naselaris, T., Prenger, R.J., Gallant, J.L., 2008. Identifying natural images from human brain activity. Nature, 452 (7185), 352–355, with permission. (B) Reproduced from Nishimoto, S., Vu, A.T., Naselaris, T., Benjamini, Y., Yu, B., Gallant, J.L., 2011. Reconstructing visual experiences from brain activity evoked by natural movies. Curr. Biol., 21 (19), 1641–1646, with permission. (C) Reproduced from Shen, G., Horikawa, T., Majima, K., Kamitani, Y., 2019. Deep image reconstruction from human brain activity. PLoS Comput. Biol., 15 (1), e1006633., CC-BY4.0. (D) Reproduced from Takagi, Y., Nishimoto, S., 2023. High-resolution image reconstruction with latent diffusion models from human brain activity. Proceedings of the IEEE/CVF Conference on Computer Vision and Pattern Recognition, 14453–14463.

Nishimoto, 2023). These contributions collectively paint a vivid picture of the ongoing evolution and potential of neural decoding as it relates to visual cognition.

Speech decoding has become another prominent area, encompassing both the decoding of spoken utterances and the decoding of auditory perception. The area of decoding spoken utterances often focuses on individuals with speech disorders. A recent study by Edward Chang converted intended speech from neural activity into synthesized speech (Anumanchipalli et al., 2019). Following the work, the decoding model is investigated to help patients with locked-in-syndrome (Metzger et al., 2023) or amyotrophic lateral sclerosis (Willett et al., 2023) regain their voices. By using surgically implanted microelectrode arrays or ECoG, electrical impulses associated with the patient's speech muscles are recorded and mapped to phonemes, the basic units of speech. These phonemes, in turn, correspond to specific configurations of the vocal tract muscles, such as lips, tongue, jaw, and larynx.

While speech has been extensively addressed in BMI due to its importance, much of the research has been conducted using multielectrode studies rather than neuroimaging, mainly because it demands higher temporal resolution compared to vision. In neuroimaging, fMRI is commonly used but doesn't offer high temporal resolution. Magnetoencephalography (MEG) and EEG, which have higher temporal resolution, have other problems such as high noise and rough spatial resolution. As a result, until recently, neuroimaging studies have mainly focused on classification (Dash et al., 2020; Farwell and Donchin, 1988; Pereira et al., 2018). However, in the decoding of speech perception, Alex Huth achieved a significant breakthrough in 2023 by successfully decoding continuous speech (Tang et al., 2023) (Fig. 6). In this study, each person listened to a few

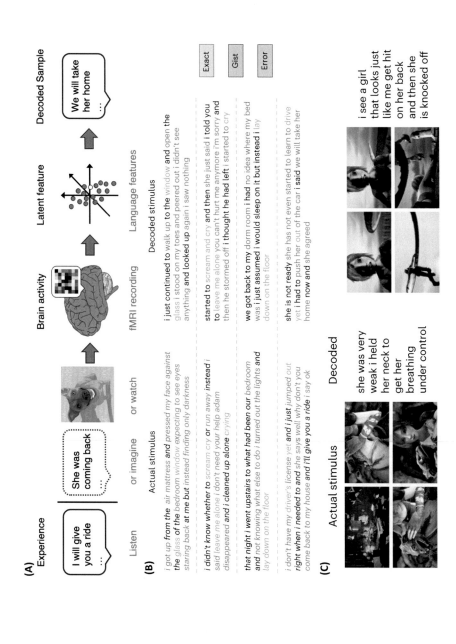

FIG. 6

(A) A schematic of semantic decoding experiments in Tang et al. (2023). (B) Four segments from test sentences are displayed together with the predictions made by the decoder for a single. (C) To examine the adaptability of the decoder across different modalities, brain activities were decoded as they viewed four mute short films (Blender Foundation; https://www.sintel.org).

Reproduced from Tang, J., LeBel, A., Jain, S., Huth, A.G., 2023. Semantic reconstruction of continuous language from non-invasive brain recordings. Nat. Neurosci., 26, 858–866, with permission.

minutes of speech for 16h. Using Generative Pre-trained Transformer 2 (GPT-2), a language model trained by a large text dataset, they trained a model to predict brain activity from GPT-2's latent representations for that sentence. Based on the predicted brain activity, the most plausible sentences were estimated, resulting in the reconstruction of sentences similar to the speech being heard. Interestingly, they further applied the decoder to brain activity during silent video viewing to test whether the decoder could be used for "mind reading." As a result, they succeeded in decoding sentences similar to the video content. Together, these pioneering efforts are opening exciting avenues in understanding both the production and perception of speech, reinforcing the comprehensive and dynamic nature of neural decoding research.

Assumptions of decoding models

Decoding models, which are intricately crafted and formulated to interpret and analyze distinct mental states, various cognitive tasks, and assorted sensory inputs, fundamentally build upon the assumption that there are reliable, consistent patterns of brain activity. These models work under the premise that each mental state or sensory input is connected to a unique and consistent pattern of neural activity within the brain. Variances or differences in these patterns can arise and might often be attributed to individual differences in physiology or other factors such as attention, learning, or the general state of the brain at a given time. These factors can introduce a degree of complexity and challenge in interpreting brain activity, but understanding them helps in refining the models.

Another assumption that is pivotal in the structure of decoding models is the principle of stationarity. This principle simplifies the relationship between brain activity and the variables being decoded as unchanging and stable over a given time. In a constantly changing and dynamic organ such as the brain, this assumption might seem oversimplified. Realistically, the relationship between brain activity and the variables can evolve and change due to various elements such as learning, fatigue, shifts in attention, or even environmental factors. Nevertheless, for short timescales or specific, constrained tasks, the assumption of stationarity can be deemed reasonable and offers practical benefits in the modeling process.

Linear relationship assumption is also commonly adopted in decoding models. This posits that the relationship between brain activity and the variables being decoded is linear in nature. Although this is a simplifying assumption, it serves as a useful approximation in many situations. This assumption is preferred for interpretations because sufficiently complex nonlinear models can decode arbitrarily complex patterns from early sensory signals (e.g., face recognition can be performed using retinal activity patterns, though this does not imply that the retina is the locus of face recognition; see Naselaris et al., 2011 for further discussion). However, it is worth noting that the real relationships might exhibit nonlinearity, and hence this assumption might not always hold. Despite this limitation, linear models are often favored due to their mathematical tractability and ease of interpretation compared to more complex nonlinear models.

The notion of independence is also a key factor in the modeling process. This assumes that the variables being decoded are independent of each other given the observed brain activity. However, in practice, this assumption is often not entirely accurate since different aspects of perception, behavior, or cognition are likely to be interconnected and interdependent. Nonetheless, assuming independence can simplify the modeling process and sometimes provide a reasonable approximation, making decoding models more tractable and computationally efficient.

Last, assumptions concerning the spatial and temporal resolution of the brain activity are made based on the specific type of neural data being utilized. For example, fMRI data provide high spatial resolution, allowing for detailed observation of brain structures, but they have relatively low temporal resolution, limiting their ability to capture rapid changes over time. Conversely, EEG data offer high temporal resolution, enabling the observation of fast neural activities, but are limited in their spatial resolution. These assumptions are typically grounded in substantial empirical evidence about the capabilities and limitations of these recording techniques, and they are instrumental in shaping the way decoding models are designed and implemented.

In summary, decoding models rely on a series of assumptions about the nature of brain activity and the relationships between different variables. While these assumptions can provide useful simplifications and approximations, they also present challenges and potential inaccuracies that must be understood and addressed in the application of these models. The careful consideration of these assumptions allows researchers to create models that can effectively interpret complex mental states, cognitive tasks, and sensory inputs, thereby enhancing our understanding of the brain and its intricate functions.

Building, testing, interpreting decoding models

Typical procedures for building, testing, and interpreting decoding models involve the following:

- **Experimental Design:** The first step is to design an experiment that will generate the necessary data. This might involve choosing which tasks or stimuli to use, determining how to measure the variables of interest, how many samples to obtain, and considering potential confounds. The design of the experiment can significantly influence the quality and interpretability of the data.
- **Data Collection:** Brain activity data are collected from subjects while they perform tasks or experience the stimuli specified in the experimental design. This collection may involve using fMRI, EEG, MEG, or other techniques. It is crucial that this data collection is executed with care and consistency to ensure the validity of the data. Proper design of spatial and temporal resolutions, as well as the coverage of the recordings, is essential to meet the specific aims of the project. Considerations

regarding data collection with repeated trials (e.g., if trials are repeated, how many times?) are vital given the noise level of the recording of interest.

- **Data Preprocessing:** The raw brain activity data are typically preprocessed to correct for noise and artifacts and to standardize the measurements. Depending on the type of data, this might involve motion correction, filtering, normalization, or other preprocessing steps. These steps may include techniques such as principal component analysis (PCA), independent component analysis (ICA), or other methods aimed at reducing data dimensionality and highlighting the most informative dimensions or components.

- **Feature Extraction:** To relate stimulus or task samples to evoked brain activity, researchers typically extract relevant features from these samples. In classical research designs, experts often manually crafted these features, drawing on insights from neurophysiology or cognitive neuroscience. However, there has been a recent shift toward utilizing deep learning for feature extraction, which automates the process and often enhances the model's performance.

- **Model Building:** The model is trained on a subset of the data, referred to as the training set. This process entails choosing a model type (e.g., linear regression, support vector machine, etc.) and tuning its parameters to optimize predictions of the variables of interest based on brain activity. Model selection should be informed by the specific research question, the data's characteristics, and considerations related to model interpretability and complexity. The model-building process might also involve generating the variables of interest from their latent forms (see Fig. 1). This generative step can take various forms, depending on the nature of the samples. Examples include vector averaging for motor command generation or employing generative machine learning models to reconstruct images or sounds.

- **Model Testing:** The trained model is tested on a separate subset of data (the test set) that the model has not seen before. This step assesses how well the model generalizes to new data. Model performance is typically evaluated using metrics such as accuracy, precision, recall, or area under the receiver operating characteristic (ROC) curve. These metrics can be quantitatively calculated by examining the distance between the decoded result and the ground truth in the feature space. Alternatively, when the decoding target is complex (e.g., images or sounds), accuracy can be assessed subjectively through human judgment. For instance, one might recruit third-party subjects and inquire how often they can correctly match the decoded results to the target.

- **Model Interpretation:** The final step is to interpret the model and its results. This might involve visualizing the model's predictions, examining the importance of different features, or testing specific hypotheses about the brain. This step is crucial for translating the results of the model into meaningful insights about brain function.

Pitfalls of decoding modeling

Decoding models in neuroscience, essential for interpreting and analyzing brain activity, come with a multitude of challenges and potential pitfalls. We will explore some of these challenges in greater detail, as follows:

- **Overfitting:** This occurs when the model fits the training data's noise rather than the underlying pattern, resulting in poor performance on new, unseen data. Essentially, the model becomes too specialized in the particular quirks and details of the training data, losing its generalizability.
  - Techniques to handle overfitting include regularization methods such as Lasso, Ridge, or Elastic Net. These approaches penalize complex models by adding constraints, thereby helping to avoid overfitting to noise. Cross-validation, in which the training data are divided into different subsets to assess the model's performance, is used to determine the optimal hyperparameters for the regularization. This involves finding a balance between error reduction and the regularization penalty, ensuring that the model generalizes well to unseen data.
- **Underfitting:** Conversely, underfitting happens when the model is too simplistic to capture the data's complexity, causing it to perform poorly on both training and new data. This reflects a model that lacks the necessary complexity or flexibility to understand the underlying structure of the data.
  - Addressing underfitting may require creating a more complex model or adding more features or data, capturing more information to understand the underlying structure of the data.
- **Attention to Data Leakage:** Data leakage, or the unintentional sharing of information between training and testing datasets, can be particularly insidious in decoding. For instance, if preprocessing steps such as normalization are applied to the entire dataset before splitting it into training and testing subsets, information from the test set could inadvertently influence the training set. This can lead to biased results and overly optimistic performance estimates, undermining the integrity of the research.
  - In decoding, where accuracy is paramount, researchers must be extremely diligent in handling data, separating datasets early in the process, and applying preprocessing steps independently to maintain the reliability of the findings.
- **Misinterpretation of Results:** A common pitfall in decoding models is misinterpreting correlations between brain activity and variables of interest as causal relationships. While these models are powerful in identifying connections, they don't directly provide causal explanations.
  - Therefore, it's crucial for researchers to exercise caution in interpretation, avoiding overinterpretation and unwarranted causal inferences that may lead to misguided conclusions or interventions.
- **Lack of Generalization:** A model that is trained and tested on one dataset may not generalize well to others. Variations in experimental design, individual differences, or differing contexts may reduce the applicability of the model across

different datasets or populations (e.g., different categories in stimulus sets, or different age groups or cultural backgrounds of subjects).

- To avoid issues with a lack of generalizability and overinterpretation of model performance, it's vital to validate models on independent and, if possible, external datasets. This practice helps ensure that the model captures essential aspects of brain function across different subjects, experimental conditions, and laboratories.

- **Challenges with Big Data and Generative AI:** Generative AI and other models trained on extensive datasets present unique challenges. The complexity of these models may render their inner workings opaque, making interpretation and understanding difficult. This "black box" problem can be exacerbated by models that are inherently inaccessible, with internal mechanics that resist examination, hindering transparency and accountability.

  - Investing in techniques such as explainable AI frameworks can provide insights into the decision-making processes of complex models, allowing for better understanding (Gunning et al., 2019). In addition, efforts to use publicly available models and codes (e.g., Stable Diffusion for image generation), that are also trained on publicly available datasets, can increase transparency and reproducibility in the field.

- **Potential Overstating of Capabilities:** As "brain-reading" or "mind-reading" studies using decoding models rise in prominence, there's a risk of exaggerating these models' capabilities. This overstatement could lead to ethical misinterpretations, such as the false belief that we can fully understand or predict someone's thoughts based on their brain activity, raising serious privacy and autonomy concerns.

  - Researchers should set realistic expectations and avoid making sweeping claims that could mislead the public or stakeholders. Articulating the limitations of a study or model and recognizing the current boundaries of the field encourages responsible and trustworthy science.

- **Privacy and Consent Issues:** With increasing model accuracy, ethical concerns such as privacy and consent become vital, especially in sensitive applications such as lie detection or personal thought analysis.

  - These concerns include maintaining subject confidentiality, obtaining informed consent, and transparently communicating potential uses and limitations of the technology. Such practices ensure ethical integrity and foster trust between researchers and subjects.

Open challenges and future directions
Open challenges

Despite advances in neuroimaging, all neuroscience models face limitations in data acquisition, analysis, and volume. Within this context, this section specifically explores the unique challenges of decoding models and their potential applications.

One pressing challenge that deserves particular attention is the issue of generalizability across different subjects. Most current decoding studies focus on gathering extensive training data from individual subjects to develop personalized brain models. While this approach allows for highly tailored decoding models, it is impractical in settings where quick setup is essential, such as in brain–computer interfaces or clinical applications. Spending hours to tailor a model to each individual is often unfeasible. On the flip side, creating a one-size-fits-all model is fraught with its own set of challenges, as individual neuroanatomy and neural responses can vary widely. Therefore, a delicate balance must be struck between individual tailoring and general applicability.

Real-time decoding, especially in clinical settings, represents an emerging frontier with implications for neurofeedback therapies and brain–computer interfaces. The urgency of real-time decoding brings with it an array of challenges, the most critical of which is the need for models capable of rapidly and accurately interpreting noisy data. Unlike traditional offline analyses where comprehensive data cleaning is feasible, real-time decoding must contend with a continuous stream of incoming data that is often fraught with artifacts and extraneous noise. Furthermore, these models must be optimized to deliver precise and reliable predictions on extremely short timescales, often within milliseconds to seconds, to be clinically or practically useful. Balancing the speed and accuracy of these predictions while maintaining robustness to noise is a formidable challenge that current methods are still striving to meet.

The difficulty of imagery decoding is a significant hurdle, especially when contrasted with perception. Imagery, the mental visualization of objects and concepts, often has a considerably lower SNR compared to direct sensory perception (Reddy et al., 2010). This makes the accurate decoding of imagery a complex task. In the context of BMIs, this issue highlights a need for subject training, where individuals must practice and refine their ability to generate consistent and detectable imagery. Though this training might offer a pathway to enhance SNR, the inherent challenges of decoding such nuanced mental visualizations require meticulous attention and innovative methodologies.

Utilizing generative AI in decoding is both promising and complex. These powerful models can create realistic outputs, but many operate as "black boxes," obscuring their internal logic. A critical challenge lies in discerning whether the generated outputs are genuinely derived from brain activity or if they have been altered or supplemented by AI. For instance, in modeling brain-generated images, it may be difficult to determine if the visualizations accurately reflect neural activity or are shaped by the AI's algorithms. This necessitates careful interpretation and validation, balancing the potential of generative AI with the need for accuracy and transparency in representing underlying neural processes.

The concept of externalizing decoded brain activity raises intriguing possibilities and challenges. Often explored in terms such as mind-uploading or the "Matrix" scenario, it refers to translating brain activity into a digital medium. This could lead to profound understanding of consciousness and new ways to preserve or replicate human cognition. However, the idea also brings up complex questions about identity, privacy, and ethics. Technological and philosophical considerations intertwine, making this a rich and multifaceted field of study that goes beyond mere technological advancement, touching on some of humanity's most profound questions.

Future directions

The future of decoding neuroimaging data is likely to be shaped by several trends and advancements. Progress in neuroimaging technology will enhance data quality through higher-resolution imaging, less noisy measurements, and more precise targeting of neural populations, facilitating better decoding.

Large-scale collaborations, similar to the Human Connectome Project (Van Essen et al., 2013) and UK Biobank (Sudlow et al., 2015), will expedite the collection of larger, more diverse datasets, addressing the challenge of inter-individual variability. In parallel, the rapid evolution of AI technologies presents advanced techniques suited for decoding tasks, including methods for handling high-dimensional data, preventing overfitting, and integrating diverse types of data.

Future research may also concentrate on developing individualized decoding models that account for each individual's unique brain characteristics, potentially incorporating personal information such as genetic data or cognitive profiles. As more types of data are collected, such as genomic, metabolic, and physiological, models will need to effectively integrate this information, leveraging techniques from multivariate statistics and machine learning.

The question of how closely AI mirrors human cognitive processes is a subject of growing intrigue and exploration. By studying and comparing the underlying mechanisms of both human brains and AI systems, researchers can probe into the similarities and differences that exist between these two complex information-processing entities. In the field of decoding, this examination becomes particularly pertinent. By using decoding models to align human brain activity with AI's internal representations, it is possible to identify corresponding functionalities between these distinct systems, even though they are constructed and operate in fundamentally different ways.

The advancement of technology necessitates the development of ethical guidelines and legal regulations to prevent misuse and protect individual privacy. These guidelines should be adaptable, evolving in tandem with the technology.

Take-home points

- Neuroscience uses decoding models to predict cognitive states based on brain activity, furthering our understanding of brain functions. These models, often built with machine learning algorithms, have diverse applications, including potential diagnostic tools for neurological disorders. Successes include predicting visual stimuli, linguistic content, and dream content using fMRI decoding models.

- However, challenges include overfitting, underestimating individual differences, and overinterpreting results, addressed through rigorous experimental design and cautious interpretation. Other hurdles involve interpreting complex models, improving generalizability, capturing temporal dynamics, integrating multimodal data, real-time decoding, and ethical and privacy concerns.
- Despite limitations such as data quality, model complexity, and ethical concerns, decoding models show promise for advancing our knowledge of the brain and developing new treatments for neurological and psychiatric disorders, with future research aiming to improve these areas.
- The interaction between generative AI and the human brain is an exciting area of study, particularly in the context of decoding. Generative AI's remarkable ability to create content, akin to human imagination, may enhance decoding performance. However, this comes with certain complexities and requires careful handling. While generative AI may hold the potential to elevate our understanding of human cognitive processes, its "black box" nature can pose challenges in interpretation and application. Careful considerations must be made to ensure that the insights gained are representative of true neural processes rather than artificial constructions.
- In conclusion, neural decoding is emerging as a powerful tool in modern neuroscience, offering profound insights into human cognition and therapeutic pathways. The challenges, while substantial, are integral to the evolutionary journey of this field, inviting innovative solutions and continuous growth. The integration of generative AI adds further complexity and excitement to the landscape, promising a future filled with deeper understanding and applications.

Further resources

- Textbooks:
  - "Pattern Recognition and Machine Learning" by Christopher Bishop (Bishop and Nasrabadi, 2006)
  - "The Elements of Statistical Learning" by Trevor Hastie (Hastie et al., 2009)
  - "Visual population codes," edited by Nikolaus Kriegeskorte and Gabriel Kreiman (Kriegeskorte and Kreiman, 2012)
- Software:
  - Nilearn: https://github.com/nilearn/nilearn
  - Himalaya: https://github.com/gallantlab/himalaya (la Tour et al., 2022)

References

Anumanchipalli, G.K., Chartier, J., Chang, E.F., 2019. Speech synthesis from neural decoding of spoken sentences. Nature 568 (7753), 493–498.
Arbabshirani, M.R., Plis, S., Sui, J., Calhoun, V.D., 2017. Single subject prediction of brain disorders in neuroimaging: promises and pitfalls. Neuroimage 145, 137–165.

Birbaumer, N., Ghanayim, N., Hinterberger, T., Iversen, I., Kotchoubey, B., Kubler, A., Perelmouter, J., Taub, E., Flor, H., 1999. A spelling device for the paralysed. Nature 398 (6725), 297–298.

Bishop, C.M., Nasrabadi, N.M., 2006. Pattern Recognition and Machine Learning. Vol. 4, Issue 4 Springer.

Blankertz, B., Dornhege, G., Schafer, C., Krepki, R., Kohlmorgen, J., Muller, K.-R., Kunzmann, V., Losch, F., Curio, G., 2003. Boosting bit rates and error detection for the classification of fast-paced motor commands based on single-trial EEG analysis. IEEE Trans. Neural Syst. Rehabil. Eng. 11 (2), 127–131.

Blankertz, B., Dornhege, G., Krauledat, M., Müller, K.-R., Curio, G., 2007. The non-invasive Berlin brain–computer interface: fast acquisition of effective performance in untrained subjects. Neuroimage 37 (2), 539–550.

Chapin, J.K., Moxon, K.A., Markowitz, R.S., Nicolelis, M.A.L., 1999. Real-time control of a robot arm using simultaneously recorded neurons in the motor cortex. Nat. Neurosci. 2 (7), 664–670.

Cox, D.D., Savoy, R.L., 2003. Functional magnetic resonance imaging (fMRI) "brain reading": detecting and classifying distributed patterns of fMRI activity in human visual cortex. Neuroimage 19 (2), 261–270.

Dash, D., Ferrari, P., Wang, J., 2020. Decoding imagined and spoken phrases from non-invasive neural (MEG) signals. Front. Neurosci. 14, 290.

Denk, T.I., Takagi, Y., Matsuyama, T., Agostinelli, A., Nakai, T., Frank, C., Nishimoto, S., 2023. Brain2Music: reconstructing music from human brain activity. ArXiv. ArXiv Preprint ArXiv:2307.11078.

Etkin, A., Wager, T.D., 2007. Functional neuroimaging of anxiety: a meta-analysis of emotional processing in PTSD, social anxiety disorder, and specific phobia. Am. J. Psychiatry 164 (10), 1476–1488.

Farwell, L.A., Donchin, E., 1988. Talking off the top of your head: toward a mental prosthesis utilizing event-related brain potentials. Electroencephalogr. Clin. Neurophysiol. 70 (6), 510–523.

Gilja, V., Pandarinath, C., Blabe, C.H., Nuyujukian, P., Simeral, J.D., Sarma, A.A., Sorice, B. L., Perge, J.A., Jarosiewicz, B., Hochberg, L.R., et al., 2015. Clinical translation of a high-performance neural prosthesis. Nat. Med. 21 (10), 1142–1145.

Güçlü, U., van Gerven, M.A.J., 2015. Deep neural networks reveal a gradient in the complexity of neural representations across the ventral stream. J. Neurosci. 35 (27), 10005–10014.

Gunning, D., Stefik, M., Choi, J., Miller, T., Stumpf, S., Yang, G.-Z., 2019. XAI—Explainable artificial intelligence. Sci. Robot. 4 (37), eaay7120.

Harrison, S.A., Tong, F., 2009. Decoding reveals the contents of visual working memory in early visual areas. Nature 458 (7238), 632–635.

Hassabis, D., Kumaran, D., Summerfield, C., Botvinick, M., 2017. Neuroscience-inspired artificial intelligence. Neuron 95 (2), 245–258. https://doi.org/10.1016/j.neuron.2017.06.011.

Hastie, T., Tibshirani, R., Friedman, J.H., Friedman, J.H., 2009. The Elements of Statistical Learning: Data Mining, Inference, and Prediction. Vol. 2 Springer.

Haxby, J.V., Gobbini, M.I., Furey, M.L., Ishai, A., Schouten, J.L., Pietrini, P., 2001. Distributed and overlapping representations of faces and objects in ventral temporal cortex. Science 293 (5539), 2425–2430.

Haynes, J.-D., Rees, G., 2005. Predicting the stream of consciousness from activity in human visual cortex. Curr. Biol. 15 (14), 1301–1307.

Hochberg, L.R., Serruya, M.D., Friehs, G.M., Mukand, J.A., Saleh, M., Caplan, A.H., Branner, A., Chen, D., Penn, R.D., Donoghue, J.P., 2006. Neuronal ensemble control of prosthetic devices by a human with tetraplegia. Nature 442 (7099), 164–171.

Horikawa, T., Kamitani, Y., 2017. Generic decoding of seen and imagined objects using hierarchical visual features. Nat. Commun. 8 (1), 15037.

Horikawa, T., Tamaki, M., Miyawaki, Y., Kamitani, Y., 2013. Neural decoding of visual imagery during sleep. Science 340 (6132), 639–642.

Horikawa, T., Cowen, A.S., Keltner, D., Kamitani, Y., 2020. The neural representation of visually evoked emotion is high-dimensional, categorical, and distributed across transmodal brain regions. Iscience 23 (5), 101060.

Huth, A.G., Nishimoto, S., Vu, A.T., Gallant, J.L., 2012. A continuous semantic space describes the representation of thousands of object and action categories across the human brain. Neuron 76 (6), 1210–1224.

Kamitani, Y., Tong, F., 2005. Decoding the visual and subjective contents of the human brain. Nat. Neurosci. 8 (5), 679–685.

Kay, K.N., Naselaris, T., Prenger, R.J., Gallant, J.L., 2008. Identifying natural images from human brain activity. Nature 452 (7185), 352–355.

Kell, A.J.E., Yamins, D.L.K., Shook, E.N., Norman-Haignere, S.V., McDermott, J.H., 2018. A task-optimized neural network replicates human auditory behavior, predicts brain responses, and reveals a cortical processing hierarchy. Neuron 98 (3), 630–644.

Kragel, P.A., Knodt, A.R., Hariri, A.R., LaBar, K.S., 2016. Decoding spontaneous emotional states in the human brain. PLoS Biol. 14 (9), e2000106.

Kriegeskorte, N., Douglas, P.K., 2019. Interpreting encoding and decoding models. Curr. Opin. Neurobiol. 55, 167–179.

Kriegeskorte, N., Kreiman, G., 2012. Visual Population Codes: Toward a Common Multivariate Framework for Cell Recording and Functional Imaging. MIT Press.

la Tour, T.D., Eickenberg, M., Nunez-Elizalde, A.O., Gallant, J.L., 2022. Feature-space selection with banded ridge regression. Neuroimage 264, 119728.

Metzger, S.L., Littlejohn, K.T., Silva, A.B., Moses, D.A., Seaton, M.P., Wang, R., Dougherty, M.E., Liu, J.R., Wu, P., Berger, M.A., et al., 2023. A high-performance neuroprosthesis for speech decoding and avatar control. Nature 620, 1037–1046.

Miyawaki, Y., Uchida, H., Yamashita, O., Sato, M., Morito, Y., Tanabe, H.C., Sadato, N., Kamitani, Y., 2008. Visual image reconstruction from human brain activity using a combination of multiscale local image decoders. Neuron 60 (5), 915–929.

Nakai, T., Nishimoto, S., 2020. Quantitative models reveal the organization of diverse cognitive functions in the brain. Nat. Commun. 11 (1), 1142.

Naselaris, T., Kay, K.N., Nishimoto, S., Gallant, J.L., 2011. Encoding and decoding in fMRI. Neuroimage 56 (2), 400–410.

Nishimoto, S., Vu, A.T., Naselaris, T., Benjamini, Y., Yu, B., Gallant, J.L., 2011. Reconstructing visual experiences from brain activity evoked by natural movies. Curr. Biol. 21 (19), 1641–1646.

Pereira, F., Lou, B., Pritchett, B., Ritter, S., Gershman, S.J., Kanwisher, N., Botvinick, M., Fedorenko, E., 2018. Toward a universal decoder of linguistic meaning from brain activation. Nat. Commun. 9 (1), 963.

Reddy, L., Tsuchiya, N., Serre, T., 2010. Reading the mind's eye: decoding category information during mental imagery. Neuroimage 50 (2), 818–825.

Rombach, R., Blattmann, A., Lorenz, D., Esser, P., Ommer, B., 2022. High-resolution image synthesis with latent diffusion models. In: Proceedings of the IEEE/CVF Conference on Computer Vision and Pattern Recognition, pp. 10684–10695.

Saarimäki, H., Gotsopoulos, A., Jaaskelainen, I.P., Lampinen, J., Vuilleumier, P., Hari, R., Sams, M., Nummenmaa, L., 2016. Discrete neural signatures of basic emotions. Cereb. Cortex 26 (6), 2563–2573.

Shen, G., Horikawa, T., Majima, K., Kamitani, Y., 2019. Deep image reconstruction from human brain activity. PLoS Comput. Biol. 15 (1), e1006633.

Stokes, M., Thompson, R., Cusack, R., Duncan, J., 2009. Top-down activation of shape-specific population codes in visual cortex during mental imagery. J. Neurosci. 29 (5), 1565–1572.

Sudlow, C., Gallacher, J., Allen, N., Beral, V., Burton, P., Danesh, J., Downey, P., Elliott, P., Green, J., Landray, M., et al., 2015. UK biobank: an open access resource for identifying the causes of a wide range of complex diseases of middle and old age. PLoS Med. 12 (3), e1001779.

Takagi, Y., Nishimoto, S., 2023. High-resolution image reconstruction with latent diffusion models from human brain activity. In: Proceedings of the IEEE/CVF Conference on Computer Vision and Pattern Recognition, pp. 14453–14463.

Tang, J., LeBel, A., Jain, S., Huth, A.G., 2023. Semantic reconstruction of continuous language from non-invasive brain recordings. Nat. Neurosci. 26, 858–866.

Taylor, D.M., Tillery, S.I.H., Schwartz, A.B., 2002. Direct cortical control of 3D neuroprosthetic devices. Science 296 (5574), 1829–1832.

Van Essen, D.C., Smith, S.M., Barch, D.M., Behrens, T.E.J., Yacoub, E., Ugurbil, K., WU-Minn HCP Consortium, et al., 2013. The WU-Minn human connectome project: an overview. Neuroimage 80, 62–79.

Velliste, M., Perel, S., Spalding, M.C., Whitford, A.S., Schwartz, A.B., 2008. Cortical control of a prosthetic arm for self-feeding. Nature 453 (7198), 1098–1101.

Wager, T.D., Atlas, L.Y., Lindquist, M.A., Roy, M., Woo, C.-W., Kross, E., 2013. An fMRI-based neurologic signature of physical pain. N. Engl. J. Med. 368 (15), 1388–1397.

Wessberg, J., Stambaugh, C.R., Kralik, J.D., Beck, P.D., Laubach, M., Chapin, J.K., Kim, J., Biggs, S.J., Srinivasan, M.A., Nicolelis, M.A.L., 2000. Real-time prediction of hand trajectory by ensembles of cortical neurons in primates. Nature 408 (6810), 361–365.

Willett, F.R., Avansino, D.T., Hochberg, L.R., Henderson, J.M., Shenoy, K.V., 2021. High-performance brain-to-text communication via handwriting. Nature 593 (7858), 249–254.

Willett, F.R., Kunz, E.M., Fan, C., Avansino, D.T., Wilson, G.H., Choi, E.Y., Kamdar, F., Glasser, M.F., Hochberg, L.R., Druckmann, S., et al., 2023. A high-performance speech neuroprosthesis. Nature 620, 1031–1036.

Wolpaw, J.R., McFarland, D.J., 2004. Control of a two-dimensional movement signal by a noninvasive brain-computer interface in humans. Proc. Natl. Acad. Sci. U. S. A. 101 (51), 17849–17854.

Yamins, D.L.K., Hong, H., Cadieu, C.F., Solomon, E.A., Seibert, D., DiCarlo, J.J., 2014. Performance-optimized hierarchical models predict neural responses in higher visual cortex. Proc. Natl. Acad. Sci. U. S. A. 111 (23), 8619–8624.

Yuste, R., Goering, S., Arcas, B.A.Y., Bi, G., Carmena, J.M., Carter, A., Fins, J.J., Friesen, P., Gallant, J., Huggins, J.E., et al., 2017. Four ethical priorities for neurotechnologies and AI. Nature 551 (7679), 159–163.

Zador, A., Escola, S., Richards, B., Ölveczky, B., Bengio, Y., Boahen, K., Botvinick, M., Chklovskii, D., Churchland, A., Clopath, C., et al., 2023. Catalyzing next-generation Artificial Intelligence through NeuroAI. Nat. Commun. 14 (1), 1597. https://doi.org/10.1038/s41467-023-37180-x.

Normative modeling for clinical neuroscience

Andre Marquand[a,b], Saige Rutherford[a,b], and Thomas Wolfers[a,c,d]

[a]*Donders Institute for Brain, Cognition and Behaviour, Radboud University Medical Centre, Nijmegen, The Netherlands,* [b]*Department of Cognitive Neuroscience, Radboud University Nijmegen Medical Centre, Nijmegen, The Netherlands,* [c]*Department of Psychiatry and Psychotherapy, Tübingen Center for Mental Health, University of Tübingen, Tübingen, Germany,* [d]*German Center for Mental Health, Tübingen, Germany*

Introduction to normative modeling

A commonly used approach to identify biological markers associated with mental disorders or neurological diseases is a statistical or machine learning paradigm, which compares a group of cases, thus individuals with a medical condition, to a group of controls on various biological variables (Thompson, 1994). This approach has been successful in detecting factors that are linked to a medical condition, and has often led to statistically significant differences between brains of patients and healthy individuals at the group level (Boedhoe et al., 2020). A biomarker can generally be defined as an objective biological measure that provides information about an individual's medical condition (Califf, 2018). The categorization of biomarkers is based on their intended purpose, with diagnostic markers being extensively studied to predict diagnoses, and prognostic markers used to predict the course or outcome of a condition. The identification of biological markers in medicine presents a significant challenge across various medical domains (Abi-Dargham and Horga, 2016). To address this challenge, researchers employ different study designs to detect biological signatures that can be utilized to enhance diagnostics, course prediction, or treatments of complex brain disorders. The diagnosis and treatment of mental disorders, which are complex phenotypes, have traditionally relied solely on symptom assessments (Kapur et al., 2012). However, this approach fails to account for individual-level differences that contribute to the development of these disorders (Foulkes and Blakemore, 2018; Kim et al., 2023a).

Computational and Network Modeling of Neuroimaging Data. https://doi.org/10.1016/B978-0-443-13480-7.00014-4

Investigations into biological factors underlying mental disorders frequently utilize observational, case–control, and cohort designs. In these types of observational studies, the scientist does not manipulate variables but observes an effect passively, usually through the application of statistics. In medicine, a clinical trial, which is a true experimental design, is often employed in treatment-effect studies. In such a true experimental design, the variable of interest is changed or manipulated by the scientist who is an active experimenter. It is usually done through random assignment of individuals to different experimental groups, one in which a variable of interest is changed, e.g., a medication is given and one group in which this is not the case. Through the process of random assignment of groups stochastic effects can be excluded, as both the experimental and the control groups are in expectation the same. If a group effect is present and sufficiently powered by a large enough sample, the experimenter can assume that an effect of the manipulation on the variable of interest is present, as this is the only factor the randomly assigned groups differ on systematically. In this scenario, a comparison of mean differences between two groups is an important statistical procedure that yields reliable and possibly causal insights.

Most studies that try to identify biological markers of mental or neurological illness are, however, passive observational. This is not surprising as brain disorders and diseases cannot be experimentally evoked in humans. While standard textbooks on applied statistics report on the caveats of interpreting effects derived from passive observational designs (Dean et al., 2017), the most dominant paradigm to detect biological factors underlying mental and neurological illness is the case–control paradigm (Thompson, 1994). These studies have generally been analyzed with standard statistics used to detect group effects, which report significance when within group differences are small and between group differences are relatively large. However, in the case of complex brain disorders, the picture is quite the opposite (Wolfers, 2019; Wolfers et al., 2018). For instance, depression can be diagnosed based on more than 1000 different combinations of symptoms (Fried and Nesse, 2015) and no reliable diagnostic markers have yet been identified (Border et al., 2019; Dinga et al., 2019). This is due to the large degree of within group heterogeneity but also overlap in terms of biology with a healthy group. With increasing sample sizes and as small effects become statistically significant, while the respective predictive value for a disorder or disease remains limited, a false certainty regarding relevant underling biology is created (Wolfers et al., 2015) and an illusion of an "average patient" can emerge. However, similar to sex differences, if we know the mental disorder of an individual, we still know very little about most other traits and certainly not much about the underlying biology. Pursuing the identification of biological markers underlying complex disorders through a comparison of cases and controls seems therefore to a certain extent misleading. The resulting biomarkers from such comparisons are generally not predictive of course, status, or outcome of these complex medical conditions resulting in little informational value for the individual patient. However, the individual is treated in medicine and should therefore matter most in translational neuroscience and medicine.

Individual-level inferences in neuroscience are crucial for understanding how the brain works. The brain is one of the most complex and intricate organs in the human body, and its study requires a multidisciplinary approach that encompasses various levels of analysis, including molecular, cellular, systems, and behavioral measurements (Bassett and Gazzaniga, 2011). Individual-level inferences are also important for developing effective treatments for brain-related disorders. Many neurological and mental disorders, such as Alzheimer's and Parkinson's disease, as well as schizophrenia, are associated with changes in the activity and connectivity of specific neural circuits (Faraone et al., 2015; Kahn et al., 2015; Poewe et al., 2017). By identifying these changes at the individual-level, researchers can develop targeted interventions that aim to restore normal neural activity and connectivity (Friston et al., 2016). For example, deep brain stimulation, a treatment that involves the application of electrical stimulation to specific regions of the brain, has been used successfully to treat movement disorders, such as Parkinson's disease, by modulating the activity of specific circuits (Lozano et al., 2019). Thus, without tools that allow for inferences at the level of the individual or subgroup, we will not be able to add significant progress in clinical practice.

As a potential answer to these problems outlined previously, with a specific focus on mapping the heterogeneity of complex mental and neurological disorders, analytical frameworks, such as normative modeling, have been introduced (Marquand et al., 2016a, 2019). This approach is based on a combination of ideas derived from classical growth charting and modern statistical and machine learning. Normative modeling enables the mapping between quantitative biological measures and various behavioral, demographic, or clinical characteristics. It provides estimates of centiles of variation across the population, allowing for the study of individual differences and the exploration of heterogeneity within cohorts (Rutherford et al., 2022a). Normative modeling provides statistical inferences at the individual-level, indicating the extent to which each person deviates from an estimated growth chart. This approach quantifies and characterizes the deviations from the expected pattern, capturing the uniqueness of each individual without requiring overlap or consistent patterns of atypicality. This approach accommodates the convergence of multiple pathological pathways leading to similar symptoms in different individuals. While case–control analyses may suffer from reduced sensitivity in detecting disorder-related effects due to a focus on group averages, normative modeling explicitly captures and quantifies the heterogeneity underlying clinical conditions at a more detailed level. It allows for the estimation of various mappings depending on the chosen variables, with a particular focus on the relationship between behavioral or demographic measures and quantitative biological readouts, frequently obtained from neuroimaging data.

Population neuroscience

Normative modeling has become feasible with the increasing availability of large-scale population studies (Casey et al., 2018; Miller et al., 2016; Thompson et al., 2020; Van Essen et al., 2013). Population neuroscience is a rapidly growing field that aims to understand how the human brain develops using a Big-Data approach with the promise to identify factors that are hidden but important in healthy development, as well as for the emergence of brain disorders or diseases. This field has benefited greatly from advances in neuroimaging and electrophysiological techniques. These techniques allowed researchers to study the brain in action and have provided new insights into how our nervous system processes information. Magnetic resonance imaging (MRI) is one of the most widely used brain imaging techniques, which allows us to understand more about the brain's structure and function in vivo (Vlaardingerbroek and Den Boer, 2003). While functional MRI (fMRI) measures the changes in blood oxygenation levels in the brain, which are related to neural activity, structural MRI measures the different magnetic properties of different tissue types. fMRI has been used to study a wide range of cognitive and perceptual processes, such as attention, memory, language, and emotion, and has provided important insights into how these processes are implemented in the brain (Posner and DiGirolamo, 2000). Another widely used technique is electroencephalography (EEG), which measures the electrical activity of the brain using electrodes placed on the scalp. EEG is particularly useful for studying the temporal dynamics of brain activity, as it can measure activity on a millisecond timescale. While early brain imaging and electrophysiological studies have been small in scale, the research today is moving toward Big-Data and population neuroscience (Bijsterbosch, 2022; Bzdok and Yeo, 2017; Frégnac, 2017).

The emergence of population-based neuroscience approaches is closely linked to the acquisition and sharing of large-scale studies in an open-science framework (Gau et al., 2021; Markiewicz et al., 2021), which became increasingly popular over the last decade. With the abundance of open-source and multicenter data, questions regarding data analytics and methods development for understanding structure in these samples become apparent. One of the early Big-Data studies, the Human Connectome Project (HCP), developed methods and analytical protocols important for the acquisition of high-quality Big-Data neuroimaging and has been of immense importance for the development of the field (Glasser et al., 2013). Samples, such as the Alzheimer's Disease Neuroimaging Imitative (ADNI), provided data regarding patients to a large community and data sharing became more prevalent (Petersen et al., 2010). Today, there are many samples available and harmonization across studies and scanners is an important research priority, with one of the most impactful data resources, the UKBiobank, being at the heart of most large-scale neuroimaging and genetic studies today (Miller et al., 2016; Sudlow et al., 2015). While it is important to bear in mind that most of these studies contain some sampling bias and do not represent the population in its entirety (Fry et al., 2017), the scale of these studies allow for training of complex algorithms. These developments set the scene for statistical and machine learning approaches that take advantage of population neuroscience to move from training in these large-scale samples toward inference at the level of the individual, with the ambition to translate insights regarding variations in the population toward clinical decision making.

From growth charts to the normative modeling

The use of growth charts, a graphical tool that displays a child's pattern of growth, dates back to the 18th century (Cole, 2012). Since then, growth charts have become an essential tool in child health screening and clinical workup. A growth chart is a reference presented as a visual display for clinical use, and it quantifies size in terms of centiles. Centile crossing on a growth chart, which can be quantified in longitudinal assessments of individual children, indicates a different relative velocity of a child's development in comparison to its reference group. The collection of anthropometry data, the statistical summary of the data, and the graphic design of the chart are the three distinct disciplines involved in the combination of growth charts and growth references, which have had huge impact in pediatrics leading up until the modern days (Cole, 2012; Borghi et al., 2006). Normative modeling is built on this and combines the insights of this classical approach with modern machine learning. It is a framework that enables the mapping between quantitative biological measures and various behavioral, demographic, or clinical characteristics and provides estimates of centiles of variation across the population, allowing for the study of individual differences and the exploration of heterogeneity within clinical groups (Marquand et al., 2019; Rutherford et al., 2022a).

In summary, case–control studies regarding the biological markers of complex brain disorders, such as mental disorders or neurological diseases, have often identified robust links to various biological variables, such as brain regions or genetic locations. However, these associations were generally weak. Given that mental as well as neurological diseases are heterogeneous, comorbid, and overlap with healthy individuals on various biological factors, the lack of markers with predictive value is not as surprising. Normative modeling combines the concepts of growth charting with modern machine learning to map individual differences between persons with the same disorder against the population reference, moving away from case–control comparisons and the average patient, toward a systematic characterization of the individual against an estimated norm.

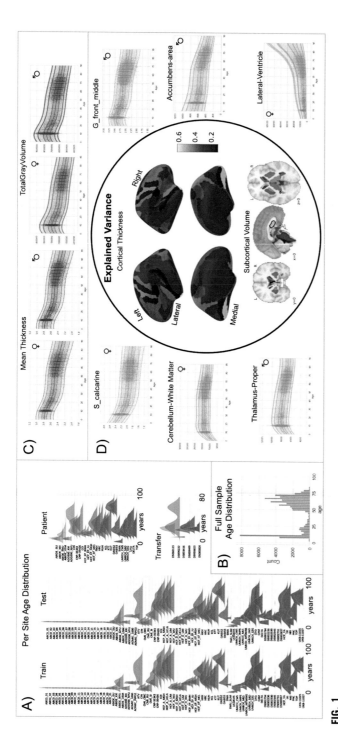

FIG. 1

Example of successful normative modeling. (A) Age density distribution (x-axis) of each site (y-axis) in the full model train and test, clinical, and transfer validation set. (B) Age count distribution of the full sample ($N = 58,836$). (C, D) Examples of lifespan trajectories of brain regions. Age is shown on x-axis and predicted thickness (or volume) values are on the y-axis. Centiles of variation are plotted for each region. In panel (C), we show that sex differences between females (red) and males (blue) are most pronounced when modeling large-scale features such as mean cortical thickness across the entire cortex or total gray matter volume. These sex differences manifest as a shift in the mean in that the shape of these trajectories is the same for both sexes, as determined by sensitivity analyses where separate normative models were estimated for each sex. The explained variance (in the full test set) of the whole cortex and subcortex is highlighted inside the circle of panel (D). All plots within the circle share the same color scale.

Courtesy from Rutherford, S., et al., 2022c. Charting brain growth and aging at high spatial precision. Elife 11, e72904.

Examples of successful normative modeling

Growth charts have been used for over a century in pediatric medicine and provide a means of quantifying individual variation against a reference population, and this idea has been generalized to clinical neuroimaging providing inferences at the individual level (Marquand et al., 2016a,b; Rutherford et al., 2022a). Normative modeling has been applied in various clinical contexts, including the development of preterm infants and the study of individuals with different brain disorders (Dimitrova et al., 2020). This approach has been used to study disorders and diseases, such as schizophrenia (Wolfers et al., 2018, 2021), bipolar disorder (Wolfers et al., 2018), Autisms (Floris et al., 2020; Zabihi et al., 2019, 2020), attention-deficit/hyperactivity disorder (Wolfers et al., 2017), Alzheimer (Verdi et al., 2021, 2023), and other dementias (Bhome et al., 2023). In one application of such approach, researchers were able to map differences in individual-level deviations across the brain. While the reference population was small, compared to more recent studies (Bethlehem et al., 2022; Rutherford et al., 2022b, 2023), the researcher could show that individuals with schizophrenia showed only a small degree of overlap in terms of extreme deviation from the estimated norm in the same brain regions, despite reliable and replicable group differences (Wolfers et al., 2018, 2021). A similar pattern was shown across various different disorders, such as Autism(s), ADHD but also neurodegenerative diseases, such as Alzheimer (Verdi et al., 2021, 2023). These findings highlight the potential of normative modeling to uncover individual-level variations and provide a more nuanced understanding of mental disorders, mapping the heterogeneity of complex brain disorders (Segal et al., 2023). In an earlier review (Marquand et al., 2019), we summarized the literature on normative modeling in a table and classified it based on (i) the choice of covariates and response variables, (ii) the extent to which the model distinguishes between different sources of variation, and (iii) the degree to which the model enables statistical inference at the individual level. We have used a similar classification in this book chapter allowing the readership a good overview of studies utilizing normative modeling (Table 1).

Recently, various well-defined studies are emerging that compile large-scale dataset for normative modeling (Bethlehem et al., 2022; Rutherford et al., 2023, 2022c; Savage et al., 2023). In these studies, various preprocessing strategies are used to extract features that are subsequently utilized for normative modeling. These samples capture the population range across the lifespan, which allows us for the first time to build a common reference frame across studies. In one such study, the researchers are interested in defining reference models for population variation to understand interindividual variability and its relationship to mental and neurological disorders. To this end, a reference cohort of neuroimaging data from 82 sites consisting of 58,836 individuals aged 2–100 was

Table 1 Overview of selected studies.

Study	Reference cohort	Target cohort	Covariates	Response variable	Algorithm
Kessler et al. (2016)	Population-based cohort	ADHD	Age	Functional connectivity measures derived from resting fMRI	Polynomial regression
Wolfers et al. (2018, 2021)	Healthy	ADHD	Age and gender	Brain volume	Gaussian process regression
Marquand et al. (2016a)	Healthy	Healthy	Delay discounting	Reward-related brain activity derived from task fMRI	Gaussian process regression
Zabihi et al. (2019b)	Typically developing	Autism	Age and gender	Cortical thickness	Gaussian process regression
Bethlehem et al. (2020)	Typically developing	Autism	Age	Cortical thickness	Local polynomial regression
Lefebvre et al. (2018)	Typically developing	Autism	Age	Alpha band brain activity derived from electro-encephalography	Local polynomial regression
Wolfers et al. (2018, 2021)	Healthy	Bipolar disorder	Age and gender	Brain volume	Gaussian process regression
Erus et al. (2015)	Healthy	Healthy	Brain volume	Age	Support vector regression
Ordaz et al. (2013)	Healthy	Healthy	Age	Task-related fMRI data	Hierarchical linear modeling

Reference	Population	Condition	Covariates / Age	Measure	Method
Huizinga et al. (2018)	Population-based	Cognitive impairment; Alzheimer	Age	Brain volume	Partial least squares and quantile regression
Ziegler et al. (2014)	Healthy	Healthy; Cognitive impairment; Dementia	Age, sex, total gray and white matter volume, total cerebrospinal fluid, MRI field strength	Brain volume	Gaussian process regression
Wolfers et al. (2018, 2021)	Healthy	Schizophrenia	Age and gender	Brain volume	Gaussian process regression
Bethlehem et al. (2022)	Population-based	Neurological diseases and mental disorders	0–100	Cortical thickness and surface	GAMLSS
Rutherford et al. (2022c)	Population-based	Neurological diseases and mental disorders	3–90	Cortical thickness and surface	BLR with warping
Rutherford et al. (2023)	Population-based	Neurological diseases and mental disorders	3–90	Cortical thickness and surface	BLR with warping
Kim et al. (2023a)	Population-based	Neurological diseases and mental disorders	3–92	Cerebellar volume	BLR with warping

assembled and normative models were estimated, which were validated against a subset of data (Rutherford et al., 2022c) (Fig. 1). By applying these models to a transdiagnostic psychiatric sample, the study showcased the models capability to quantify variability across multiple disorders. Furthermore, researchers were able to delineate individual differences among patients sharing the same diagnosis. Recently, this approach was extended to include resting-state data (Rutherford et al., 2023) as well and currently researchers are developing those models across cerebellar voxels (Kim et al., 2023a). There are many follow-up studies that apply those models to clinical cohorts trying to parse individual differences. Recent findings have demonstrated that normative models can effectively eliminate scanner effects, establish a common reference space, and enhance prediction accuracy when normalizing data against a population reference (Kia et al., 2022; Rutherford et al., 2023; Bayer et al., 2021; Kia et al., 2020).

In summary, early normative modeling studies were small in scale estimating the population range within sample. Recently, large-scale studies became prevalent, which open the possibility using estimated norms as common reference frame to perform studies across different samples including various sites and scanners. We can determine that individual differences are prevalent across neurological diseases, as well as mental disorders, and that the mapping of these differences is essential if we want to make progress toward an individual-level understanding of illness. Normative modeling has been instrumental in this process and is influencing both neurological and psychiatric research toward parsing individual differences.

Assumptions of normative modeling

The first important decision in normative modeling is to select a *reference cohort*, which seems like a trivial question but involves considerations regarding the sex, age-range, socio-economic background, race, cofounds such as scanner effects, and many other factors (Benkarim et al., 2022; Greene et al., 2022; Li et al., 2022). It is often assumed that we estimate normative models in a representative way for a specific population; however, most current samples are not representative. Therefore, typical approaches in the field usually combine samples across scanners and sites to capture more of the typical variance across brains in the general population. Due to the constraints with respect to currently available samples, the available models do not apply to all backgrounds equally but are generally and predominantly derived from WEIRD (White, Educated, Industrialized, Rich, and Democratic) samples, with the exceptions, such as Chinese Human Connectome Project, Color Nest Project, and various data sharing initiatives, such as ENIGMA, and there are recent efforts to assemble and scan populations across the globe, which will make these models more applicable to different racial and ethnic backgrounds (Benkarim et al., 2021; Kopal et al., 2023; Ricard et al., 2023). However, the scientists employing normative modeling have to be aware of the importance of the selected reference. The way the estimated models behave is largely dependent on this and investigations into appropriate reference given specific clinical questions are still in its infancy. The second class of assumptions revolves around the algorithm that is selected. Normative models can be calculated based on various different machine learning algorithms and all these algorithms have specific prerequisites, which, for correct inference, have to be met and generally tested. Therefore, when building normative models, it is of pivotal importance, next to paying attention to the selection of the reference and its specific confounds, to also check the respective machine learning algorithm on whether all assumptions are met.

Building, testing, interpreting normative models

The first step of normative modeling involves the selection of an *appropriate reference* and a vast amount of data need to be *prepared and preprocessed*, the modeling starts, in the second step, by specifying *the mapping* that the normative model should be estimated with respect to and importantly the *algorithm* it is based on. In the past, the mapping usually involved a brain read out, as well as age and sex (Wolfers et al., 2018; Bethlehem et al., 2018; Zabihi et al., 2019). However, the framework is more flexible than this and can be estimated across various different covariate and response variable configurations. For practical reasons, many reference samples are different in terms of the acquired variables, their coding and or acquisition procedure, and only demographic variables are consistently available. Therefore, normative models are generally still based on age and sex, other mappings are in the minority (Kjelkenes et al., 2022). The *machine learning algorithm* that is used within the normative modeling framework is of great importance. It can be of varying complexity, starting with simple linear models leading up to complex machine learning architectures (Rutherford et al., 2022a; Ge et al., 2023). The statistical model has to capture the variance in a response variable (also known as the target or dependent variable) based on clinically relevant covariates (also known as predictor or independent variables) across the reference cohort. As stated earlier, and in contrast to case–control analyses that primarily focus on group means and seek a consistent pattern of atypicality, normative modeling explicitly models heterogeneity by considering individual variation around the mean. In addition, normative models can capture the higher-order moments of the reference distribution (such as skew and kurtosis). In the third step, the *performance* of the estimated machine learning model is evaluated, by calculating the accuracy with which the response variable can be estimated. This is typically done by reporting performance metrics, such as mean-squared error or explained variance in a left-out validation set, and it is also important to quantifying goodness of fit to the shape of the distribution (Fraza et al., 2021; Dinga et al., 2021). To ensure reliable estimates of generalizability, this assessment is performed on withheld data. In case the estimated model needs to be transferred to new scanner setting, a *transfer learning* can be performed, which allows for application of the estimated norms to unseen scanners and samples and which has been a research priority in the last years (Fraza et al., 2021; He et al., 2022; Rutherford et al., 2020). Finally, the developed model can be applied to *quantify the deviations* of individuals in a target cohort (such as a clinical cohort) compared to the normative models trained on the reference. These deviations are generally quantified in terms of *Z-scores,* which can feed into *various post-normative modeling statistical tests* to quantify or to make sense of these individual-level deviations. Here, the guiding questions is usually how those deviations can be decerned that are linked to illness from those that are mere chance, a consequence of illness or entirely unrelated to the illness. Many research endeavors have recently been started to quantify heterogeneity in more comprehensive ways, and in future, it will be interesting to combine this approach with frameworks developed under the banner of *causal machine learning* (Schölkopf, 2022; Schölkopf and von Kügelgen, 2022).

Various regression models have been proposed for normative modeling, including Gaussian process regression, hierarchical linear models, polynomial regression, quantile regression, support vector regression, generalized additive model for location, scale and shape (GAMLSS), and autoencoders, among others (Bethlehem et al., 2022; Marquand et al., 2019; Rutherford et al., 2022a; Fraza et al., 2021; Dinga et al., 2021; Marquand et al., 2016b; Wolfers et al., 2019; Kia et al., 2022; Bayer et al., 2022; Zabihi et al., 2021). As with classical growth charts, the regression models must address specific demands, such as precise estimation of outer centiles in sparse data regions, smooth variation of centiles with covariates without crossing, and the ability to estimate deviations for individual samples (e.g., Z-scores). For the calculations of such Z-scores, it is important that the model estimates aleatoric (irreducible) and epistemic (modeling) uncertainties interpedently, which then can be used to determine Z-statistics based on irreducible and thus variance naturally present in the population independent of modeling uncertainty (Fig. 2).

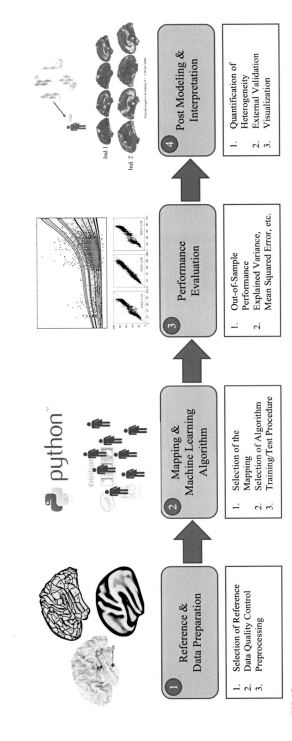

FIG. 2

Overview of normative modeling framework. The workflow consists of four stages: (1) Selection of the Reference, Data Quality Control, and Preprocessing; (2) Variable Mapping and the Machine Learning Algorithm; (3) Performance Evaluation; and (4) Post Modeling and Interpretation.

Pitfalls and limitations of normative modeling

Normative modeling is a machine learning framework; thus, all limitations that apply to the specific model employed within this framework also apply to that specific normative modeling application. Here, we can specifically address the fact that as a bottom-up approach, it is designed to capture variations and should not be regarded as a substitute for hypothesis testing or top-down theory-driven approaches. Instead, a preferable approach is to integrate the strengths of both methods. One approach to achieving this goal is by integrating top-down, theory-driven techniques with normative models. We are currently exploring this integration within the context of integrating reinforcement learning with normative modeling (Kumar et al., 2024). This integration allows for mapping individual variations based on a theory-driven model.

Moreover, it is important to recognize that normative models do not directly indicate the clinical relevance of observed deviations. These deviations might have biological significance but may not be associated with psychopathology. It is hard to identify causal relationships of biology to mental disorders (Leonardsen et al., 2023; Young, 2015). Therefore, it is important to incorporate recent developments on causality into a normative modeling framework, by, for instance, running causality estimators that are specifically geared to capture the relationships of individuals to each other based on their genetic links (Taschler et al., 2022). Furthermore, it is crucial to externally validate the identified deviations using additional measures, such as symptoms, genotype, and environmental factors. Rigorous data preprocessing and thorough validation procedures play a vital role in this process of understanding extreme deviations. However, the most pressing question in normative modeling is still unanswered, how do we understand if a particular deviation from a biological readout or biological or cognitive process is disease relevant? While this can in parts be answered in the context of true experimental designs with the manipulation of single putatively causal variable, for large-scale normative models, we require considerable methodological and experimental advancements, so that complex disorders can eventually be reconceptualized at the level of the individual or subgroup.

Open challenges and future directions

Normative modeling provides a framework for detecting deviations from anticipated patterns at the individual-level. This technique has been instrumental in understanding individual differences in brain development, aging, and diverse cognitive domains across various illnesses (Table 1). By analyzing cognitive measurements and brain organization, it allows us to investigate individual variations within a multidimensional framework, in line with initiatives, such as RDoC (Research Domain Criteria) (Cuthbert and Insel, 2013; Insel et al., 2010; Michelini et al., 2021; Sanislow, 2020). Nevertheless, there is further progress to be made in integrating these approaches and better understanding the clinical relevance of individual variability. Specifically, we must comprehend the circumstances, timing, and reasons behind how individual differences can manifest as vulnerability or resilience.

The field is clearly still in its infancy and many current developments are going to shape the future of neuroscientific research on complex brain disorders. The principles developed for neuroscience, however, are applicable more broadly to other medical fields, proposing a transferable and clinical relevant machine learning framework, for parsing complex disorders and diseases. We can identify several innovations in the context of normative modeling today. First, normative models are being developed for different modalities (Rutherford et al., 2023, 2022c; Savage et al., 2023; Winter et al., 2022). Second, the models are increasing in scale. Third, different algorithms, either in the process of constructing norms or building features, are being employed (Ge et al., 2023; Dinga et al., 2021; Kia et al., 2022; Kumar, 2021). While normative modeling in neuroimaging has only been recently established with large-scale studies being conducted in the last years, other data modalities are left out. Recently, a paper was published on norms of aging for different organs of the human body (Tian et al., 2023a,b). This development indicates a trend toward estimating norms across various biological variables. Further, different demographics and ethnic groups are considered within the normative modeling framework, making such models compatible with different backgrounds. More complex algorithms are being established that can be deployed to estimate and construct explainable and interpretable features from neuroimaging data, which can then be charted in the context of the normative modeling framework (de Boer, 2022; Frotscher et al., 2023; Leonardsen et al., 2022, 2024; Yang et al., 2023).

An essential goal is to create techniques that can stratify individuals based on deviations from normative models. One potential method is training clustering algorithms using these deviations from multiple normative models, for example, using clustering algorithms or latent variable models. However, traditional "hard" clustering algorithms assign each person to a single cluster, disregarding the possibility of multiple overlapping mechanisms in different individuals, which are important to understand heterogeneity and comorbidity of such illnesses. Therefore, recently developed methodology to better parse these layers of complexity and using more flexible clustering methodology (Zabihi et al., 2020) is important. Furthermore, the models became more precise with respect to spatial precision; for instance, models developed on the cerebellum contain more than 140,000 features (Kim et al., 2023a,b).

In summary, normative modeling is a valuable tool for investigating individual differences. However, additional research is necessary to successfully integrate these approaches and understand the clinical implications of individual variability. Key areas of emphasis include developing methods to parse individuals and understand structure in normative space and explicitly incorporating spatial information to enhance our ability to identify abnormal patterns across brain regions adaptively and flexibly. Furthermore, normative models need to be developed for higher spatial precision, which can also capture different modalities. We require machine learning methods that can be adapted to construct features for normative models, which ideally incorporate interpretability. Exploring recent developments in causal machine learning and applying those approaches to data that is normalized against the population reference is an interesting research endeavor as well.

Take-home points

1) The field is clearly still in its infancy but had considerable success and exerted a large influence on how we are dealing with problems pertaining heterogeneity and comorbidity of complex brain disorders and diseases.

2) Many studies have shown a remarkable adaptability of this framework, which can be effectively employed to estimate variations in brain growth and establish connections between various aspects of behavior, cognition, and biology.

3) The key advantage of this framework lies in its capacity to predict individual-level outcomes based on a normative pattern in reference to population samples, providing the opportunity to translate insights from population neuroscience into clinical practice. Consequentially, shifting the focus from average effects to comprehending individual differences has been the biggest contribution of this work thus far.

4) The normative modeling framework bridges the gap between theoretical considerations such as RDoC and the study of individual variations, thereby presenting a promising pathway toward achieving more precision in psychiatry, neurology, and other medical disciplines. However, this ambitious goal requires many more conceptual, methodological, and experimental contributions that go beyond the current state-of-the-art in normative modeling.

References

Abi-Dargham, A., Horga, G., 2016. The search for imaging biomarkers in psychiatric disorders. Nat. Med. 22, 1248–1255.

Bassett, D.S., Gazzaniga, M.S., 2011. Understanding complexity in the human brain. Trends Cogn. Sci. 15, 200–209.

Bayer, J.M.M., et al., 2021. Accommodating site variation in neuroimaging data using normative and hierarchical Bayesian models. bioRxiv. https://doi.org/10.1101/2021.02.09.430363. 2021.02.09.430363.

Bayer, J.M., et al., 2022. Accommodating site variation in neuroimaging data using normative and hierarchical Bayesian models. Neuroimage 264, 119699.

Benkarim, O., et al., 2021. The cost of untracked diversity in brain-imaging prediction. bioRxiv. https://doi.org/10.1101/2021.06.16.448764.

Benkarim, O., et al., 2022. Population heterogeneity in clinical cohorts affects the predictive accuracy of brain imaging. PLoS Biol. 20, e3001627.

Bethlehem, R.A., Seidlitz, J., Romero-Garcia, R., Dumas, G., Lombardo, M.V., 2018. Normative age modelling of cortical thickness in autistic males. Bioarchive.

Bethlehem, R.A.I., et al., 2020. A normative modelling approach reveals age-atypical cortical thickness in a subgroup of males with autism spectrum disorder. Commun. Biol. 3, 486.

Bethlehem, R.A.I., et al., 2022. Brain charts for the human lifespan. Nature 604, 525–533.

Bhome, R., et al., 2023. A neuroimaging measure to capture heterogeneous patterns of atrophy in Parkinson's disease and dementia with Lewy bodies. Neuroimage Clin. https://doi.org/10.1101/2023.08.01.23293480.

Bijsterbosch, J., 2022. Piggybacking on big data. Nat. Neurosci. 25, 682–683. https://doi.org/10.1038/s41593-022-01058-w.

Boedhoe, P.S., et al., 2020. Subcortical brain volume, regional cortical thickness, and cortical surface area across disorders: findings from the ENIGMA ADHD, ASD, and OCD working groups. Am. J. Psychiatry 177, 834–843.

Border, R., et al., 2019. No support for historical candidate gene or candidate gene-by-interaction hypotheses for major depression across multiple large samples. Am. J. Psychiatry 176, 376–387.

Borghi, E., et al., 2006. Construction of the World Health Organization child growth standards: selection of methods for attained growth curves. Stat. Med. 25, 247–265.

Bzdok, D., Yeo, B.T.T., 2017. Inference in the age of big data: future perspectives on neuroscience. Neuroimage 155, 549–564.

Califf, R.M., 2018. Biomarker definitions and their applications. Exp. Biol. Med. (Maywood) 243, 213–221.

Casey, B.J., et al., 2018. The adolescent brain cognitive development (ABCD) study: imaging acquisition across 21 sites. Dev. Cogn. Neurosci. 32, 43–54.

Cole, T.J., 2012. The development of growth references and growth charts. Ann. Hum. Biol. 39, 382–394.

Cuthbert, B.N., Insel, T.R., 2013. Toward the future of psychiatric diagnosis: the seven pillars of RDoC. BMC Med. 11, 126.

de Boer, A.A.A., et al., 2022. Non-Gaussian normative modelling with hierarchical Bayesian regression. Imaging Neuroscience. https://doi.org/10.1101/2022.10.05.510988.

Dean, A., Voss, D., Draguljić, D., 2017. Design and Analysis of Experiments. Springer International Publishing, https://doi.org/10.1007/978-3-319-52250-0.

Dimitrova, R., et al., 2020. Heterogeneity in brain microstructural development following preterm birth. Cereb. Cortex 30, 4800–4810.

Dinga, R., et al., 2019. Evaluating the evidence for biotypes of depression: methodological replication and extension of Drysdale et al. (2017). NeuroImage Clin. 22, 101796.

Dinga, R., et al., 2021. Normative modeling of neuroimaging data using generalized additive models of location scale and shape. bioRxiv. https://doi.org/10.1101/2021.06.14.448106.

Erus, G., et al., 2015. Imaging patterns of brain development and their relationship to cognition. Cereb. Cortex 25, 1676–1684.

Faraone, S.V., et al., 2015. Attention-deficit/hyperactivity disorder. Nat. Rev. Dis. Primers. 1, 15020.

Floris, D.L., et al., 2020. Atypical brain asymmetry in autism—a candidate for clinically meaningful stratification. Biol. Psychiatry Cogn. Neurosci. Neuroimaging. https://doi.org/10.1016/j.bpsc.2020.08.008.

Foulkes, L., Blakemore, S.J., 2018. Studying individual differences in human adolescent brain development. Nat. Neurosci. 21, 315–323. https://doi.org/10.1038/s41593-018-0078-4.

Fraza, C.J., Dinga, R., Beckmann, C.F., Marquand, A.F., 2021. Warped Bayesian linear regression for normative modelling of big data. Neuroimage 245, 118715.

Frégnac, Y., 2017. Big data and the industrialization of neuroscience: a safe roadmap for understanding the brain? Science 358, 470–477.

Fried, E.I., Nesse, R.M., 2015. Depression is not a consistent syndrome: an investigation of unique symptom patterns in the STAR*D study. J. Affect. Disord. 172, 96–102.

Friston, K., Brown, H.R., Siemerkus, J., Stephan, K.E., 2016. The dysconnection hypothesis (2016). Schizophr. Res. 176, 83–94.

Frotscher, A., Kapoor, J., Wolfers, T., Baumgartner, C.F., 2023. Unsupervised anomaly detection using aggregated normative diffusion. arXiv preprint arXiv:2312.01904.

Fry, A., et al., 2017. Comparison of sociodemographic and health-related characteristics of UK biobank participants with those of the general population. Am. J. Epidemiol. 186, 1026–1034.

Gau, R., et al., 2021. Brainhack: developing a culture of open, inclusive, community-driven neuroscience. Neuron 109, 1769–1775.

Ge, R., et al., 2023. Normative modeling of brain morphometry across the lifespan using CentileBrain: algorithm benchmarking and model optimization. bioRxiv. https://doi.org/10.1101/2023.01.30.523509.

Glasser, M.F., et al., 2013. The minimal preprocessing pipelines for the human connectome project. Neuroimage 80, 105–124.

Greene, A.S., et al., 2022. Brain–phenotype models fail for individuals who defy sample stereotypes. Nature 609, 109–118.

He, T., et al., 2022. Meta-matching as a simple framework to translate phenotypic predictive models from big to small data. Nat. Neurosci. 25, 795–804. https://doi.org/10.1038/s41593-022-01059-9.

Huizinga, W., et al., 2018. A spatio-temporal reference model of the aging brain. Neuroimage 169, 11–22.

Insel, T., et al., 2010. Research domain criteria (RDoC): toward a new classification framework for research on mental disorders. AJP 167, 748–751.

Kahn, R.S., et al., 2015. Schizophrenia. Nat. Rev. Dis. Primers. 1, 15067.

Kapur, S., Phillips, A.G., Insel, T.R., 2012. Why has it taken so long for biological psychiatry to develop clinical tests and what to do about it. Mol. Psychiatry 17, 1174–1179.

Kessler, D., Angstadt, M., Sripada, C., 2016. Growth charting of brain connectivity networks and the identification of attention impairment in youth. JAMA Psychiatry 73, 481–489.

Kia, S.M., et al., 2020. Hierarchical Bayesian regression for multi-site normative modeling of neuroimaging data. In: Martel, A.L., et al. (Eds.), Medical Image Computing and Computer Assisted Intervention—MICCAI 2020. Springer International Publishing, pp. 699–709, https://doi.org/10.1007/978-3-030-59728-3_68.

Kia, S.M., et al., 2022. Closing the life-cycle of normative modeling using federated hierarchical Bayesian regression. PLoS One 17, e0278776.

Kim, M., et al., 2023a. Machine learning for parsing individual differences. Biol. Psychiatry 93, S53.

Kim, M., Leonardsen, E., Rutherford, S., Selbæk, G., Persson, K., Steen, N.E., Smeland, O.B., Ueland, T., Richard, G., Beckmann, C.F., Marquand, A.F., Alzheimer's Disease Neuroimaging Initiative (ADNI), Andreassen, O.A., Westlye, L.T., Wolfers, T., Moberget, T., 2023b. Mapping cerebellar anatomical heterogeneity in mental and neurological illnesses. bioRxiv 2023.11.18.567647. https://doi.org/10.1101/2023.11.18.567647.

Kjelkenes, R., et al., 2022. Mapping normative trajectories of cognitive function and its relation to psychopathology symptoms and genetic risk in youth. Biol. Psychiatry Glob. Open Sci. 3, 255–263.

Kopal, J., Uddin, L.Q., Bzdok, D., 2023. The end game: respecting major sources of population diversity. Nat. Methods 20, 1122–1128. https://doi.org/10.1038/s41592-023-01812-3.

Kumar, S., 2021. NormVAE: normative modeling on neuroimaging data using variational autoencoders. arXiv.

Kumar, P., Dayan, P., Wolfers, T., 2024. From complexity to precision-charting decision-making through normative modeling. JAMA Psychiatry 81 (2), 117–118. https://doi.org/10.1001/jamapsychiatry.2023.4611.

Lefebvre, A., et al., 2018. Alpha waves as a neuromarker of autism Spectrum disorder: the challenge of reproducibility and heterogeneity. Front. Neurosci. 12, 662.

Leonardsen, E.H., Peng, H, Kaufmann, T., Agartz, I., Andreassen, O.A., Celius, E.G., Espeseth, T., Harbo, H.F., Høgestøl, E.A., Lange, A.M., Marquand, A.F., Vidal-Piñeiro, D., Roe, J.M., Selbæk, G., Sørensen, Ø., Smith, S.M., Westlye, L.T., Wolfers, T., Wang, Y., 2022. Deep neural networks learn general and clinically relevant representations of the ageing brain. Neuroimage 256, 119210. https://doi.org/10.1016/j.neuroimage.2022.119210. Epub 2022 Apr 21.

Leonardsen, E.H., et al., 2023. Genetic architecture of brain age and its causal relations with brain and mental disorders. Mol. Psychiatry 28, 3111–3120. https://doi.org/10.1038/s41380-023-02087-y.

Leonardsen, E.H., Persson, K., Grødem, E., et al., 2024. Constructing personalized characterizations of structural brain aberrations in patients with dementia using explainable artificial intelligence. NPJ Digit. Med. 7, 110. https://doi.org/10.1038/s41746-024-01123-7.

Li, J., et al., 2022. Cross-ethnicity/race generalization failure of behavioral prediction from resting-state functional connectivity. Sci. Adv. 8, eabj1812.

Lozano, A.M., et al., 2019. Deep brain stimulation: current challenges and future directions. Nat. Rev. Neurol. 15, 148–160.

Markiewicz, C.J., et al., 2021. The OpenNeuro resource for sharing of neuroscience data. Elife 10, e71774.

Marquand, A.F., Rezek, I., Buitelaar, J., Beckmann, C.F., 2016a. Understanding heterogeneity in clinical cohorts using normative models: beyond case control studies. Biol. Psychiatry 80, 552–561.

Marquand, A.F., Wolfers, T., Mennes, M., Buitelaar, J., Beckmann, C.F., 2016b. Beyond lumping and splitting: a review of computational approaches for stratifying psychiatric disorders. Biol. Psychiatry Cogn. Neurosci. Neuroimaging 1, 433–447.

Marquand, A.F., et al., 2019. Conceptualizing mental disorders as deviations from normative functioning. Mol. Psychiatry. https://doi.org/10.1038/s41380-019-0441-1.

Michelini, G., Palumbo, I.M., DeYoung, C.G., Latzman, R.D., Kotov, R., 2021. Linking RDoC and HiTOP: a new interface for advancing psychiatric nosology and neuroscience. Clin. Psychol. Rev. 86, 102025.

Miller, K.L., et al., 2016. Multimodal population brain imaging in the UK biobank prospective epidemiological study. Nat. Neurosci. 19, 1523–1536.

Ordaz, S.J., Foran, W., Velanova, K., Luna, B., 2013. Longitudinal growth curves of brain function underlying inhibitory control through adolescence. J. Neurosci. 33, 18109–18124.

Petersen, R.C., et al., 2010. Alzheimer's Disease Neuroimaging Initiative (ADNI). Neurology 74, 201–209.

Poewe, W., et al., 2017. Parkinson disease. Nat. Rev. Dis. Primers. 3, 17013.

Posner, M.I., DiGirolamo, G.J., 2000. Cognitive neuroscience: origins and promise. Psychol. Bull. 126, 873–889.

Ricard, J.A., et al., 2023. Confronting racially exclusionary practices in the acquisition and analyses of neuroimaging data. Nat. Neurosci. 26, 4–11.

Rutherford, S., Angstadt, M., Sripada, C., Chang, S.-E., 2020. Leveraging big data for classification of children who stutter from fluent peers. bioRxiv. https://doi.org/10.1101/2020.10.28.359711. 2020.10.28.359711.

Rutherford, S., et al., 2022a. The normative modeling framework for computational psychiatry. Nat. Protoc. 17, 1711–1734.

Rutherford, S., et al., 2022b. Evidence for embracing normative modeling. Elife. https://doi.org/10.1101/2022.11.14.516460. 2022.11.14.516460 Preprint at.

Rutherford, S., et al., 2022c. Charting brain growth and aging at high spatial precision. Elife 11, e72904.

Rutherford, S., et al., 2023. Evidence for embracing normative modeling. Elife 12, e85082.

Sanislow, C.A., 2020. RDoC at 10: changing the discourse for psychopathology. World Psychiatry 19, 311–312.

Savage, H.S., et al., 2023. Unpacking the functional heterogeneity of the Emotional Face Matching Task: a normative modelling approach. bioRxiv. https://doi.org/10.1101/2023.03.27.534351. 2023.03.27.534351 Preprint at.

Schölkopf, B., 2022. Causality for Machine Learning. pp. 765–804, https://doi.org/10.1145/3501714.3501755.

Schölkopf, B., von Kügelgen, J., 2022. From statistical to causal learning. arXiv. https://doi.org/10.48550/arXiv.2204.00607. Preprint at.

Segal, A., et al., 2023. Regional, circuit and network heterogeneity of brain abnormalities in psychiatric disorders. Nat. Neurosci. 26, 1613–1629.

Sudlow, C., et al., 2015. UK biobank: an open access resource for identifying the causes of a wide range of complex diseases of middle and old age. PLoS Med. 12, e1001779.

Taschler, B., Smith, S.M., Nichols, T.E., 2022. Causal inference on neuroimaging data with Mendelian randomisation. Neuroimage 258, 119385.

Thompson, W.D., 1994. Statistical analysis of case-control studies. Epidemiol. Rev. 16, 33–50.

Thompson, P.M., et al., 2020. ENIGMA and global neuroscience: a decade of large-scale studies of the brain in health and disease across more than 40 countries. Transl. Psychiatry 10, 100.

Tian, Y.E., et al., 2023a. Heterogeneous aging across multiple organ systems and prediction of chronic disease and mortality. Nat. Med. 29, 1221–1231.

Tian, Y.E., et al., 2023b. Evaluation of brain-body health in individuals with common neuropsychiatric disorders. JAMA Psychiatry 80, 567–576.

Van Essen, D.C., et al., 2013. The WU-Minn human connectome project: an overview. Neuroimage 80, 62–79.

Verdi, S., Marquand, A.F., Schott, J.M., Cole, J.H., 2021. Beyond the average patient: how neuroimaging models can address heterogeneity in dementia. Brain. https://doi.org/10.1093/brain/awab165.

Verdi, S., et al., 2023. Personalising Alzheimer's disease progression using brain atrophy markers. medRxiv. https://doi.org/10.1101/2023.06.15.23291418. 2023.06.15.23291418 Preprint at.

Vlaardingerbroek, M.T., Den Boer, J.A., 2003. Magnetic Resonance Imaging. Springer, https://doi.org/10.1007/978-3-662-05252-5.

Winter, N.R., et al., 2022. Quantifying deviations of brain structure and function in major depressive disorder across neuroimaging modalities. JAMA Psychiatry 79, 879–888.

Wolfers, T., 2019. Towards Precision Medicine in Psychiatry. 250.

Wolfers, T., Buitelaar, J.K., Beckmann, C.F., Franke, B., Marquand, A.F., 2015. From estimating activation locality to predicting disorder: a review of pattern recognition for neuroimaging-based psychiatric diagnostics. Neurosci. Biobehav. Rev. 57, 328–349.

Wolfers, T., et al., 2017. Refinement by integration: aggregated effects of multimodal imaging markers on adult ADHD. J. Psychiatry Neurosci. 42, 386–394.

Wolfers, T., et al., 2018. Mapping the heterogeneous phenotype of schizophrenia and bipolar disorder using normative models. JAMA Psychiatry 11, 1146–1155.

Wolfers, T., et al., 2019. From pattern classification to stratification: towards conceptualizing the heterogeneity of Autism Spectrum Disorder. Neurosci. Biobehav. Rev. 104, 240–254.

Wolfers, T., et al., 2021. Replicating extensive brain structural heterogeneity in individuals with schizophrenia and bipolar disorder. Hum. Brain Mapp. 42, 2546–2555.

Yang, H.C., Andreassen, O., Westlye, L.T., Marquand, A.F., Beckmann, C.F., Wolfers, T., 2023. Learning cortical anomaly through masked encoding for unsupervised heterogeneity mapping. arXiv preprint arXiv:2312.02762.

Young, G., 2015. Causality in psychiatry: a hybrid symptom network construct model. Front. Psych. 6, 164.

Zabihi, M., et al., 2019. Dissecting the heterogeneous cortical anatomy of Autism Spectrum Disorder using normative models. Biol. Psychiatry Cogn. Neurosci. Neuroimaging 4, 567–578. https://doi.org/10.1016/j.bpsc.2018.11.013.

Zabihi, M., et al., 2020. Fractionating autism based on neuroanatomical normative modeling. Transl. Psychiatry 10, 1–10.

Zabihi, M., et al., 2021. Explanatory latent representation of heterogeneous spatial maps of task-fMRI in large-scale datasets. bioRxiv.

Ziegler, G., Ridgway, G.R., Dahnke, R., Gaser, C., 2014. Individualized Gaussian process-based prediction and detection of local and global gray matter abnormalities in elderly subjects. Neuroimage 97, 333–348.

Index

Note: Page numbers followed by *f* indicate figures, *t* indicate tables and *b* indicate boxes.

Printed in the United States
by Baker & Taylor Publisher Services